Springer Series in
Experimental Entomology

Thomas A. Miller, Editor

Springer Series in Experimental Entomology
Editor: T. A. Miller

Insect Neurophysiological Techniques
By T. A. Miller
Neurohormonal Techniques In Insects
Edited by T. A. Miller
Sampling Methods In Soybean Entomology
By M. Kogan and D. Herzog
Cuticle Techniques In Arthropods
Edited by T. A. Miller

Neuroanatomical Techniques

Insect Nervous System

Edited by
N. J. Strausfeld and Thomas A. Miller

With Contributions by
J. S. Altman · J. Bacon · L. G. Bishop · A. D. Blest
C. B. Boschek · P. S. Davie · H. E. Eckert · G. E. Gregory
B. Hengstenberg · R. Hengstenberg · N. Klemm
G. A. Nevmyvaka · M. Obermayer · M. P. Osborne
M. O'Shea · A. A. Panov · S. I. Plotnikova
F.-W. Schürmann · M. K. Shaw · N. J. Strausfeld
N. M. Tyrer

Springer-Verlag
New York Heidelberg Berlin

N. J. Strausfeld
European Molecular Biology Laboratory
D-6900 Heidelberg, F.R.G.

Thomas A. Miller
Department of Entomology
University of California
Riverside, California 92521 U.S.A.

With 172 figures

Library of Congress Cataloging in Publication Data
Main entry under title:
Neuroanatomical techniques.
  (Springer series in experimental entomology)
  Includes index.
  1. Neuroanatomy—Technique. 2. Nervous system—
Insects. 3. Insects—Anatomy. I. Strausfeld, Nicholas
James, 1942–   II. Miller, Thomas A.   III. Series.
QL927.N49   595.7′04′8   79-10145

9 8 7 6 5 4 3 2 1

ISBN 0-387-90392-5 Springer-Verlag New York Heidelberg Berlin
ISBN 3-540-90392-5 Springer-Verlag Berlin Heidelberg New York

Preface

Most neurobiological research is performed on vertebrates, and it is only natural that most texts describing neuroanatomical methods refer almost exclusively to this Phylum. Nevertheless, in recent years insects have been studied intensively and are becoming even more popular in some areas of research. They have advantages over vertebrates with respect to studying genetics of neuronal development and with respect to studying many aspects of integration by uniquely identifiable nerve cells.

Insect central nervous system is characterized by its compactness and the rather large number of nerve cells in a structure so small. But despite their size, parts of the insect CNS bear structural comparisons with parts of vertebrate CNS. This applies particularly to the organization of the thoracic ganglia (and spinal cord), to the insect and vertebrate visual systems and, possibly, to parts of the olfactory neuropils. The neurons that make up these areas in insects are often large enough to be impaled by microelectrodes and can be injected with dyes. Added to advantages of using a small CNS, into which the sensory periphery is precisely mapped, are the many aspects of insect behaviour whose components can be quantitized and which may find both structural and functional correlates within clearly defined regions of neuropil. Together, these various features make the insect CNS a rewarding object for study.

This volume is the first of two that describe both classic and recent methods for neuroanatomical research on insect CNS. Most techniques

v

are derived from methods first used on vertebrates; but there are some notable exceptions, such as methylene blue methods, special stains for synapses, and the fluorescent dye and cobalt marking techniques. The methods described in this volume are mostly biased towards correlating structure with function. Others, however, are purely descriptive. But without these even the best and most fluorescing neuron would be lost for a context in which to fit.

The editors would like to thank all the contributors for their labours, and also for their patience during the preparation of this volume which was initiated in 1976. And, lastly, we invite them all to join with us in dedicating this book to six colleagues who have played a major role in neurobiological research because they have introduced important and powerful techniques. They are, A. O. W. Stretton and E. A. Kravitz; R. M. Pitman, C. D. Tweedle and M. J. Cohen; and W. Stewart: The methods are familiar to all of us.

N. J. Strausfeld
Thomas A. Miller
September 1980

Note to the reader: A companion volume to this book is presently being written by an expert group of authors and edited by N.J. Strausfeld. Topics to be treated include: Electronmicroscopy (EM) of Golgi and cobalt-silver stained cells; EM resolution of transsynaptic cobalt and horse radish peroxidase; marking cells with Cytochrome C; double marking techniques for EM; High voltage EM; Interpretation of EM of freeze fracture replicas of neuropil; combined reduced silver and cobalt methods; methylene blue methods; Lucifer yellow histology; localization of functional activity by radioactive deoxyglucose; radioactive amino-acid mapping of sensory pathways; biochemistry and immunological characterization of proctolin containing neurons; histochemical distinction between octopamine, dopamine, noradrenalin and 5HT; immunocytochemical methods for identifying peptides in insect CNS; methods for studying the developing nervous system; uses for computer-graphics in structural analysis.

Call to Authors

Springer Series in Experimental Entomology will be published in future volumes as contributed chapters. Subjects will be gathered in specific areas to keep volumes cohesive.

Correspondence concerning contributions to the series should be communicated to:

Thomas A. Miller, Editor
Springer Series in Experimental Entomology
Department of Entomology
University of California
Riverside, California 92521
USA

Contents

Chapter 16
The Use of Horseradish Peroxidase as a Neuronal Marker in the Arthropod Central Nervous System
HENDRIK E. ECKERT and C. BRUCE BOSCHEK. With 16 Figures.

Chapter 17
Cobalt Staining of Neurons by Microelectrodes
MICHAEL O'SHEA. With 3 Figures.

Chapter 18
Nonrandom Resolution of Neuron Arrangements
J. BACON and N. J. STRAUSFELD. With 7 Figures.

Chapter 19
Filling Selected Neurons with Cobalt through Cut Axons
J. S. ALTMAN and N. M. TYRER. With 11 Figures.

Chapter 20
**Silver-Staining Cobalt Sulfide Deposits within Neurons of
Intact Ganglia**
M. OBERMAYER and N. J. STRAUSFELD.   With 8 Figures.

Chapter 21
**Intensification of Cobalt-Filled Neurons in Sections (Light and
Electron Microscopy)**
N. M. TYRER, M. K. SHAW, and J. S. ALTMAN.   With 7 Figures.

List of Contributors

J. S. Altman   Department of Zoology, The University of Manchester, Manchester M13 9PL, England

J. Bacon   Max Planck Institut für Verhaltensphysiologie, D-8131 Seewiesen über Starnberg, F.R.G.

L. G. Bishop   Biology Department, University of Southern California, Los Angeles, California 90007 U.S.A.

A. D. Blest   Department of Neurobiology, Research School of Biological Sciences, The Australian National University, Canberra City, A.C.T., Australia

C. B. Boschek   Institute für Virologie, Fachbereich Humanmedizin der Justus Liebig Universität, Frankfurter Strasse 107, D-6300 Giessen, F.R.G.

P. S. Davie   Department of Zoology, University of Canterbury, Christchurch, New Zealand

H. E. Eckert   Department of Animal Physiology, Ruhr University Bochum, P.O. Box 102148, D-4630 Bochum-Querenburg, F.R.G.

G. E. Gregory   Insecticides and Fungicides Department, Rothamsted Experimental Station, Harpenden, Herts. AL5 25Q, England

B. Hengstenberg   Max Planck Institut für Biologische Kybernetik 38, Spemannstrasse, 7400 Tübingen, F.R.G.

R. Hengstenberg   Max Planck Institut für Biologische Kybernetik 38, Spemannstrasse, 7400 Tübingen, F.R.G.

N. Klemm   Fakultät für Biologie, Universität Konstanz, Postfach 5560, D-7750 Konstanz 1, F.R.G.

Plate 2

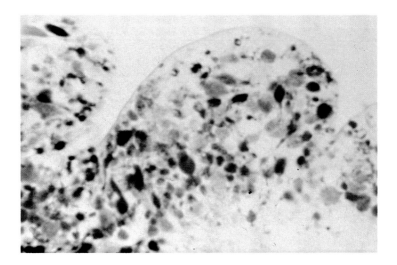

Fig. 4-34. *Rhodnius prolixus,* imago; NSC of the pars intercebralis: *blue* = A cells, rich in strong acid groups after KMnO₄ oxidation; *purple* = C cells, rich in weak acid groups after KMnO₄ oxidation. PTh-PF, ×720.

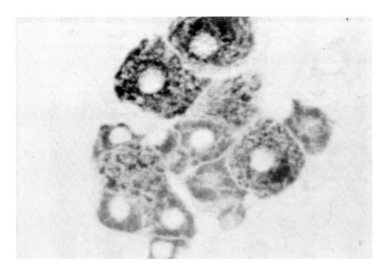

Fig. 4-35. *Locusta migratoria,* imago; storage lobe of the corpus cardiacum: *blue* = NSM of the brain A1 cells rich in strong acid groups after KMnO₄ oxidation; *purple* = NSM of the A2 cells of the pars intercerebralis rich in weak acid groups after KMnO₄ oxidation. PTh-PF, ×720.

Plate 3

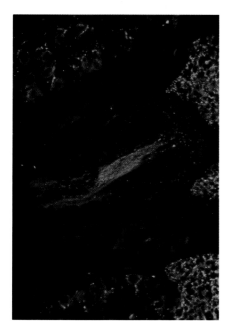

Fig. 5-11. Oblique horizontal section of the brain of adult *Calliphora vomitoria* L. CA fluorescence. In the middle of the picture is the ellipsoid body. Kodakchrome KX.

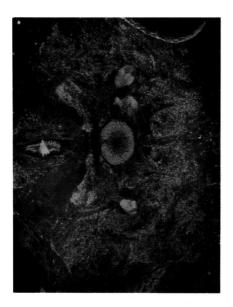

Fig. 5-12. Occipital ganglion of the adult cricket *Acheta domesticus* L. IA-induced fluorescence in the middle of the ganglion is bordered by CA-containing fibers. Ektachrome High Speed, developed for ASA 400 (27 DIN).

Plate 4

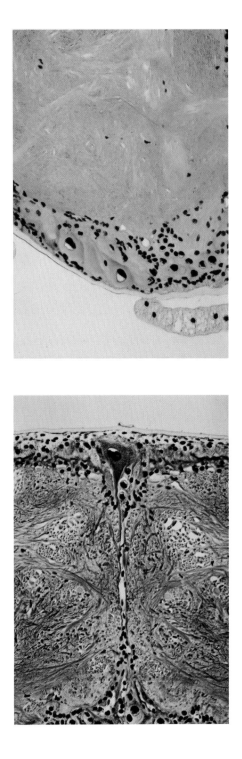

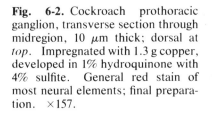

Fig. 6-1. Cockroach mesothoracic ganglion, longitudinally oblique frontal section through posteroventral region, 10 μm thick; anterior at *top*. Impregnated with 1.3 g copper/65 ml 2% protargol, developed in 1% hydroquinone with 4% sodium sulfite. Yellow/brown colors after second impregnation and development; neural lamella and fat body (at *bottom*) unstained. Photographed in distilled water during wash after development. ×157.

Fig. 6-2. Cockroach prothoracic ganglion, transverse section through midregion, 10 μm thick; dorsal at *top*. Impregnated with 1.3 g copper, developed in 1% hydroquinone with 4% sulfite. General red stain of most neural elements; final preparation. ×157.

Plate 5

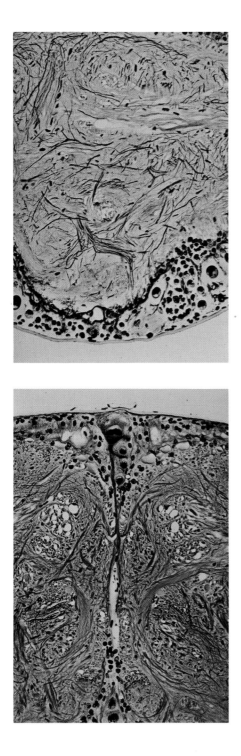

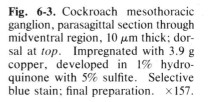

Fig. 6-3. Cockroach mesothoracic ganglion, parasagittal section through midventral region, 10 μm thick; dorsal at *top*. Impregnated with 3.9 g copper, developed in 1% hydroquinone with 5% sulfite. Selective blue stain; final preparation. ×157.

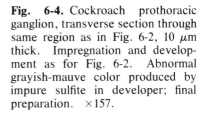

Fig. 6-4. Cockroach prothoracic ganglion, transverse section through same region as in Fig. 6-2, 10 μm thick. Impregnation and development as for Fig. 6-2. Abnormal grayish-mauve color produced by impure sulfite in developer; final preparation. ×157.

Plate 6

Fig. 7-7

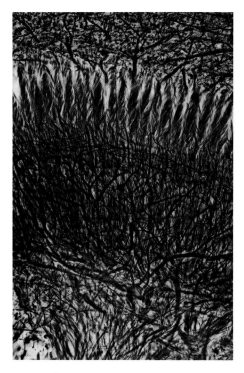

Fig. 7-8

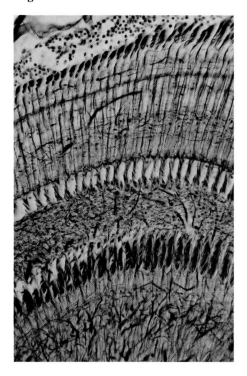

Figs. 7-7–7-10. *Calliphora.* The results of basic Holmes-Blest routines of reduced silver impregnation following fixation in Gregory's Dubuscq-Brasil, and using either pyridine (**Fig. 7-7,** lobula), 2,6-lutidine (**Fig. 7-8,** lobula and medulla), or 2,4,6-collidine (**Fig. 7-9,** lobula) in the impregnating bath. Pyridine, in these methods, can give a very wide range of colors, from the reddish-purple shown here, to blue-black. Lutidines characteristically yield blue-gray fibers, with those in fine neuropil black in the more successful preparations. Collidine gives brownish impregnations. [Preparations courtesy of Dr. N. J. Strausfeld.]

Fig. 7-9   Plate 7

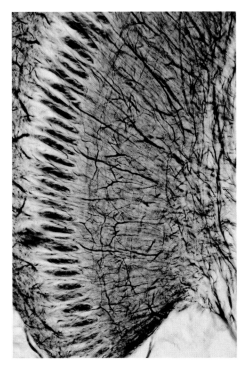

Fig. 7-10

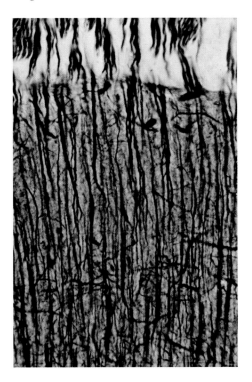

Fig. 7-10: The effect of preceding a lutidine Holmes-Blest impregnation by mercury-cobalt mordanting. The outer medulla of *Apis* shown is refractory to most reduced silver techniques. Typically, this routine yields dark blue impregnations for which the adjustment of toning needs to be very precise.

Plate 8

Fig. 7-11

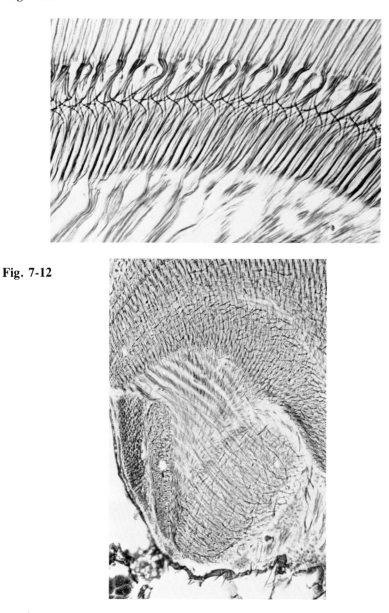

Fig. 7-12

Figs. 7-11–7-14. Shown are the order of resolution obtained from modified Ungewitter impregnations and the differences between results obtained after fixation in Duboscq-Brasil solution and in FBP. **Fig. 7-11:** Vertical section of the lamina of *Calliphora erythrocephala* (Diptera). Basic method with pyridine after fixation in Duboscq-Brasil. **Fig. 7-12:** *Calliphora.* Lobula, lobula plate and proximal medulla, processing as in Fig. 7-11.

Plate 9

Fig. 7-13

Fig. 7-14

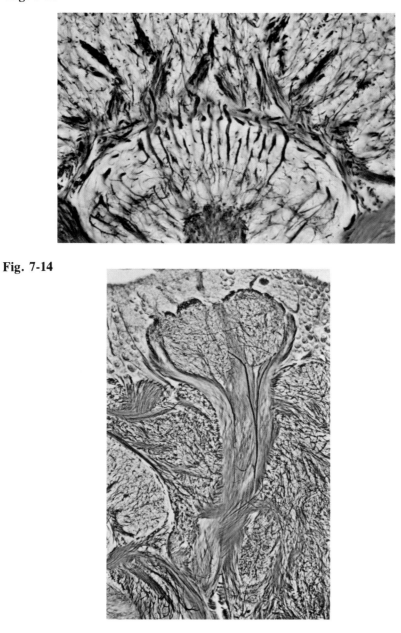

Fig. 7-13: *Calliphora.* Horizontal section of a portion of the central body, following 1-h fixation in FBP at 20°C and the basic method using 2,6-lutidine. Large fibers in tracts are typically red or crimson after this routine, and contrast with fine fibers in neuropil, which are gray or black, and contrast with a clear matrix.
Fig. 7-14: *Calliphora.* Corpus pedunculatum in longitudinal section; method as for Fig. 7-13.

Plate 10

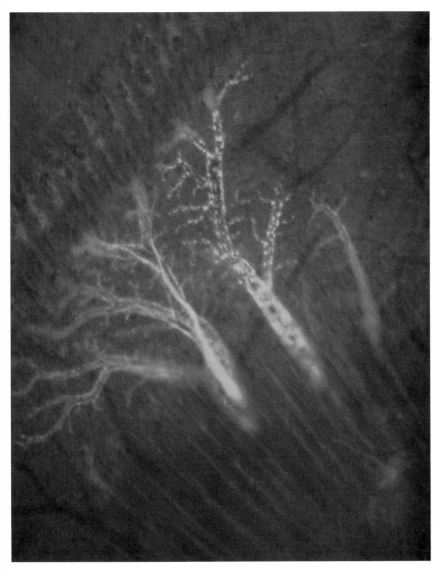

Fig. 14-1. Photograph of a 12-μm frontal section of a "giant" horizontal directionally selective movement detector in the lobula plate of the fly optic lobe. This cell was filled with Procion yellow by the application of 20 nA for approximately 10 s, fixed in FAA, and embedded in paraplast.

Plate 11

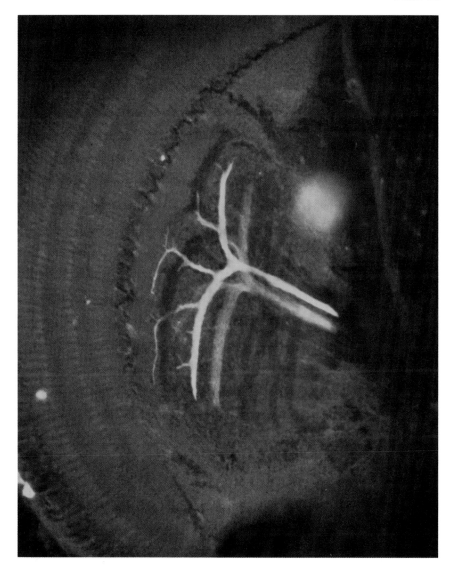

Fig. 14-3. Photograph of a 12-μm frontal section of two large vertical cells in the lobula plate of the fly optic lobe. These cells were injected with Procion yellow separately. Note that unstained cells are outlined by the autofluorescence of the tissue. This effect can be enhanced by fixation in glutaraldehyde. This tissue was fixed in FAA.

Plate 12

Fig. 15-1, A–H. 10-μm frontal sections through the brain and optic lobes of the fly *Calliphora erythrocephala,* and parts of different stained visual interneurons. Scale bars are 100 μm. (A) Fragment of the dendritic arborization of HS1 in the anterior surface layer of the lobula plate. Notice the circular dark zone, where tissue autofluorescence has been partially bleached. (B) Fine dendrites of the neuron H2 demonstrate that even very thin (<1 μm), long, and extensively branched fibers are uniformly stained up to their very ends. Many profiles appear blurred because the depth of focus is less than 0.5 μm at this magnification. (C) Side branches of main dendrites of VS3 neuron appear "spiny." (D) Telo-dendritic arborizations of a VCH-cell are distinctly "blebbed" or "varicose." (E) Crude penetration and excessive injection current result in local swelling and dye leakage. (F) Even in perfect penetrations dye leakage occurs with unduly large injection currents and/or overfilling. (G) Sometimes stainings appear "grainy," i.e., stained clumps adhere to the membrane, whereas the fiber lumen appears empty. Presumably these cells died before fixation. (H) When several fibers in a bundle are stained successively, and the amount of displaced charge is sufficiently different (x, 1 μC; y, 2.3 μC), cells can be correlated with physiological measurements by their respective fluorescence intensity.

Plate 13

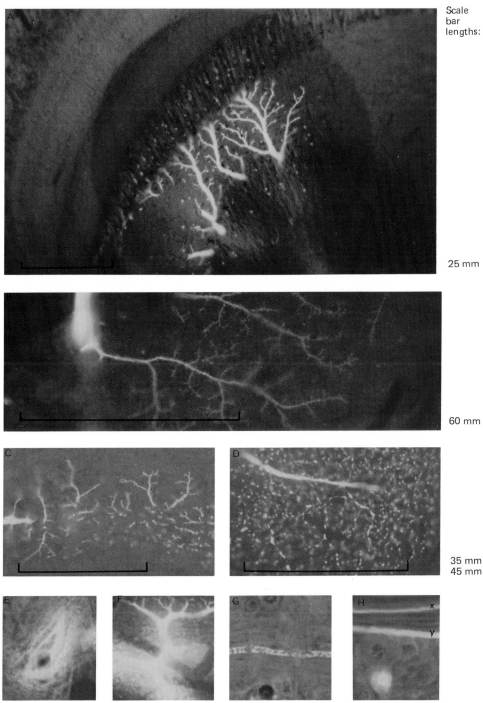

Scale
bar
lengths:

25 mm

60 mm

35 mm
45 mm

Fig. 15-1A–H. Legend on facing page.

Plate 14

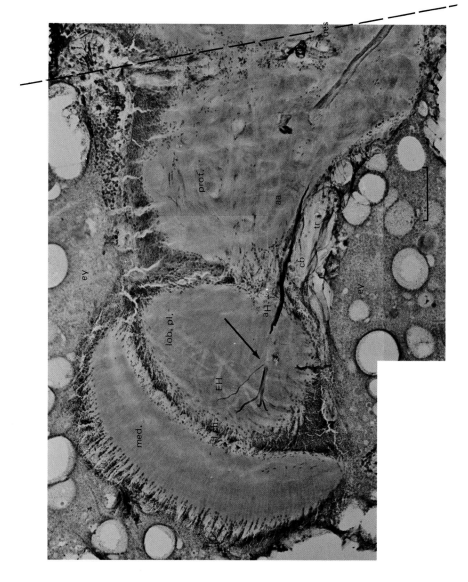

Fig. 16-1. Slightly oblique, frontal section through the brain of the blowfly, *Calliphora*. The retina and first optic neuropil are omitted. The dark brown profiles are parts of HRP-marked neurons belonging to a horizontal cell (*EH*) and the ventral main dendrite of a vertical cell (*V*). Part of the dendrite in the lobula plate, the axon, cell body, and the axonal arborization of the horizontal cell are contained in this section. The cresyl violet counterstain results in a light blue to purplish color of the unmarked neuropil tissue and a dark blue color of the cell bodies of unmarked neurons. The *dashed line* corresponds to the midline of the animal. *aa* axonal arborization; a_H axon of EH; *cb* cell body of EH; *ey* egg yolk embedding medium; *i.ch.* inner chiasma; *lob.pl.* lobula plate; *med.* medulla; *oes.* esophagus; *prot.* protocerebrum; *tr* trachea. The *arrow* points to the injection site. Marker: 100 μm.

Plate 15

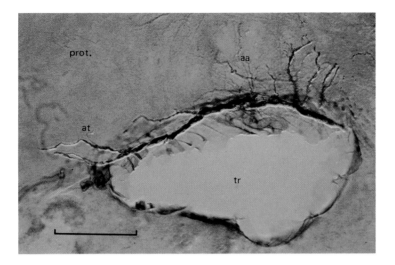

Fig. 16-2. Protocerebral axon part of a horizontal cell showing bulbous endings of axon branches. Interference contrast microscopy. Abbreviations as in Fig. 16-1. Marker: 100 μm.

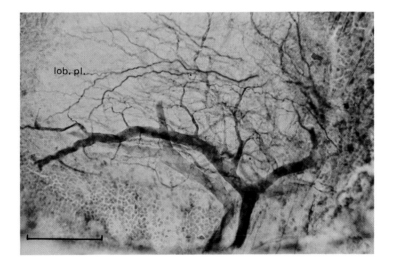

Fig. 16-3. En masse staining of small- to large-diameter cells in the lobula plate. Note the two closely adjacent dendrites of two vertical cells. Note cell bodies of unmarked fibers in the lower left-hand corner. *Calliphora*. Marker: 100 μm.

Plate 16

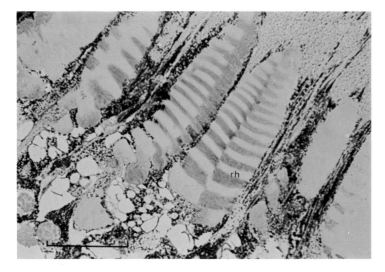

Fig. 16-4. Longitudinal section through the retina demonstrating the extracellular space. Microvillar layers of rhabdom (*rh*) cut perpendicularly to their long axes appear most heavily stained. *Astacus* (Decapoda). Marker: 100 μm. [Courtesy of Drs. W. Schröder, H. Stieve, and I. Claassen-Linke.]

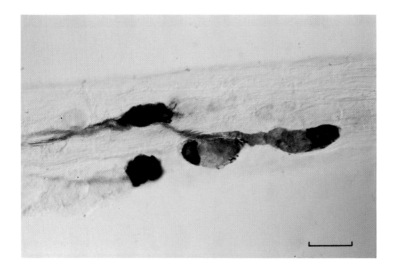

Fig. 16-5. Photoreceptor cells of the ventral nerve of the horseshoe crab, *Limulus* (Xiphosura), marked by intracellular injection of horseradish peroxidase. Marker: 100 μm. [Courtesy of Drs. W. Schröder, H. Stieve, and G. Maaz.]

Chapter 1

The Methylene Blue Technique: Classic and Recent Applications to the Insect Nervous System

S. I. Plotnikova
G. A. Nevmyvaka

The Sečenov Institute of Evolutionary
Physiology and Biochemistry
U.S.S.R. Academy of Sciences
Leningrad, U.S.S.R.

Historical Review

At the end of the last century Ehrlich (1886) discovered that injection of a methylene blue solution into an animal's circulatory system could stain nerve cells and their processes blue. Thereafter histologists began to use methylene blue in preference to any other method of staining, particularly for the peripheral nervous system of insects (Monti, 1893, 1894; Holmgren, 1895, 1896; Rath, 1896; Duboscq, 1898).

Their findings were, however, contradictory since these authors were unable to undertake a detailed examination of their preparations: The dye proved to be unstable and disappeared soon after it was applied.

Dogiel (1902) was the first to develop the technique of fixing and dissecting preparations, which he stabilized in a solution of ammonium molybdate. This method enabled retention of the blue coloration, and the preparations have not lost their quality for three-quarters of a century. Even today these original preparations are used for research at the University of Leningrad's cytology and histology department.

Of all its users, Zawarzin obtained superior results, applying the method to insect cerebral nervous systems. His studies, which were begun in the first decade of this century in Dogiel's laboratory, have made

1

him especially well known among neurohistologists. In 1911 his first work of the series "Histological Studies on Insects" was published under the title "The structure and innervation of the heart of the larval *Aeschna.*" This was followed by "The sensory nervous system of the larval *Aeschna*" in 1912 (1912a), "The sensory nervous system of the larval *Melolontha vulgaris*" in the same year (1912b), "The optic ganglia of the larval *Aeschna*" in 1913, and, in 1924, "The histological composition of the unpaired abdominal nerve in insects" (1924a) and "The abdominal brain of insects (1924b)."

With respect to the study of *in toto* preparations, stained *in vivo* with methylene blue, Zawarzin comprehensively described features of the peripheral nervous system, including the sensory innervation of lip fibrils, the mandibuli, jaws, labium, abdomen, chordotonal organs, and chemosensory apparatus. These studies made the simple distinction between two major classes of sensory cells. The first kind is a bipolar element that lacks peripheral ramifications and simply invests or inserts into the base of the sensory apparatus, whereas the second kind gives rise to apical processes and forms tree-like terminal ramifications. The latter are almost always located at articulations, where they innervate the hypoderma of the membrane, that is, the softest and most delicate part of the chitinous skeleton.

The usefulness of the methylene blue technique for staining insect nervous system was, however, best demonstrated by Zawarzin's studies on the abdominal chain of the larval *Aeschna,* which he began in 1914 and worked on for the following eight years.

The characteristic feature of the methylene blue method is that in each preparation only single elements of the whole ganglionic mass can be identified. In order to collect his results Zawarzin systematically recorded every element in all the ganglia that had taken up the stain. He claimed that usually no novel elements were revealed after the 400th preparation. In all, Zawarzin prepared and examined ventral cords of about 750 animals and ten times that many ganglia.

He found that neuronal elements that belong to the same ganglion are characterized by their astonishingly constant forms and numbers, and because of this, specific cell types can be easily recognized, even in preparations that are only partially stained. One essential feature of a neuron is the position of its cell body and processes. Zawarzin carefully described the position of each single element in the horizontal and dorsoventral planes and so revealed the precise topography of neurons, from which he was then able to speculate about their possible functional significance. He showed, for example, that there was a dorsoventral differentiation within the ganglia. This comprised special loci for inter- and intraganglionic conduction pathways that could be distinguished from sensory and motor regions and the central bulk of neuropil.

Zawarzin described the neuronal composition of the ganglia more comprehensively and more precisely than any other contemporary author, and it is claimed that for some ganglia he identified more than 75% of the constituent neurons. Although his extensive and complex investigations recognized cardinal features of each thoracic and abdominal ganglion, his publications describe only two of these in detail, namely, the second (thoracic) and fourth (1st abdominal) ganglia. It was presumably his intention at some stage to describe the others but he was unable to complete this study. This work has been continued in Zawarzin's tradition by Tsvileneva, who has since published several extensive works on this and other arthropod nervous systems (see, for example, Tsvileneva, 1950, 1951, 1970).

Zawarzin's studies on the insect visual system precede descriptions of the ventral nerve cord. At that time little was known about the interrelationships between various elements of the visual neuropil, and terminology was not even defined. Zawarzin referred to the three neuropil masses that subserve the retina as the first, second, and third optic "ganglia." [1] By using both reduced silver techniques and methylene blue preparations, he was able to reveal both the general structures of these neuropils and the course of many single fibers between them.

Zawarzin was also the first to make detailed comparison between the fiber interrelationships in the visual systems of insects, vertebrates, and cephalopods and to draw specific analogies between them. He was struck by the similarity of organization between the visual neuropils of these three groups of animals, and although some similar comparisons had attracted the attention of other authors, such as Radl (1912), their studies did not take into account specific neuronal arrangements and were derived from gross preparations alone.

Zawarzin's analysis, however, demonstrated that throughout the visual system cell processes are organized as stratifications in which synaptic specializations ought to be located. He proposed that these layered structures should be termed "screening centers" and that they were invariably located at the same geometric position in the neuropil: Similar elements were arranged across each layer.

He suggested that the structure of the retina, as well as that of the cerebral cortex, midbrain, and cerebellum of all higher vertebrates, contains layered arrangements that correspond to those found in the insect visual system. Zawarzin then proposed that many centers, including the cortex, may find reference to a model system of development represented by the stratified vertebrate retina.

---

[1] The first, second, and third optic ganglia are, respectively, the lamina, medulla, and lobula, by modern terminology. Zawarzin used the term "ganglion" for the synaptic part of the nervous system, not considering whether this part is anatomically similar to the segmental ganglia. [Eds.].

As a consequence of such comparative studies he formulated the hypothesis of parallelism in neurohistologic structures, and he went on to find support for his thesis by investigating the thoracic-abdominal nervous system of *Aeschna*.

In retrospect, the ganglia of *Aeschna* seem to be an ideal choice that is made even more logical if one considers that both vertebrates and insect cords supply axons to striated musculature that moves jointed limbs and appendages.

In comparing details of the abdominal nervous system of larval *Aeschna* with the spinal cord of vertebrates, Zawarzin found that these systems are surprisingly similar. In vertebrates, the gray matter of the spinal cord, containing the majority of the dendrites and terminal ramifications, is located as a core lined by a peripheral sheath of axon fasciculi that projects down the cord (its white matter). The griseum of the cord is further subdivided into characteristic regions of neuroarchitectures that contain specific types of neurons. In insects, peripheral areas of the thoracic and abdominal ganglia are occupied by arborizations from dorsal and ventral regions that Zawarzin claimed to be exactly analogous to dendritic and terminal ramifications of elements that pass to, and project from, the spinal cord. Zawarzin's thesis also suggests that the central portions of the insect ganglia are occupied by elements that are entirely analogous to those in vertebrate gray matter.

The conduction pathways of the abdominal ganglia of insects are roughly grouped above and below the central neuropil and form dorsal and ventral groups of commissural fibers. An exception is found in insects whose neuropil is less profoundly compartmentalized into discrete regions, such as in some Orthoptera. In these species some longitudinal fasciculi invade central neuropil regions. However, as far as their relative locations are concerned, dorsal and ventral commissural bundles correspond to the white matter of the spinal cord. Zawarzin went on to state that structural similarities between insect ganglia and spinal cord became even more evident with respect to the arrangements of single neurons. In both insects and vertebrates, motor neurons are generally located at specific superficial regions that are geometrically opposite to those containing sensory terminals. Neurons of the central neuropil regions lie between these. In insects, ventral neuropils of ganglia receive sensory fibers and comprise sensory neuromeres, whereas dorsal neuropils comprise the motor roots. This is topologically identical to the situation in vertebrates but rotated through 180°. Zawarzin makes the simple analogy between the ventral horns of the spinal cord and dorsal effector (motor) roots of insect ganglia and, likewise, the ventral roots of the insect ganglia and the dorsal horns of the spinal cord. Analogies can be further extended to include neuropil subdivisions between these areas.

Zawarzin published these ideas in 1925, setting down his concept of principles of parallelism of tissue structures. Comparison of insect and vertebrate nervous systems led him to the conclusion that there is an evolutionary tendency for congruence of neural organization in nervous tissue that subserves functionally analogous organs. However, similarities between nervous systems depend on the level of their phylogenetic development, so comparisons need to be made between animals that are similarly equipped with appendages, and so forth. Even though the principal structure of the brain stem, or ventral cord, shows specific features that persist in both archaic and recent forms, the associated neuropil structures become more complex the greater the number and repertoire of effector organs and the greater the number of central neurons.

At the time when Zawarzin was formulating these concepts, he also turned his attention to the structure of the unpaired ventral nerves of insects, which at that time was regarded as the analog of the vertebrate sympathetic nerve. Using the methylene blue technique Zawarzin demonstrated that this nerve was the major stem of the vegetative nervous system, and he demonstrated the cellular composition of the unpaired ventral nerve as well as its course between effector organs and centers in the ganglia.

Zawarzin's colleague, Orlov, continued these studies on several arthropods. His remarkable work on the structure of the stomatogastric system of the larval coleopteran *Oryctes nasicornis,* and of the crayfish, wholly revised contemporary knowledge about vegetative nervous systems in arthropods. It is fair to say that Orlov's investigations did much to contribute to a clear understanding of sympathetic nervous systems in general, including those of vertebrates. His works between the years 1922 and 1930 (Orlov, 1922, 1924, 1925, 1927, 1929, 1930) do much to demonstrate the beauty of the methylene blue procedure. It is worthwhile listing a short summary of the kinds of studies that have been inspired by, and are derived from, Zawarzin and his pupils.

Phylum Platyhelminthis; class Turbellaria (Kolmogorova, 1959, 1960)

Phylum Annelida; class Polychaeta (Fominych, 1977; Lagutenko, 1978); class Oligochaeta (Newmywaka, 1947a,b, 1948, 1952, 1956, 1966; Lagutenko, 1977), class Hirudina (Lagutenko, 1974, 1975a,b, 1976).

Phylum Arthropoda; class Arachnida (Tsvileneva, 1964); class Crustacea (Tsvileneva, 1970, 1972; Fedosova, 1974; Tsvileneva *et al.,* 1975, 1976; Tsvileneva and Fedosova, 1975); class Insecta (Rogosina, 1928; Plotnikova, 1949, 1967a,b, 1969, 1975, 1977, 1978, 1979; Tsvileneva, 1950, 1951, 1970; Knyazeva, 1969, 1970a,b, 1976; Knyazeva *et al.*, 1975; Svidersky, 1967, 1973; Karelin, 1974). The last two mentioned works are methylene blue studies in conjunction with electrophysiology.

Studies by Zawarzin and his immediate co-workers, which set down his views on the evolution of the nervous system ("Survey of the evolutionary histology of the nervous system"), remain unknown to many biologists because this book was first published in Russian in 1941 and reprinted once in 1950.

Methods

Zawarzin's method for *intra vitam* staining with methylene blue has the reputation of being a difficult process that requires persistence and patience.   The present authors usually are unable to achieve more than a 20% success rate, and often the staining process must be repeated in order to obtain neurons that are distinctly stained.   In our experience acceptable results depend on several rather abstruse and difficult-to-define factors such as the season and, thus, the developmental stage of the insect. Larval *Aeschna* are best stained during the spring, summer, and early autumn when their metabolism is particularly intense.   However, "domestic" insects that live in constant environmental conditions, such as the cockroach, are susceptible to methylene blue staining throughout the year.

The simplest staining method is to inject the methylene blue solution into the body cavity through a microsyringe.   The fluid should be injected under pressure so that the body distends.   Zawarzin injected up to 1.5 ml of the solution[2] into the body of the dragonfly larva.   We inject as much as 2.5 ml (minimum 1.5 ml) into male and female locusts, respectively. To avoid the solution leaking out of the puncture made by the syringe needle, we inject the dye into an extremity of the meta- or mesothorax and then apply a ligature.   To resolve elements of the abdominal chain, final dissection is made between 1 and 1.5 h after initial injection.   To resolve the peripheral nervous system, final dissection is made between 0.5 and 1 h after injection.   During this time animals are kept at a temperature of 20–25°C.

Staining conditions must be varied so that they are appropriate for the species of insect and the tissue that is to undergo examination.   Variations of the duration of the staining process and temperature are also relevant, as are the concentrations of the methylene blue and sodium chloride solutions.

Staining does not occur synchronously in all neurons.   The first elements to be revealed are those that reside near the neuropil surface or

---

[2] In none of the classic accounts is the exact concentration of methylene blue solution given, possibly because the compound stain from different sources contains various amounts of the dye [Eds.].

those that are not specially protected by the neurolemma sheath. Neurons at some depth in the tissue are stained next. If the duration of staining is extensive, elements that have already taken up the stain may bleach. We observe that whenever the central regions of the nervous system have taken up the stain, peripherally placed elements can be barely detected and vice versa. It is thus necessary to make a great number of preparations, using a wide range of staining periods.

To stain the nervous systems of adult insects, it is advisable to treat them with methylene blue within the first 48 h after the last molt or eclosion. After the required staining period, the neural tissue is fixed in a solution of ammonium molybdate.

The Staining Schedule

1. Prepare a 0.25% or 0.5% solution of methylene blue in double distilled water. This is effectively a saturated solution that should be warmed, but not boiled, in order to achieve maximum solubility. Of great importance is the quality of the dye. If the results are unsatisfactory one may try other brands.
2. Prepare a 4% solution of sodium chloride in double distilled water.
3. Prepare a 12% solution of ammonium molybdate in distilled water and warm this in order to dissolve it completely. Then filter and cool.

The stock solutions should always be prepared on the day of their use, and receptacles and instruments should be perfectly clean.

The concentrations of methylene blue and NaCl in the working solution can be varied. Zawarzin used solutions with a methylene blue concentration of 0.08% or 0.16% for the peripheral nervous system and 0.8% up to 0.125% for ganglia. The concentration of NaCl was 0.75%. It is also possible to use an appropriate Ringer's solution. To stain elements of the Orthoptera central nervous system we used a solution whose methylene blue concentration was 0.06% and NaCl concentration was 2.25%; to stain the peripheral nervous system we increased the concentration of methylene blue and decreased the concentration of sodium chloride. To reveal sensory elements that are enclosed by a cuticular sheath, some authors use a methylene blue concentration of as much as 1% in distilled water.

Increasing the concentration of methylene blue leads to coarser staining and decreases the distinct resolution of single fibers. Raising the concentration of NaCl gives rise to deformations of cell bodies.

Subsequent dissection after staining must be cursory: To expose the abdominal ganglionic chain, the time for dissection, before fixation, should not exceed half a minute. If dissection demands a longer time than this, then the animal should be cut open, fixed in toto, and the final dissection can be later made after fixation, under distilled water during the washing procedure.

Fixation is carried out in a solution of ammonium molybdate. We recommend that this be done in a refrigerator set at 4°C for 7 to 16 h depending on the size and density of the neuropil. A diluted solution of osmium tetroxide (2–4 drops of 0.25% OsO_4 solution in distilled water added to 50 ml ammonium molybdate) improves fixation and assists the resolution of axons and their ramifications. However, in our experience, osmium tetroxide also has some negative effects on the staining of some neural components. After fixation and dissection the preparation is suspended in a glass flask and washed thoroughly with distilled water, after which any further dissection is completed. Dissected ganglia are then carefully spread out on a coverslip, or better, on a thin sheet of translucent mica, and excess water is gently removed with a strip of filter paper. The object, spread out on the glass or mica support, is then dehydrated in three changes of absolute alcohol (surplus fluid is also removed by filter paper). The preparation should not be kept longer than half a minute in the first change of alcohol, because nonabsolute alcohol extracts methylene blue. The preparation is then placed in xylene until it is entirely cleared, whereafter it is mounted in dammar-xylol, which, in the case of thick preparations, should be viscous. It is necessary to add dammar-xylol regularly beneath the cover glass in the case of large preparations. On no account should Canada balsam be used as the mounting medium, because this will cause the blue pigment to bleach out. It should be noted that preparations that have been insufficiently washed after fixation gradually become opaque.

Recent Applications of the Method

Needless to say, knowledge of the technique is not the only requirement for obtaining meaningful results, and total preparations can be intelligently analyzed only when some general principles of ganglionic organization are realized. Some examples of these are their divisions into strata and regional zonations, and we are obliged to refer to, and recommend, Zawarzin's classic accounts (1924a,b, 1941) based on the larval dragonfly *Aeschna*. These works will doubtlessly be of value for future studies.

In our experience whole-mount preparations of methylene blue-treated ganglia are not susceptible to photography at higher magnifications simply because of their considerable thickness. To study the most slender ramifications and to map out their positions, it is necessary to reconstruct elements in total preparations through their various levels so as to distinguish many similarly colored neurons that would otherwise be obscure in photomicrographs. In practice, all nervous elements in any given part of the neuropil have to be reconstructed by a camera lucida device, after which they are mapped onto a standard outline of the ganglion. One such example is shown in Fig. 1-1 which depicts neuronal elements of the motor and sensory neuropils of the prothoracic ganglion of *Locusta migratoria* (L.).

Figure 1-1 gives a general impression of the location of juvenile motor center nerves 1, 3, 4, and 6 in the first ganglion (Fig. 1-1; *N1, Nm, N3,*

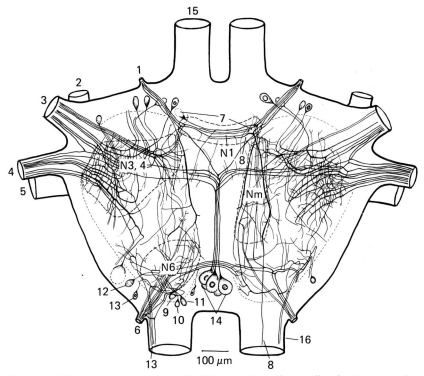

Fig. 1-1. Motor area of the neuropil of the prothoracic ganglion in *Locusta migratoria* L. Methylene blue. *1*, first nerve pair; *2*, sensory roots of the second nerve pair; *3*, anterior; *4*, posterior motor roots of the second nerve pair; *5*, 3rd nerve pair; *6*, nerve bundle intermingling with the first nerve of the mesothoracic ganglion; *N1* = motor nucleus of the first nerve pair; *N3, 4* = motor nucleus of the second nerve pair; *Nm* = medial motor nucleus; *N6* = motor nucleus of the sixth nerve; *7, 8* = neurons whose axons pass to the periphery via the first nerve; *9–12* = neurons whose axons pass into the sixth nerve; *13*, a neuron whose axon passes first to the connective and then into the first nerve of the mesothoracic ganglion; *14*, mediodorsal neurons; *15*, anterior; *16*, posterior connectives.

N4, N6), omitting the nucleus of the fifth nerve, which lies more ventrally in the neuropil and is not resoluble with the others depicted here.

Axons of neurons that are identical to those described by Neville (1963; see also Bentley, 1970; Tyrer and Altman, 1974) as motor neurons of the dorsal longitudinal muscles, project to the periphery via the first nerve (Fig. 1-1, *7*), and similar neurons exist in all the ganglia of the abdominal chain. Shown with these elements is a ramifying neuron whose axon projects to the periphery in parallel with the first nerve and whose secondary process projects through the posterior connective into the mesothoracic ganglion (Fig. 1-1, *8*). Axons of neurons are shown that pass through the sixth nerve but that also intermingle with the first nerve of the mesothoracic ganglion (Fig. 1-1, *9, 10, 11, 12*). We are also able to identify with methylene blue a neuron whose dendritic ramifications are located in a fashion identical to nerve cells described by Neville (1963), Bentley (1970), and Tyrer and Altman (1974), which also innervate dorsal longitudinal muscle (Fig. 1-1, *13*) via the first nerve. Also illustrated in Fig. 1-1 are ramifications of dendrites of nerve cells, destined for both the first and sixth nerves, which form a discrete region of neuropil in the medial part of the ganglion (Fig. 1-1, *Nm*). Likewise, neurons of the third and fourth nerves form a common nucleus (neuropil region or center) whose dendritic ramifications are located laterally in motor neuropil.

A further group of neurons, derived from particularly large perikarya, are located mediodorsally in the posterior zone of all the thoracic ganglia (Plotnikova, 1969; see also Fig. 1-1, *14*), Their processes are directed anteriorly, and they bifurcate to give rise to characteristic patterns of short dendrites: Between 2 and 4 such axons enter into the first pair of motor nerves or into the symmetric third and fourth nerves. Seabrook (1968) was the first to describe a neuron with this kind of ramification in the last abdominal ganglion of the locust. Later these neurons were described by Crossman *et al.* (1971, 1972). The present account briefly describes only a few of the neurons in the ganglia; those remaining are shown in the illustrations.

Figure 1-2, which is based upon many methylene blue preparations, depicts sensory neuropil of the metathoracic ganglion of the locust, which consists of one thoracic and three abdominal neural segments. Sensory fibers enter through the first pair of nerves of the third (*7 III*) and fourth (*7IV*) segments (*9*) where they are directed into frontal connectives. In each segment they give rise to collateral branches. Sensory fibers enter the second pair of nerves and pass medially through the ganglion. Sensory fibers derived from the tympanic organ enter the ganglion through the first nerve of the second neural segment (Fig. 1-2, *10*) and terminate in a frontal expansion of the neuropil, here termed the "first sensory nucleus" (Fig. 1-2, *11*). Terminal fibers of the first nerve, derived from the wing,

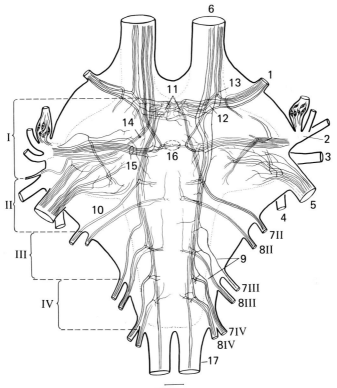

100 µm

Fig. 1-2. Sensory area of the neuropil of the metathoracic ganglion in *Locusta migratoria* L. Methylene blue. *I*—first, *II*—second, *III*—third, *IV*—fourth neural segment; *1–5* = segmental nerves of the first neural segment (the same terms as in Fig. 1-1); *6*, anterior connectives; *7II*, first (acoustic); *8II*, second nerve of the second neural segment; *7III*, first; *8III* second nerve of the third neural segment; *7IV*, first; *8IV*, second nerve of the fourth neural segment; *10*, cord of sensory fibers of the tympanal organ; *11*, first sensory nucleus; *12, 13* cords of sensory fibers of the first nerve; *14, 15*, cords of the sensory root of the second nerve; *16*, second sensory nucleus; *17*, posterior connectives.

enter into the same region, and some of the fibers in fact end here (Fig. 1-2, *12*), whereas others give rise to collateral branches that project into the mesothoracic ganglion (Fig 1-2, *13*). There they enter into an equivalent nucleus and finally terminate. Some sensory fibers of the second nerve also invest this region: Fibers of one strand of the second nerve project as collaterals into the frontal connectives (Fig. 1-2, *14*), whereas others pass posteriorly into interganglionic connectives (Fig. 1-2, *15*). Collateral branches from these fibers form the second sensory nucleus (Fig. 1-2, *16*), whereas the sensory nuclei of the fifth nerve project to a separate location in the neuropil.   To avoid overloading the figure, some of these fiber pathways are depicted only on one side of the ganglia and some major cord tracts are omitted.

The methylene blue method occasionally reveals many neural elements within the same region of the neuropil.   Figure 1-3 illustrates such an example: In this illustration two tracts of fibers are distinctly identified, medially and laterally (Fig. 1-3, *1,2*).   They appear to diverge from the supraesophageal ganglion and can be distinctly resolved in all three thoracic ganglia due to the characteristic branching patterns of their collaterals. Methylene blue reveals a prominent fiber that we identify as being identical to the descending contralateral movement detector axon (DCMD), which was first identified by cobalt iontophoresis in *Schistocera gregaria*

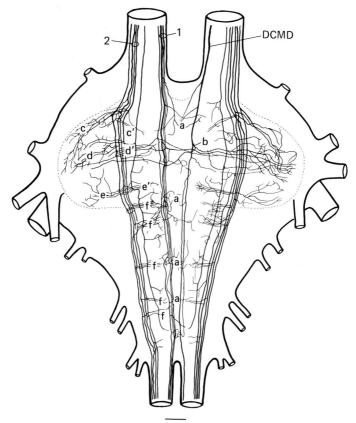

100 μm

Fig. 1-3. Dorsal connective area of the neuropil of the metathoracic ganglion in *Locusta migratoria* L. Methylene blue. *1*, medial fiber bundle; *a*, medial; *b*, great lateral collateral branches; *DCMD*, descending contralateral movement detector; *2*, lateral fiber tract; *c*, *c′*, *d*, *d′*, *e*, *e′*, *f* collateral branches.

by O'Shea *et al.* (1974) whereby its relationship with a number of more peripheral motor neurons (among them inhibitory neurons) was identified.   These also included its excitatory relationship with the fast extensor tibia neuron, and the anterior, common, and posterior inhibitor neurons (Burrows and Rowell, 1973).   It is unlikely that the DCMD neuron is exclusively connected to these latter elements, and we have been able to resolve several other branches (Fig. 1-3, *a*) that terminate alongside medial branches of thin fiber tracts in the dorsal zone of central neuropil where together they form common centers.

Lateral tracts to the thoracic ganglia give rise to collateral branches that extend medially into the ganglia and outward toward the edge, in particular within the anterior, mid, and posterior regions of the pro- and mesothoracic ganglia and within the first thoracic neural segment of the metathoracic ganglion (see Fig. 1-3, *c, c', d, d', e, e'*).   Descending lateral tracts that pass into abdominal ganglia only give rise to lateral, rather than medial, collaterals (Fig. 1-3, *f*).   It seems from these distributions that descending "command" elements may invest a great number of motor neurons.

Further Developments of the Method

Methylene blue can provide an overview of the structures in ganglionic neuropil where it is possible to distinguish sensory, motor, and interneurons and to reveal some possible candidates for reflex arcs that may be associated with appendages.   However, it is hardly possible to attach specific integrative roles to ascending and descending pathways since it is not yet possible to differentiate neuronal subsystems that mediate a particular function.   For this we depend upon the neurophysiologist who is better equipped to obtain insight into the function of a single cell by combined electrophysiology and structural identification.

However, it is common knowledge to the users of the present method that when the nervous system is excited by an electric current some neurons take up the stain better than others.   For example, when an electric current is passed through a crustacea's ventral cord, prior to staining, the subsequent uptake by neurons of methylene blue was enhanced (Tsvileneva *et al.*, 1976).   Similar experiments have been attempted on insects, and Karelin (1974) attempted to inject methylene blue into the stump of the dorsaltegumentary nerve by electrophoresis.   Although collateral fibers of this nerve were not revealed, the main fibers could be traced over considerable distances.   Further improvements of staining have also been achieved by adjusting the methylene blue solutions to achieve a pH and ionic balance that corresponds to the internal milieu of the tissue.

The ideal collaboration for the morphologist is to work closely with electrophysiologists that employ single-cell identification with, for example, cobalt chloride iontophoresis. However, structural studies with methylene blue at least give us some topographic information about arrangements in neuropil that serve as a basis for functional study. By combining these methods physiologists have, for example, revealed the course of wind-sensitive receptors on the locust's head, which are presumed responsible for maintaining the animals flight, to neurons of the thoracic ganglia that directly control wing muscle activity (Svidersky, 1967, 1973). In this study the sensory cells were revealed by methylene blue staining (Svidersky and Knyazeva, 1968) and the pathway of these cells through the dorsal tegumentary nerve was traced by iontophoretically injected cobalt chloride, Procion yellow, and methylene blue (Karelin, 1974). This combination of methods we like to term "electrophysiologic anatomy" (see also Svidersky and Varanka, 1973; Varanka and Svidersky, 1974a,b). The general practice is thus to examine topographic relationships of whole-mount preparations with methylene blue and, once having prepared a schema of their identification, to examine them by electrophysiologic techniques.

In summary, it is fair to say that the method of intra vitam methylene blue has by no means lost its importance even though new techniques have since been developed for studying the structure of neuropil. Combinations of the methylene blue technique and novel methods of investigation are a prerequisite for obtaining optimal results.

The original formula specified "Santomerse No. 3" (Monsanto Chemical Company, 800 N. Lindbergh Blvd., St. Louis, Missouri 63166) as the detergent, but "Teepol" (Shell Chemicals) has proved to be a satisfactory substitute. Probably many household liquid detergents would work equally well. The amount does not seem to be critical, as it acts simply as a wetting agent. Without it the stain penetrates only into the large tracheal trunks in the cockroach and not into the tracheae of the CNS. The other reagents have been of analytical grade purity wherever possible. The solubility of barium chloride ($BaCl_2 \cdot 2H_2O$) is 35.7 g/100 ml "cold" water, or 58.7 g/100 ml "hot" water (Weast, 1975).

Practical Procedure

The dye solution is placed in a suitable vacuum-resistant flask or bottle having a tube through the stopper fitted with a stopcock and connected to some form of vacuum pump. Some means for the readmission of air is also needed. This can be either a second tube and stopcock fitted through the stopper, as in Hagmann's original description, or, more simply, a three-way stopcock inserted in the vacuum tubing or an air-bleed screw integral with the vacuum pump. In Wigglesworth's (1950) technique a more elaborate apparatus for filling the tracheae is described. Sufficient dye solution must be used to completely submerge the insects to be stained. It should be degassed before use, to prevent it from bubbling too much subsequently. Whole, live insects are killed with chloroform (probably other killing agents would be equally suitable) and suspended in the flask above the dye solution by a thin but fairly stiff wire, the end of which is gripped in the partly opened exit stopcock, so that air can still pass through. They must be weighted in some way to prevent them from floating when they are later dropped into the dye solution. Small insects can be enclosed in some form of cage or basket, as suggested by Hagmann, but larger ones, such as cockroaches, can be simply weighted with, for instance, a strip of stainless steel gauze wrapped loosely round the body.

The air is then evacuated from the flask, though not to the extent that the dye solution bubbles excessively—a vacuum gauge is helpful in controlling this—and the vacuum maintained for a time. For insects the size of adult cockroaches, 20 min is usually about right, though smaller insects may require a shorter time. The exit stopcock is then opened fully, releasing the end of the wire and allowing the insect or insects to drop into the dye solution. The vacuum is held for a further 5–10 min, after which air is allowed to reenter the flask gently. In this way the atmospheric pressure forces the dye solution into the tracheal system. Staining at atmospheric pressure should be continued for a time—about 15 min for adult cockroaches—after which the insect is withdrawn from the flask by the attached wire. It can be blotted briefly to remove excess dye from the body surface and is then placed in the fixative.

Large insects, such as cockroaches, are best injected with some of the fixative first, using a hypodermic needle inserted between the abdominal segments, until the body becomes slightly distended, before they are immersed in the fixative. They should be weighted down in some way if they tend to float. Fixation can conveniently be completed overnight at room temperature (20–25°C), though smaller insects may not need to be fixed for as long as this.

After fixation the insect is washed for 1–2 min in distilled water, and the appendages—legs, wings, and antennae—can usefully be removed. The body is then transferred to a low grade of ethanol, say 20%, in which it can be dissected and the required parts of the CNS removed to a bath of 50% ethanol. Here unwanted tissue such as fat body can be cleaned away. Dehydration is continued through 90% and absolute ethanol, about 10 min in each being sufficient for cockroach ventral nerve cords. The ganglia can be further cleaned up in the 90% ethanol bath, peripheral nerves being cut short and tracheae removed from the outside of the nerve cord if desired.

The cord can finally be cleared in xylene or a similar agent. It can then be mounted in balsam, using glass or celluloid strips to support the cover glass if necessary, to yield a permanent whole mount. Alternatively, it may be stored, according to Hagmann, for an indefinite period in either 70% ethanol, xylene, cedar oil, or clove oil; or it may be infiltrated with paraffin wax and sectioned in the usual way, the sections being dewaxed and either covered without further treatment or lightly counterstained— with, say, eosin, or some other suitable dye.

Results and Comments

The tracheal branches within the CNS are stained an intense dark blue, with the background completely clear if no counterstain has been used (Fig. 2-1). The degree of staining of the tracheoles has not been investigated to any extent, because it is the larger branches that are of importance as neuroanatomical landmarks, but in cockroach ganglia a vast number of very fine branches, down to 1 μm or so in diameter, are filled. The method thus shows the ganglia to be very highly tracheated.

The chief weakness of the technique is the relatively poor preservation of the surrounding nerve tissue, the quality of which does not approach that given by alcoholic Bouin (Duboscq-Brasil) (see Chap. 6 of this vol-

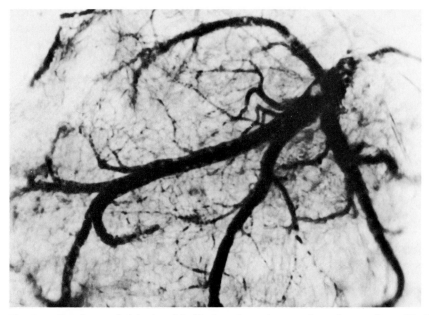

Fig. 2-1. Tracheae of right-hand half of cockroach mesothoracic ganglion, filled with trypan blue.   Ventral view of whole mount, midline at bottom, anterior at right.   × 152.

ume).   A better method of fixation would therefore be a useful improvement.   However, for general observations on the arrangement of the larger tracheal branches the method as it stands is very effective.   By its nature, of course, the procedure cannot be used, as noted by Hagmann himself, to fill the closed (apneustic) tracheal systems of many aquatic and endoparasitic larval forms (such as in the Ephemeroptera, Odonata, Trichoptera, Diptera, Hymenoptera, and Coleoptera).   However, the method has the advantages over other techniques of being reliable, simple, and relatively quick to perform, and requiring only easily obtainable reagents.   For the purpose described here it can be thoroughly recommended.

Chapter 3

Toluidine Blue as a Rapid Stain for Nerve Cell Bodies in Intact Ganglia

J. S. Altman

The University of Manchester
Manchester, England

The distribution of nerve cell bodies can be demonstrated in whole insect ganglia by staining with toluidine blue, making lengthy sectioning and reconstruction procedures (e.g., Cohen and Jacklett, 1967) unnecessary. This rapid and reliable method was developed as an adjunct to cobalt chloride staining of neurons (see Chaps. 17 and 19–21) for examining the relationship of cobalt-filled neurons to other nerve cells (Altman and Bell, 1973). It is also useful for locating positions of somata for intracellular recording, for comparative studies of neuron numbers and distribution, and for developmental work. It is excellent for preparing demonstration material for teaching and for practical classes in neurobiology.

Stain

1 g   toluidine blue
6 g   borax ($Na_2B_4O_7$)
1 g   boric acid
100 ml distilled water

pH should be 7.6–9.0

Freshly dissected ganglia are immersed in warmed stain. Staining time depends on the temperature, the age of the stain, and the organism, particularly on the size of the ganglion, and the optimum must be determined experimentally for each species. With fresh stain, 15 min in an oven at 50°C is a good starting point but if the time is too long or the temperature too high overstaining will result and it will be impossible to differentiate completely. Alternatively, ganglia can be placed in cold stain and left on a hot plate at 50°C for up to 1 h. Stain that has been aged for a month or more appears to give a crisper result.

Differentiator and Fixative

Bodian's No. 2 fixative (Bodian, 1937):

> 5 ml formalin
> 5 ml glacial acetic acid
> 90 ml 80% ethanol

Transfer the ganglia directly from the stain to the fixative, which should be changed several times in the first 5 min as it becomes saturated with stain. At this point the ganglia appear dark blue with a metallic sheen. Continue to differentiate until the cell bodies are clearly visible under a dissecting microscope, and nerve roots and neuropil areas are almost white. Large ganglia, such as those of the locust, may take 15–20 min to reach this stage, depending on the intensity of the initial staining, but ganglia from small insects such as Diptera and Hymenoptera may only require 5 min for complete differentiation. Some stain is lost during dehydration, so differentiation should be stopped just before the desired appearance is reached.

Dehydration and Mounting

Dehydrate in two changes of 90% ethanol (5 min each), and two changes of absolute ethanol (10 min each). Clear in methyl benzoate and mount in neutralized Canada balsam.

Results

Cell-body cytoplasm stains bright blue and the nucleus pale blue, with a very dark nucleolus. All other tissue should be colorless (see Fig. 3-1,

Plate 1). Both neuron and glial cell bodies will be stained by this procedure but the glial cells, being smaller, lose their stain more rapidly than the neurons and so can be eliminated by staining for a shorter time and differentiating longer. Similarly, large neurons can be demonstrated selectively by manipulating staining and differentiation times.

An alternative method is to stain after fixation (Levi-Montalcini and Chen, 1971; Truman and Reiss, 1976), but in my experience differentiation is not so effective on prefixed material, and the ganglia lack the clarity of material stained before fixation.

Special Applications

1. *Toluidine blue staining of cobalt-filled ganglia.* Ganglia are treated with ammonium sulfide to precipitate cobalt sulfide (Chap. 19) and washed well with saline solution before toluidine blue staining. Because ammonium sulfide is a reducing agent, the stain is initially decolorized and the cell bodies appear blue only during fixation and dehydration. This makes differentiation somewhat difficult to control, and after dehydration it may be necessary to return to fixative for further differentiation.

2. *Species with pigment in the ganglionic sheath.* In some species of Orthoptera, the details of cell bodies in whole mounts are obscured by heavy pigment deposits in the ganglionic sheath. This may be bleached before toluidine blue staining by a 5% sodium hypochlorite solution or a dilute solution of a commercial bleach (e.g., Clorox or Parazone) until it is no longer visible. This bleach too has a reducing action, and the same problem with differentiation is encountered as in (1) above.

3. *Toluidine blue as a block stain for wax sections.* Ganglia are stained by the usual method but differentiation is stopped well before the end point, because stain tends to be lost during subsequent processing. The specimen is transferred to fresh fixative with acetic acid omitted (5 ml formalin, 95 ml 80% ethanol) for 1–2 h, dehydrated, wax embedded, and sectioned conventionally. Sections can be counterstained with eosin, to give a result similar to a hematoxylin-eosin stain, or with the same toluidine blue stain heated on the slide for a few seconds, followed by brief differentiation in

> 2.5 ml formalin
> 2.5 ml glacial acetic acid
> 95 ml 80% ethanol

This has a metachromatic effect, staining cell body cytoplasm greenish-blue, nuclei pale blue, nucleoli greenish-yellow, and neuropil pinkish-purple. It gives much better resolution of neuropil structure than counter-staining with eosin. Wax sections of unstained ganglia can also be stained with toluidine blue by this method to give a useful, though less dramatic, general picture of neuropilar organization with reasonable definition of larger fibers.

Acknowledgment

Much of the original work on this method was done by my collaborator, Elizabeth Bell, to whom I owe many thanks.

Chapter 4

Demonstration of Neurosecretory Cells in the Insect Central Nervous System

A. A. Panov

Institute of Evolutionary Morphology
and Ecology of Animals
U.S.S.R. Academy of Sciences
Moscow, U.S.S.R.

The neurosecretory cells (NSC) of insects were first described more than 40 years ago as *drüsenartige Nervenzellen* in the pars intercerebralis of the honey bee *Apis mellifera* (Weyer, 1935). However, the study of insect neurosecretion progressed slowly during the years 1930–1950. Only after the introduction of Gomori's methods in the 1950s [first Gomori's chrome hematoxylin by Bargmann (1949) and then paraldehyde-fuchsin by Gabe (1953, 1955)] was the study of the insect neurosecretory system (NSS) intensified. Extensive studies of this period, which were later summarized in Gabe's (1966) monograph, led to the concept of the universal occurrence of the NSC in insect nervous systems and allowed the establishment of a generalized scheme of their distribution in the brain and, to a lesser extent, in the thoracic and abdominal ganglia.

In the 1960s and 1970s, several histochemical, electron-microscopic, radioautographic, immunohistochemical, and other experimental methods were introduced specifically for the study of insect neurosecretion. The characteristic investigations of this period are, for example, those by Highnam (1961) and by Girardie and Girardie (1967) on locusts, by Schooneveld (1970) on the Colorado beetle, by Kind (1964, 1968) on several Lepidoptera, as well as by Raabe (1965) on thoracic and abdominal neurosecretion in several insects.

However, detailed studies of the NSS in specimens of a single species were somewhat negative in their contribution to the idea of an essential peculiarity of the NSC composition in each of the species studied. Thus, the concept of the insect NSS was in contrast to that of other organ systems of insects, which were characterized by a certain degree of generality in their organization within a given class.

At present, extensive comparative studies of the NSS of differently related species appear to be necessary to solve this problem. This comparative method of investigation is of great value because it not only allows the investigator to establish common characters of the NSS of a given insect group; it also permits a more objective study of the composition of the NSC in each species, sharply distinguishing age, sex, physiologic, and other species-specific characters.

A significant discrepancy often occurring in the data of different authors on the composition of the NSS (not only between related species but also within one and the same) seems to a great extent to be determined by differences in their staining methods. That is why the main task of this chapter is to report the known staining methods, to compare them, and to recommend those that allow the investigator to reveal the maximum diversity of NSC composition in secretory products so that they may be a basis for comparing the data obtained by different authors.

Methods for NSC Detection

In Vivo Observations

Neurosecretory elements can be revealed as bluish-white or yellowish-white opalescent spots when the NSC or neurohemal organs contain large amounts of the neurosecretory material (NSM) and when the ganglion sheath is thin and weakly pigmented.

The first in vivo observations on the perikarya of NSC were made by Thomsen (1948) on the imago blowfly (Calliphoridae) and by Bounhiol (1955) on *Bombyx*. The white-blue opalescence of the cardiac bodies was first discovered in the cockroach *Blatta* by Scharrer and Scharrer (1954).

The NSC become more distinct under oblique or dark-field illumination (Thomsen and Thomsen, 1954). In some instances, in vivo observations allow the apparent distinction between particular NSC types. In each half of the pars intercerebralis of *Calliphora* females, for example, between 7 and 8 NSC were found to have a bluish-white opalescence, and 4 NSC, a yellowish-white opalescence (Thomsen and Lea, 1968). According to their number and position, the first 7–8 NSC seem to correspond to A2 cells, and the second quartet to A1 cells, as described in stained preparations (Panov, 1976).

Supravital staining of NSM improves significantly the visibility of the NSC. Of the different fluorochroms tested, acridine orange was found to be the most satisfactory (Beattie, 1971). Pieces of nerve tissue were excised in insect saline solution and stained in acridine orange (0.1 mg in 1 ml of 0.9% NaCl) for 1 min. They were then transferred to a drop of saline solution on the slide and observed under a cover glass by fluorescent microscopy. The NSM appears as red fluorescing granules against the pale green cytoplasm of nonsecretory cells and axons. The nuclei of the glial and nervous cells were intensively green with distinct nucleoli. The cytoplasm of the glial cells did not stain. The red fluorescence was unstable and in the case of permanent irradiation bleached after 5 min. Using this method all NSM types were stained. Therefore it is not clear whether staining with acridine orange is dependent on the chemical structure of the NSM or associated with the special type of membrane structure of the neurosecretory granules. Lysosomes found in the perikarya of nonsecretory neurons also fluoresce red, but they were fewer and larger than neurosecretory granules.

Observations on Fixed Tissue

Large amounts of NSM can be seen in preparations stained with common histologic methods applied for the study of neurosecretion prior to the development of special "neurosecretory" methods. For example, an affinity of the NSM to acid stains (Day, 1940) (Figs. 4-1 and 4-2), to iron hematoxylin (Rehm, 1950), and to different components of Heidenhain's method (Figs. 4-3) had been established.

Most special methods for NSM staining are complex and include staining with several stains. Because of this, they serve not only to improve the selectivity of NSM staining but also to distinguish different kinds of NSM. Oxidation is an obligatory step followed by staining with a basic and then, if necessary, with an acid stain. A counterstain of cytoplasmic and nuclear structures is possible, provided that the color contrast is obtained between them and the NSM.

Staining of Sectioned Material

Gomori's Chrome Hematoxylin-Phloxine (GChHPh)

Staining with GChHPh, introduced for the study of neurosecretion by Bargmann (1949), was of great importance some time ago. However, it was shown (e.g., Schooneveld, 1970) that in insect material, this method frequently gives unsatisfactory results, not being sufficiently sensitive and selective. At present, GChHPh has apparently fallen out of favor, because it can be substituted by other staining methods that give more defi-

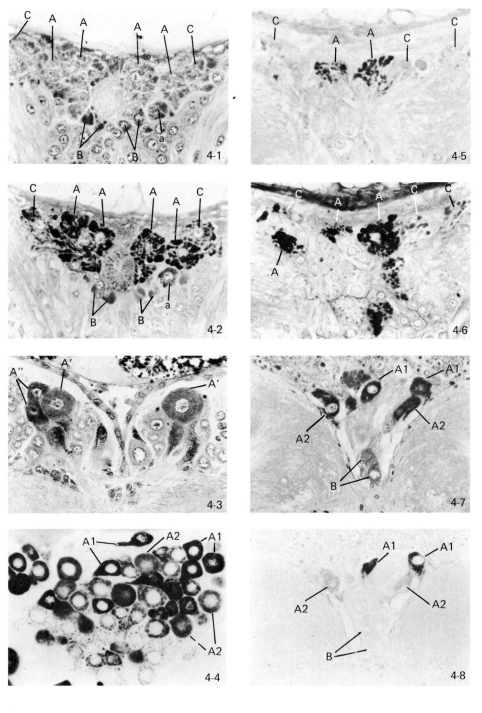

nite results.   Thus, the GChHPh staining procedure described in detail by Gabe (1966) is omitted here.

Paraldehyde-Fuchsin (PF) by Gomori-Gabe

Despite all its limitations (see below), staining with PF has turned out to be very convenient and sensitive.   The data obtained with this method were exactly those that have allowed the development of the concept that neurosecretory elements in invertebrates are commonly and universally distributed.

The PF method has several modifications with respect to preparation of the stain and the staining procedure.   The original mode of preparation of the staining solution by Gomori has largely fallen into disuse because it is unstable and does not permit the user to obtain comparable results after staining with solutions prepared at different times before their application.   A stable staining solution of PF can be obtained from previously prepared or commercially available dry dye.   Two methods of preparations of PF in the dry form are known.

Paraldehyde-fuchsin by Gabe (1953).   Add 1 g of basic fuchsin (C.I. 677) to 200 ml of boiling water, boil 1 min, cool, filter.   To the filtrate add 2 ml each of concentrated HCl and paraldehyde.   Leave stoppered at room temperature.   Each day after preparation of this solution withdraw a drop and place it on a filter paper.   The fuchsin will be found to decrease and the precipitate to increase in quantity.   When the solution has lost its reddish fuchsin color (3–4 days), filter it and discard the filtrate.   Dry the precipitate on the filter paper at 50°C.   The dry crystals can then be removed from the filter paper and stored for years in a reagent bottle.   The yield is about 1.9 g.

Paraldehyde-fuchsin by Rosa (1953).   Add 2 ml of paraldehyde and 2 ml of concentrated HCl to 200 ml of a solution of basic fuchsin (1 g) in 70% alcohol [Meola (1970) obtained good results using fuchsin from Allied Chemical Co., C.I. 42500, whereas the fuchsin of E. Gurr Ltd. was wor-

---

Figs. 4-1 through 4-24, the NSC of the pars intercerebralis, fixed in Bouin's fluid, are shown × 307.   *A1, A'₁, A'₂, A''*, *a:* corresponding types of NSC, NSM that possess mainly strong acid groups after permanganate oxidation; *A2, C:* corresponding types of NSC, NSM that possess mainly weak acid groups after permanganate oxidation; *B:* B-NSC, NSM that remain acidophilic despite permanganate oxidation.

Fig. 4-1. *Eurygaster integriceps,* imago; phloxin without oxidation.
Fig. 4-2. The same, the next section; PThPh.
Fig. 4-3. *Lymantria dispar,* mature larva; Heidenhain's azan.
Fig. 4-4. *Schistocerca gregaria,* imago; PF.
Fig.4-5. *Eurygaster intergriceps,* imago; AB at pH 1.0.
Fig. 4-6. The same, the next section; AB at pH 2.5.
Fig. 4-7. *Calliphora vicina,* imago; PThPh.
Fig. 4-8. The same, the next section; AB-AY.

se]. Allow it to "ripen" at room temperature for 3 days or so, after which, presumably, the concentration of the paraldehyde fuchsin reaches a maximum. Add 100 ml of the original Gomori mixture to 50 ml of chloroform in a separation funnel and then add an additional 200 ml of distilled water. Shake briefly and allow the formed precipitate to settle (for at least 30 min). Drain off the contents of the separation funnel, which contains the suspended precipitate of dye, and run this through filter paper without suction. Filtration usually requires several hours. Allow the moist dye precipitate to dry at 50°C. When the precipitate is dry, scrape it from the filter paper and store in a closed or stoppered vial. The yield is about 0.3 g. According to Meola (1970), Rosa's method of preparing the dry PF dye yields superior results compared with that of Gabe.

The working solution of PF by Gabe (1953) can be prepared by dissolving 0.125 g of PF crystals in 50 ml of 70% alcohol and adding 1.0 ml of glacial acetic acid. This solution keeps its staining properties at least 6 months. Ewen (1962) also recommends acidification of the PF working solution, whereas Cameron and Steele (1959) found no difference between sections stained with and without acid.

Meola's (1970) method is to prepare a 0.5% solution of PF and add 1% (v/v) of concentrated HCl. This solution is allowed to ripen one week and is then filtered once prior to use so as to remove contaminants. The NSM stains most selectively with PF solution at pH 1.6–1.7. Meola's PF working solution is twice as concentrated as Gabe's, and because of this, duration of staining with the former must be accordingly shorter.

There are contradictory opinions about the advantages of different methods of preparations of PF solutions. Thus, Meola (1970) found his PF working solution to be significantly more selective and sensitive in staining the mosquito's NSC than that of Gabe. On the contrary, Credland and Scales (1976) studied midges of the family Chironomidae (related to Meola's Culicidae) and obtained better results with the staining procedure according to Ewen (1962) (working solution of PF after Gabe), whereas Meola's PF solution did not reveal the NSC at all.

For many years, in our laboratory, good results were obtained from different insects using PF prepared according to Gabe's method from basic fuchsin by E. Merck, A. G. (Darmstadt, West Germany). Excellent pictures were also obtained with PF made by "Chroma Gesellschaft" (Stuttgart, West Germany).

The pH of the working solution is of great importance for obtaining satisfactory results. We found 0.25% solutions of PF from different origins to have a pH between 2.5 and 2.9. At this pH, the NSM stains intensively, but some batches of PF also cause relatively strong background coloration. Adding a few drops of concentrated HCl brings down the solution acidity to pH 1.4–1.6, at which point certain kinds of basophilic NSM stain fairly strongly, whereas the background remains colorless. A decrease of pH to 1.0–1.1 significantly diminishes the intensity of staining of the NSM. At pH 0.75, only the NSM rich in SS- and SH-groups stained distinctly in our preparations.

Thus, to reach the maximum sensitivity of staining with PF, the working solution needs acidification with drops of concentrated HCl until the background coloring disappears whereas that of the NSM remains intense. These requirements are achieved by some PF batches without acid, but by reducing the acidity of the working solution of pH 2.0–2.5, it is possible to guarantee the success of the staining procedure. As mentioned previously, Meola (1970) found pH 1.6–1.7 to be optimal. Above and below this pH the stain did not differentiate as well.

Staining procedure after Ewen (1962). (1) Fix in aqueous Bouin's fluid or in Helly's mixture, dehydrate, and embed in paraffin. (2) Remove paraffin and hydrate sections in usual way. Hydrolyze sections fixed in Helly in $1\,N$ HCl at 60°C for 6 min according to Wendelaar Bonga (1970). (3) Oxidize in freshly prepared mixture of equal volumes of 0.6% $KMnO_4$ and 0.6% H_2SO_4 for 1 min at 20–22°C. (4) Rinse in distilled water. (5) Bleach in 2.5% sodium metabisulfite for 20–40 s. (6) Wash in running tap water for 5 min or more. (7) Pass through distilled water, 30% to 70% alcohol. (8) Stain in PF for 2 min. (9) Wash in 2 or 3 changes of 96% alcohol. (10) Differentiate, if the background is stained, in 0.5% HCl in absolute alcohol for 10–30 s. (11) Wash in 96% alcohol.

In case only the staining of PF-positive NSM is needed (e.g., to reveal the thinnest neurosecretory fibers), dehydrate sections in 2 changes of acetone, clear in 2 changes of xylene, and mount in Canada balsam.

RESULTS: The NSM turned basophilic after permanganate oxidation stain to show up in different shades of purple (Fig. 4-4).

If more complete information is desired, hydrate sections in the usual way and counterstain. Staining with a 0.1%–0.5% solution of light green is the simplest one. It stains the whole section green and the acidophilic NSM that stains red with GChHPh is more intense.

Halmi's mixture is also often used (Halmi, 1952): distilled water, 100 ml; light green CF yellowish, 0.2 g; orange G, 0.1 g; chromotrope 2R, 0.2 g; glacial acetic acid, 1.0 ml. The mixture keeps indefinitely. Immerse hydrated sections for 10 min in the mixture containing 4% phosphotungstic acid and 1% phosphomolybdic acid, rinse briefly in distilled water, stain in Halmi's mixture [to 1 h according to Ewen (1962)], rinse briefly in 0.2% acetic acid in 95% alcohol, dehydrate in absolute alcohol, clear in xylene, mount in Canada balsam.

RESULTS: Acidophilic NSM is green; cytoplasm, yellowish green; nuclei, orange.

Counterstaining of the nuclei with Groat's hematoxylin or with iodine green, as well as the whole sections with picroindigocarmine (picrofuchsin) was occasionally recommended (Gabe, 1955).

Paraldehyde-Thionin-Phloxine (PThPh) (Panov, 1964)

Paraldehyde-thionin (PTh) was initially found to be a precise method for demonstrating β cells of the pancreatic islets and certain cells in the anterior pituitary (Paget and Eccleston, 1959).

Staining solution of PTh (from Paget and Eccleston, 1959). Dissolve 0.5 g thionin in 91.5 ml 70% alcohol, add 7.5 ml paraldehyde, and 1.0 ml concentrated HCl. Allow the solution to ripen in a tightly stoppered bottle for about 2 weeks. Paget and Eccleston tested many brands of thionin of British origin. The most consistently satisfactory was thionin "Revector" (Hopkin and Williams, Ltd.). A sample of thionin from E. Merck, A. G. (Darmstadt, West Germany) was also satisfactory. Huber (1963) obtained good results using only thionin E. Merck, A. G. and thionin Chroma Gesellschaft (Stuttgart, West Germany). In our laboratory, we used successfully an old sample of thionin "Grübler" and thionin Georg T. Gurr. PTh prepared from thionin BDH (U.K.) did not stain the NSC.

To demonstrate the NSC we adapted the staining procedure with GChHPh in which ChH was replaced with PTh. This technique has been the usual method of NSC staining in our laboratory for many years.

Staining procedure. (1) Fix in aqueous Bouin's fluid, in Zenker, or in Helly, but not in alcohol-containing mixtures, and then embed in paraffin. (2) Dewax and hydrate sections. Preparations fixed in sublimate-containing mixtures are hydrolized in 1 N HCl at 60°C for 6 min (Wendelaar Bonga, 1970). Wash in water. (3) Oxidize in freshly prepared mixture of equal volumes of 0.6% $KMnO_4$ and 0.6% H_2SO_4 for 1 min. (4) Rinse in water and bleach in 2.5% potassium metabisulfite for 20–40 s. (5) Wash in running tap water for 5 min or more. (6) Rinse in distilled water. (7) Stain with PTh for 10 min. (8) Rinse in 2 or 3 changes of distilled water. (9) Stain with 0.2% aqueous solution of phloxine for 5 min. Phloxine Geigy, Phloxine BDH and Cyanosin Apolda (GDR) were equally satisfactory. (10) Rinse in tap water. (11) Immerse in 5% phosphotungstic acid for 1 min. (12) Wash in running tap water for 5 min. (13) Differentiate in 95% alcohol under microscopic control until the neuropil turns pale rose. If it is certain that the phloxinophilic NSC exist in sections, differentiate until they become distinct. (14) Dehydrate in 2 changes of acetone, clear in xylene, mount in Canada balsam.

RESULTS: The NSM turned basophilic after permanganate oxidation stain in different shades of blue; the NSM remained acidophilic—bright pink; nucleoli—bright pink; cytoplasm—rose (Figs. 4-2, 4-7, 4-9, 4-13, 4-15, 4-17, 4-19, 4-21, 4-23; Fig. 4-33, Plate 1)

Performic Acid-Alcian Blue (PFAAB) (Adams and Sloper, 1956)

In sections of the corpora cardiaca of a cockroach, *Nauphoeta cinerea,* Sloper (1957) found a great similarity in the distribution of substances stained with GChH and of those stained with Alcian blue after peracid oxidation. This was one reason to propose the PFAAB method for demonstrating insect NSM. It was suggested that protein-bound cystine of the NSM oxidizes by peracid to cysteic acid that then reacts selectively with Alcian blue at pH below 1.0.

Performic acid. This is prepared each day by adding to 40 ml of 98% formic acid 4 ml of fresh H_2O_2 and 2–3 drops of concentrated H_2SO_4. Allow to stand at least 3 h; stir to remove gas bubbles before using.

Staining solution. A 1% solution of Alcian blue 8GS (G. Gurr) in 2 N H_2SO_4 is prepared by adding the dye to the acid. The solution, heated to 70°C and filtered, is freshly prepared each day (pH 0.2).

Staining procedure. (1) Dewax and bring sections to water in usual way. (2) Rinse successively first in 70% and then in absolute performic acid. (3) Oxidize in fresh solution of performic acid for 5 min. (4) Wash in distilled water for about 5 min. (5) Since by this treatment, sections are often loosened from the slide, they are warmed at 60°C until they are dry and adhere again to the slide. (6) Wash sections again in running tap water for 15 min. (7) Stain in Alcian blue for 1 h. (8) Wash in water for about 5 min. (9) Dehydrate in alcohol and mount in Canada balsam.

RESULTS: NSM containing large amounts of cystine stain in different shades of blue.

Humbertsone's Performic Acid-Victoria Blue (PFAVB)
(Dogra and Tandan, 1964)
This technique is analogous to the preceding one, but Alcian blue is replaced by an iron-resorcin lake of Victoria blue, which according to the authors, gives a more distinct picture.

Performic acid. See the PFAAB method.

Staining solution. Mix in a flask: distilled water, 200 ml; dextrine, 0.5 g; Victoria blue 4R, 2.0 g; resorcin, 4.0 g. Bring to boil. When boiling briskly add 25 ml of boiling 29% ferric chloride. Boil for another 3 min; cool. A heavy precipitate is formed. Filter and dry the precipitate in an oven at 50°C. Dissolve the whole precipitate in 400 ml of 70% alcohol and then add 4 ml of concentrated HCl and 6 g phenol. For better results use 2 weeks later. The stain keeps for months.

Staining procedure. (1) Fix the dissected brain in 10% formaldehyde-saline solution, embed in paraffin wax and section. (2) Bring the sections fixed to the slide to xylene and then to absolute alcohol; allow them just to dry. (3) Place the slide on matchsticks in a petri dish and drop performic acid solution on the dry sections; oxidize for 5 min. Because the sections detach easily, it is recommended that the slides be washed in boiling water, to which sodium laurylsulfate is added. Then rinse and keep them in a refrigerator, in distilled water. Dry before using. Ganagarajah and Saleuddin (1970) recommend oxidation of the whole tissue. It is possible (however, with worse results) to oxidize the sections in acidified $KMnO_4$ as

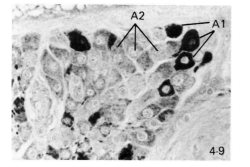

4-9

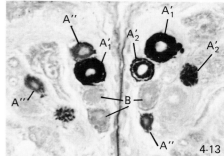

4-13

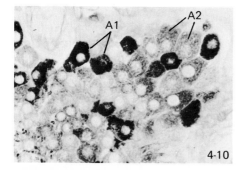

4-10

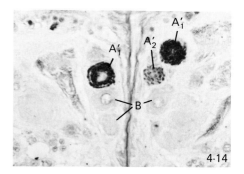

4-14

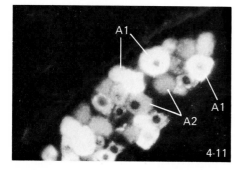

4-11

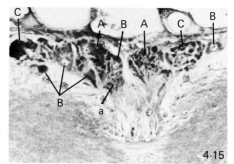

4-15

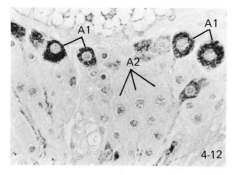

4-12

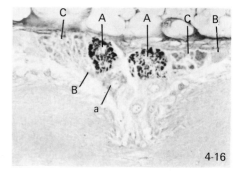

4-16

described previously. However, the specificity of staining SS- and SH-groups is then lost. (4) Wash in distilled water for 15 min. (5) Rinse in 70% alcohol. (6) Stain in staining solution for 12 h. (7) Wash in 70% alcohol. (Here one may rinse in tap water and counterstain in 0.001% safranin in 1% acetic acid for 5 min. Afterward rinse again in tap water.) (8) Dehydrate in graded alcohols, clear in xylene, mount in Canada balsam.

RESULTS: NSM rich in cystine and/or cysteine stain in blue.

Alcian Blue-Phloxine (ABPh)

In routine histologic practice, however, staining with Alcian blue following permanganate oxidation is more convenient. At an appropriate pH of the working solution, Alcian blue demonstrates practically the same NSM after both peracid and permanganate oxidation (Herlant, 1958). Sections are then stained with phloxine, revealing NSM that retains its acidophilia despite oxidation.

Staining procedure. (1) Fix preferably in aqueous Bouin's fluid. (2) Dewax and hydrate sections in usual way. (3) Oxidize in freshly prepared mixture of equal volumes of 0.6% $KMnO_4$ and 0.6% H_2SO_4 for 1–2 min. (4) Bleach in 2.5% potassium metabisulfite for 20 s. (5) Wash in running tap water for 5 min or more. (6) Stain in 0.5% solution of Alcian blue in 3% acetic acid (pH 2.5) for 30 min. (7) Rinse in distilled water. (8) Stain in 0.5% aqueous solution of phloxine for 3–5 min. (9) Immerse in 5% phosphotungstic acid for 1 min. (10) Wash in running tap water for 30 s–5 min. (11) Differentiate in 95% alcohol under microscope control until the neuropil turns pale pink. (12) Dehydrate in 2 changes of acetone, clear in 2 changes of xylene, and mount in Canada balsam.

RESULTS: NSM converted to the basophilic after oxidation stains in different shades of blue; NSM that retain their acidophily stain bright pink; the nucleoli are bright pink, chromatin and cytoplasm, grayish pink.

The pH of the working solution of Alcian blue is extremely important for staining results. At pH 1.0 or below, the same NSM stains as with PFAAB method. At pH 2.5, this NSM begins to stain more intensely, and other kinds of basophilic NSM start to take up the color (see Figs. 4-5, 4-6). At pH 3.5, selectivity of staining disappears due to the strong background coloration.

Fig. 4-9. *Gryllus bimaculatus,* imago; PThPh.
Fig. 4-10. The same; PTh-PF.
Fig. 4-11. The same; PIC.
Fig. 4-12. The same; AB-AY.
Fig. 4-13. *Dendrolimus pini,* mature larva; PThPh.
Fig. 4-14. The same, the next section; AB-AY.
Fig. 4-15. *Eurygaster integriceps,* imago; PThPh.
Fig. 4-16. The same, the next section; AB-AY.

The following brands of Alcian blue were found to be satisfactory: Alcian blue 8GS Fluka, Alcian blue BDH, Alcian blue Gee Lawson, Alcian Blau Kepec. Alcian Blau 6B (Schuchardt, München) was unsatisfactory because of background staining at pH 2.5.

Alcian Blue-Alcian Yellow (AB-AY)

The AB-AY method was developed to distinguish mucosubstances rich in sulfo or in carboxy groups (Ravetto, 1964). At pH 1.0 or below, Alcian blue reacts only with sulfo groups that ionize at this pH. Alcian yellow is used in a solution at pH 2.5 at which carboxy groups also ionize and bind Alcian yellow. As a result, substances rich only in sulfo groups stain blue, those rich in carboxy groups stain yellow, and substances containing both sulfo and carboxy groups stain in different shades of green depending on the ratio of acid groups.

Peute and van de Kamer (1967) used the AB-AY method after permanganate oxidation for demonstrating various kinds of NSM in the hypothalamo-neurohypophyseal system of the frog, *Rana temporaria*. There they found that NSM stained blue, green, or yellow and suggested that acidified potassium permanganate oxidizes SS- and/or SH-groups to sulfo groups and hydroxy groups to carboxy groups.

After the first step of staining with Alcian blue at pH 1.0, all of the reacted substances stain blue, but they generally turn green after the second step of staining with Alcian yellow at pH 2.5. This change in color may be of a different origin (Sorvari and Sorvari, 1969). The substances 'stained with Alcian blue at pH 1.0 may also possess weak acid radicals in addition to strong ones. The former ionize at pH 2.5 and react with Alcian yellow. Or, the color change may be an artifact giving the wrong information on the ratio of strong and weak acid groups. This may result from (1) the substitution of Alcian blue by Alcian yellow or (2) the incomplete saturation of strong acid radicals with Alcian blue during short staining and their subsequent reaction with Alcian yellow.

To avoid potential errors, Sorvari and Sorvari (1969) recommend increasing the time in Alcian blue at pH 1.0 up to 18 h when a complete saturation of strong acid groups with Alcian blue occurs and the substitution of Alcian blue by Alcian yellow diminishes. We found, however, that even in this case a partial substitution of Alcian blue by Alcian yellow takes place. To avoid the latter, we advise converting Alcian blue in insoluble monastral fast blue in an alkaline bath according to Lillie (1965).

Staining procedure (Wendelaar Bonga, 1970; modified by Panov, 1978). (1) Fix in aqueous Bouin's fluid and embed in paraffin. (2) Dewax and hydrate sections in the usual way. (3) Oxidize in a mixture of equal volumes of 0.6% H_2SO_4 and 0.6% $KMnO_4$ for 2 min. (4) Rinse in water and bleach in 2.5% sodium (potassium) metabisulfite (up to 30 to 40 s). (5) Wash in running tap water for 5 min or more. (6) Rinse in distilled water and then with pH 0.2 or 1.0 buffer. (7) Stain with 0.5% Alcian blue 8GS at pH 0.2 for 1 h or at pH 1.0 for 30 min. (8) Rinse in corresponding buffer (pH 0.2 or 1.0). (9) Wash in distilled water for 5 min. (10) Transfer through 40% to 70% alcohol. (11) Treat with 0.5% borax in 80% alcohol for 30 to 40 min. (12) Rinse in 70% alcohol, transfer to a second portion of 70% alcohol, and rinse in 40% alcohol and then in distilled water; finally, rinse in pH 2.5 buffer. (13) Stain with 0.5% Alcian yellow GXS at pH 2.5 for 40 min. (14) Rinse with pH 2.5 buffer and wash in distilled water for 2 to 3 min. (15) Dehydrate in two changes of acetone, clear in two changes of xylene, and mount in Canada balsam.

RESULTS: Neurosecretory material turning basophilic after permanganate oxidation stains blue, green, or yellow depending on the ratio of the resulting strong and weak acid groups (Figs. 4-8, 4-12, 4-14, 4-16).

After step 14, sections can be stained with 0.1% nuclear fast red for 1 min or with phloxine (as described under Paraldehyde-Thionin). Staining with phloxine reveals which NSM remain acidophilic after permanganate oxidation.

All of the used batches of Alcian blue were sufficiently water soluble, whereas many batches of Alcian yellow were weakly soluble in water. Gabe (1972) could not obtain solutions of Alcian yellow above 0.1% using samples of Alcian yellow Michrome No. 636 E. Gurr, Alcian yellow G. Gurr, and Alcian Gelb GXS Chroma Gesellschaft. We used the latter brand of Alcian yellow and also could not obtain a 0.5% solution. On the other hand, Wendelaar Bonga (1970) prepared a 0.5% solution of Alcian blue GXS, and Sorvari and Sorvari (1969) prepared 0.1–5% solutions. Unfortunately, the sources of Alcian yellow they used were not indicated.

Victoria Blue (Alcian Blue)-Paraldehyde-Fuchsin
(Ganagarajah and Saleuddin, 1970)
Oxidize fixed ganglia with performic acid according to PFAAB method, embed in paraffin, and stain sections first with Victoria blue according to PFAVB method (or Alcian blue according to PFAAB method) and then with paraldehyde-fuchsin according to PF method.

RESULTS: NSM stains blue and purple.

Paraldehyde-Thionin-Paraldehyde-Fuchsin (PTh-PF) (Panov, 1976)
The method of a double staining by Ganagarajah and Saleuddin (1970) has two main defects. First, because the whole ganglion oxidizes, the degree of tissue oxidation is depth dependent. Second, the staining procedure is time consuming. Therefore, we proposed a simple and time-saving method of double staining, first with paraldehyde-thionin and then with paraldehyde-fuchsin.

Staining procedure. (1) Fix tissues in aqueous Bouin's fluid, embed in paraffin. (2) Dewax and hydrate sections in the usual way. (3) Oxidize in a freshly prepared mixture of equal volumes of 0.6% $KMnO_4$ and 0.6%

H_2SO_4 for 1 min. (4) Rinse in water. (5) Bleach in 2.5% potassium or sodium metabisulfite for 20–40 s. (6) Wash in running tap water for 5 min or more. (7) Rinse in distilled water. (8) Stain with PTh for 10 min. (9) Rinse twice in distilled water, pass through 30% and 70% alcohol. (10) Stain with PF at pH 2.5 for 2 min. (11) Rinse 2–3 times in 95% alcohol. (12) Dehydrate in acetone, clear in xylene, and mount in Canada balsam.

RESULTS: Different NSM, turned basophilic after permanganate oxidation, stain blue, purple, or a mixture of colors (Fig. 4-10; Figs. 4-34 and 4-35, Plate 2).

Pseudoisocyanin Method (PIC) (Sterba, 1961)

A fluorescent microscopic method with demonstrated NSM was developed from the observation that the hypothalamic NSM stains metachromatically with pseudoisocyanin after oxidation with acidified potassium permanganate (Schiebler, 1958). Sterba (1961) showed that the metachromatically stained product fluoresces in ultraviolet light. He also found the PIC method to be significantly more sensitive than the PF one (Sterba, 1964).

Staining procedure. (1) Fix in aqueous Bouin's fluid, in Held's fluid, buffered Lillie's formaldehyde, Zenker or Helly; embed in paraffin. (2) Section at 3–4 μm. (3) Dewax quickly in benzene and hydrate. (4) Oxidize in a freshly prepared mixture of equal volumes of 0.5% $KMnO_4$ and 1.0% H_2SO_4 for 3 min. (5) Bleach in 1.5% potassium (or sodium) metabisulfite or in 3% oxalic acid for 10–20 s. (6) Wash in running tap water for 5 min or more. (7) Rinse in distilled water. (8) Stain with N, N'-diethylpseudoisocyanin chloride (14.5 mg in 100 ml of distilled water) for 1–3 min. (9) Bring to ammoniacal water (2 drops of concentrated NH_4OH in 100 ml of water) to avoid depolymerization of the dye and cover with a cover glass. Because sections detach readily from slides in ammoniacal water, Dr. D. A. Sakharov (personal communication) recommends the use of a solution of medinal at pH 8. Embedding in a permanent medium diminishes fluorescence intensity (Sterba, 1961). (10) Examine sections under a fluorescent microscope using excitation (Schott) filter UG 1/1.5 and ocular barrier filter OG 1.

RESULTS: NSM exhibits an intense orange fluorescence. Neural lamella, "gliosomes" occurring in glial cells of some insects, as well as granules of hemocytes also fluoresce orange. Other structures show a very weak unspecific fluorescence (Figs. 4-11, 4-18, 4-20, 4-22, 4-24). Background is dark red.

Pseudoisocyanin according to Coalson (1966)

Schooneveld (1970) found that staining with PIC by Sterba (1961) may sometimes give unsatisfactory results because of detachment of sections from slides in strong alkaline solution and a rapid image fading by the ultraviolet light. He recommended the use of the technique by Coalson (1966), which was developed originally for demonstrating insulin-containing cells in islets of Langerhans.

Staining procedure (adapted from Coalson, 1966). (1) Fix in aqueous Bouin's fluid, embed in paraffin, section at 3–5 μm, and affix to slides with albumen. (2) Dewax and hydrate sections in usual way. (3) Oxidize in freshly prepared mixture of equal volumes of 1.2% $KMnO_4$ and 0.6% H_2SO_4 at 28–29°C for 30–45 s. This oxidizing step is critical and appears to be temperature sensitive. (4) Rinse in tap water. (5) Bleach in 5% oxalic acid for 30 s. (6) Wash in running tap water for 5 min or more. (7) Wash in 2 changes of distilled water for 3–5 min each. (8) Stain in refrigerator in cold N,N'-diethylpseudoisocyanin chloride solution (35 mg/100 ml of distilled water) for 20 min. This staining solution can be used immediately and, if refrigerated, appears to be stable for more than a year. (9) Wash in cold running tap water for about 5 min until the albumen adhesive is decolorized and the sections appear bright red. (10) Cover with cover glass and examine in fluorescent microscope as in PIC method.

RESULTS: These were the same as those for the PIC method.

Heidenhain's Azan Method after Prolonged Fixation in Helly's Fluid
Although Heidenhain's azan method is usually used to obtain survey preparations in histology, it is a very sensitive technique for demonstrating some kinds of NSM provided that the tissue fixation with Helly's mixture is prolonged (Raabe, 1963).

Staining procedure. (1) Fix with Helly's mixture for 20–24 h; wash in running tap water for 24 h, pass through graded iodinized alcohols, clear in chloroform, embed in paraffin. (2) Dewax sections, pass to water through graded iodinized alcohols to remove residual sublimate deposits. (3) Remove iodine with 0.25% sodium thiosulfate for 5–10 min. (4) Wash in running tap water for 5 min, rinse in distilled water. (5) Stain with 0.1% azocarmine G at 55–60°C for 15 min. (6) Cool to room temperature. (7) Rinse briefly in distilled water and differentiate in 0.1% aniline in 90% alcohol until nuclei and NSM stand out sharply against their pale pink background. (8) Rinse in 1% acetic acid in 95% alcohol. (9) Immerse in 5% phosphotungstic acid for 1 h. (10) Stain for 1–2 h in the following mixture: water-soluble aniline blue, 0.5 g; orange G, 2.0 g; glacial acetic acid, 8 ml; distilled water, 100 ml. Dissolve in boiling distilled water, cool, filter, and dilute with an equal volume of distilled water. (11) Differentiate in 96% alcohol until cells appear red, blue, and orange. (12) Dehydrate in absolute alcohol, clear in 2 changes of xylene, and mount in Canada balsam.

RESULTS: Various kinds of NSM stain bright red, pink, or orange against a polychrome background.

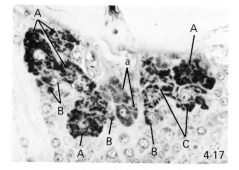

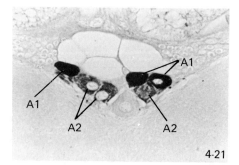

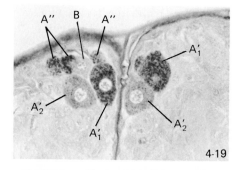

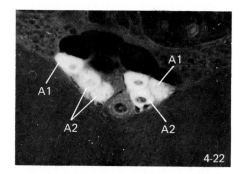

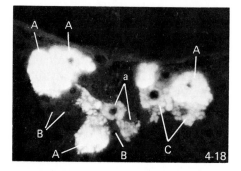

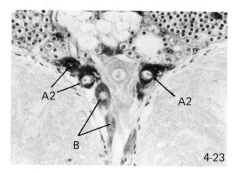

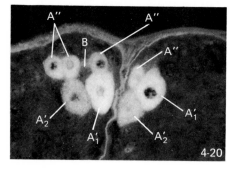

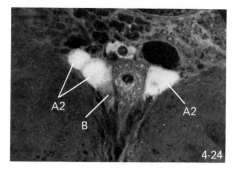

In Toto Demonstration of the Neurosecretory Elements

The minute dimensions of insect ganglia as well as their relatively large NSC and their superficial position gave rise to the possibility of demonstrating the neurosecretory elements in situ. A staining technique of the whole ganglia was first developed by Dogra and Tandan (1964) who adopted some of the above-mentioned staining methods after increasing the duration of oxidation and staining.

Performic Acid-Victoria Blue (Dogra and Tandan, 1964)
(1) Expose the brain or other organ of an insect in a physiologic solution, wash off the hemolymph, clean the surface of the ganglion of trachea, fat bodies, etc. Fix in situ in 10% formaldehyde-saline solution (physiologic saline, 90 ml; commercial formalin, 10 ml). After 2–3 h dissect out the organs required and fix in fresh fixative for 24–36 h. The neurilemma can be removed in order to see the NSC more distinctly. Wash well in tap water for 2–3 h; next in distilled water for 10–20 min; blot off the water with strips of filter paper. (2) Oxidize the organs in performic acid until transparent (5 min or more). (3) Blot off excess oxidant with filter paper. (4) Wash repeatedly in distilled water for 20–30 min. (5) Transfer through 30% to 70% alcohol. (6) Stain with Victoria blue staining solution for 12–18 h, the period depending on size of material. (7) Quickly blot off excess stain with filter paper. (8) Differentiate in 70% alcohol; change the alcohol repeatedly, until no more superfluous stain is given off. (9) Dehydrate; clear in cedarwood oil for 2–4 h; remove the cedarwood oil with xylene for some minutes, mount in Canada balsam.

Paraldehyde-Fuchsin for Total Staining of Ganglia
(Dogra and Tandan, 1964)
Staining procedure. (1) Expose an organ in a physiologic solution, wash off the hymolymph, fix in Bouin's fluid for 12–24 h, carefully wash off the picric acid with lithium salt in 70% alcohol, bring to water. (2) Oxidize in a mixture of equal parts of 0.6% $KMnO_4$ and 1.1% H_2SO_4 for 2–3 min. (3) Rinse in water. (4) Bleach in a 4% solution of sodium or potassium metabisulfite under the microscope until the tissue becomes completely white. (5) Wash in distilled water for 5–10 min. (6) Pass through 30% and 70% alcohol for 10–15 min in each. (7) Stain in PF solution for 2–

Fig. **4-17.** *Eurygaster integriceps,* imago; PThPh.
Fig. **4-18.** The same, the next section; PIC.
Fig. **4-19.** *Hyalophora cecropia,* larva; PThPh.
Fig. **4-20.** The same, the next section; PIC.
Fig. **4-21.** *Calliphora vicina,* imago; PThPh.
Fig. **4-22.** The same, the next section; PIC.
Fig. **4-23.** The same, a more caudal section; PThPh.
Fig. **4-24.** The same, the next section; PIC.

10 min.   (8) Differentiate in 95% alcohol until no more superfluous stain is given off.   If there is some stained precipitate on the surface of the brain, wash it off in 70% alcohol.   (9) Dehydrate, clear in cedarwood oil, wash off the oil with xylene, and mount in Canada balsam.

For staining with PF in toto, we used the same reagents as described in the PF method, but oxidized the ganglia for about 15 min (bleaching under microscopic control) and stained them for about 30 min.   By differentiating in acidified alcohol it is possible to decolorize the background (Figs. 4-25–4-32).

Performic Acid-Resorcin-Fuchsin (PFARF)
(Ittycheriah and Marks, 1971; Ittycheriah, 1974)
Resorcin-fuchsin, like PF, was originally used for staining elastic fibers (Romeis, 1932).   The use of resorcin-fuchsin for NSC staining was rejected earlier by Gabe (1953) on the grounds that the stain was difficult and dangerous to prepare, that staining time was excessively long, and that tissue coloration was poorly preserved.   The recent commercial availability of the dye as a dry powder has eliminated difficulties in stain preparation.

McGuire and Opel (1969) found resorcin-fuchsin to be a simple and reliable vertebrate neurosecretory stain with characteristics similar to those of PF except that it stains NSM black. Ittycheriah and Marks (1971) preferrèd PFARF to other staining methods for in situ demonstration of insect NSC because of sharpness of the NSC and clearness of the background.

Performic acid.   Mix 4 ml of 97–100% formic acid, 1 ml of fresh 30% H_2O_2 and 0.2 ml of concentrated H_2SO_4.   Allow to stand for 2 h before use.   The solution keeps well for 24–36 h.   According to Ittycheriah (1974), this oxidant gives better results than the PFAAB method.

Resorcin-fuchsin staining solution.   Dissolve 0.5 g of resorcin-fuchsin [Resorcin-fuchsin No 11430 Chroma Gesellschaft (Stuttgart, West Germany) was used successfully] in 100 ml of 70% alcohol and add 2 ml of concentrated HCl.   The solution can be used immediately and keeps for several months.

Staining procedure.   (1) Dissect the brain and associated endocrine organs in an appropriate saline solution, fix in modified Bouin's fluid (0.5% trichloroacetic acid replaces the 5% acetic acid) for 4–25 h.   (2) Thoroughly wash in water containing a few crystals of lithium carbonate. (3) Oxidize in oxidant solution for 5 min or more (one can also use a mixture of equal volumes of 0.6% $KMnO_4$ and 0.6% H_2SO_4).   (4) Blot off excess oxidant with filter paper. (5) Wash repeatedly in distilled water for 20–30 min.   (6) Transfer through 30% to 70% alcohol (10–15 min each).   (7) Stain in staining solution for 1–30 min, depending on the size of the ganglion.   (8) Blot off excess stain with filter paper.   (9) Differentiate in 70%

alcohol.   Change alcohol repeatedly until no more superfluous stain is evident.   If necessary, use acid alcohol to remove stain from sheath and then wash thoroughly in 70% alcohol until the NSC stand out prominently.   (9) Dehydrate in 95% and 100% alcohol; clear in cedarwood oil for 4 h or more; remove cedarwood oil with xylene and mount in Permount or Euparol.

Comparisons of Present-Day Neurosecretory Staining Methods

As shown above, all known methods of staining of the NSM can be subdivided into two groups: the empirical methods (PTh, PF, GChHPh, PFARF, AB after permanganate oxidation, Heidenhain's azan) and histochemical ones (PFAAB, PFAVB).   At pH 1 and below the latter stains react with the sulfo groups of cysteine acid, which is obtained after oxidation of cystine and/or cysteine with performic acid (Adams and Sloper, 1956).

Among the empirical methods of NSM staining, the histochemical basis of the PF technique has been the most carefully investigated.   It has been shown that PF staining has no specificity in the histochemical sense insofar as it reacts with both aldehydes and various acid radicals (Bangle, 1954; Konečný and Pličzka, 1958; Sumner, 1965).   Furthermore, it has yet to be revealed in which naturally occurring substances these groups can be formed by oxidation with acidified potassium permanganate and by subsequent bleaching with potassium metabisulfite (Gabe, 1966).   However, a far-reaching similarity in results of peracid and permanganate oxidation speaks in favor of the similarity in their oxidizing action on thiol groups (Gabe, 1966, 1972).   In addition, the histochemical unspecificity of the PF staining is also determined by the fact that PF itself seems to be a mixture of several substances possessing differing staining properties (Elftman, 1956; Schooneveld, 1970).

All data given above suggest a dual limitation of PF and other similar staining methods.   On the one hand, they apparently cannot reveal all possible existing kinds of insect NSM, e.g., the so-called azocarmine-positive NSC do not often stain with PF but were observed by Heidenhain's azan method in many insects (Raabe, 1963).

On the other hand, the widespread occurrence of substances that can react with PF or with other "neurosecretory" stains demonstrates the existence of a variety of tissue and cell components that can be determined by these methods.   In insects, in addition to NSM, PF will, for example, stain the hemocytic granules, the neural lamella, the striated border of some epithelia, the lysosomes, and other structures.   This gives rise to certain difficulties for interpreting the origin of the PF-positive substances occurring outside the NSC and incites speculations about NSM migration through the insect body.

In the cells of the central nervous system, PF also reveals some nonsecretory structures. In the glial cells of the cockroach brain, "gliosomes" that under the electron microscope resemble lysosomes, were stained with PF (Pipa, 1962). Therefore, the occurrence of a few minute PF-positive granules in the cytoplasm of the nerve cell does not always prove its neurosecretory nature (cf. Gabe, 1966).

Finally, the occurrence of "typical NSM" in the cytoplasm of the nerve cell also does not necessarily indicate its neurosecretory nature, and other evidence is required to prove it (Bern, 1962). It is not excluded that some NSM-containing nerve cells are nonsecretory neurons with a particular chemistry. The widespread occurrence of a diffuse network of nerve fibers containing PF-positive material, e.g., in the brain of some bugs (Johansson, 1958; Panov, 1969), favors this idea.

As shown by several authors, there is no uniform PF-positive NSM (commonly termed type A NSM) in insects (Drawert, 1966; Tandan and Dogra, 1966; Schooneveld, 1970). In the Colorado beetle, Schoonveld (1970) distinguished a number of PF-positive NSM stained in different shades of purple. Comparison of staining with PF and PFAAB (Schooneveld, 1970), with PF and PFAVB (Tandan and Dogra, 1966), or with PF and DDD (Drawert, 1966) showed that the proteins rich in thiol groups are only partially responsible for staining with PF. Namely, they form the NSM of typical A cells, which stains deep purple with PF. Other kinds of PF-positive NSM either give a weak reaction to SS- and SH-groups (*Leptinotarsa decemlineata:* Schooneveld 1970) or do not react at all (*Agrotis segetum:* Drawert, 1966).

The comparison of NSM staining with PF, PTh-PF, and AB-AY that we carried out on alternate sections of the brains of various insects allowed, to some extent, the elucidation of the nature of staining of different NSM with PF. Thus, for example, Table 4-1 includes the data on the comparative staining of two types of PF-positive NSC (A1 and A2 cells) found in the pars intercerebralis of *Calliphora* females (see Figs. 4-7 and 4-8).

Table 4-1. Color of the NSM of Different NSC Types in the Pars Intercerebralis of *Calliphora vicina* Stained by Different Methods

	Cell type	
Staining method	A1	A2
PF	Deep purple	Purple
AB-AY	Greenish blue	Yellowish green
PTh-PF	Deep blue	Purple
PThPh	Deep blue	Purple

Staining with AB-AY shows that the A1-NSM of *Calliphora* is rich in SS- and/or SH-groups and poor in OH-groups if Peute's and van de Kamer's (1967) supposition on the mode of oxidizing action of acidified potassium permanganate is correct. On the other hand, the NSM of the A2 cells is rich in OH-groups and poor in thiolic groups (SS- and/or SH-groups). Since the NSM of the A2 cells reacts intensely with PF after oxidation, the ability of PF also to bind carboxy groups is apparently realized here (Konečný and Pličzka, 1958).

PF that is used first only after oxidation does not permit a precise distinction of substances such as those of A1 and A2 cells of *Calliphora,* both in perikarya (see, e.g., Fig. 4-4) and especially in neurosecretory fibers. Our experience shows that staining with AB-AY is also insufficient to distinguish these NSM in routine histologic practice because of the low-intensity AY color (see Figs. 4-8, 4-12, 4-16). With AB-AY techniques only large aggregations of the A2-NSM can be revealed.

These inconveniences can be eliminated using sequential staining with PTh and then with PF. The PTh-PF technique was initially used to analyze the composition of the PF-positive NSC in the pars intercerebralis of *Calliphora* and *Lucilia* imagos (Panov, 1976), and later in numerous mecopteroid and neuropteroid insects (Panov and Davydova, 1976). As mentioned above, the PTh-PF method is a more convenient analog to the PFAVB-PF technique (Ganagarajah and Saleuddin, 1970) that has not yet been widely used.

PTh first used after oxidation reacts mainly with the A1-NSM as AB in the AB-AY method, whereas the A2-NSM either remains colorless or stains weakly pale blue. During the next step of staining, the A2-NSM reacts with PF. Due to the great intensity of the PF color, even minute granules of the A2-NSM stand out sharply against a clear background and can be well distinguished from the A1-NSM stained blue (Fig. 4-10; Figs. 4-34 and 4-35, Plate 2).

A comparison of the color of certain NSC after staining with PTh and then after subsequent staining with PF shows, however, that the phenomenon of binding of the stain to the NSM is apparently much more complex than was indicated above. It seems to include the different affinities of various NSM to PTh or PF. Thus, for example, when stained with PTh, the NSM of the NSC that compose the groups M2 and M3 of the larval brain of *Chironomus* becomes equally intense blue. However, following staining with PF, the NSM of M2 cells remains blue whereas that of M3 cells turns purple. The binding of the M3-NSM to PTh appears to be unstable, and PTh is substituted by PF.

According to Sterba (1964), the secondary fluorescence of the NSM stained with PIC is determined by a polymerization of the dye on closely located (about 4 Å) sulfo groups arising from SS- and/or SH-groups after oxidation. Consequently, Sterba considers the PIC method to be a spe-

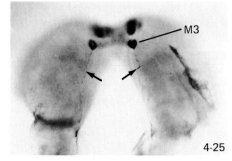

4-25

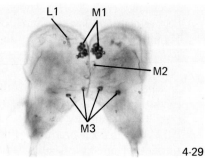

4-29

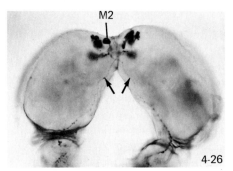

4-26

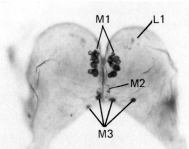

4-30

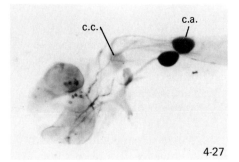

4-27

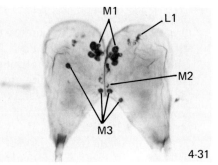

4-31

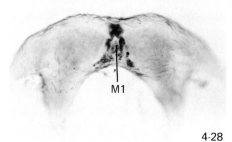

4-28

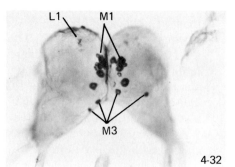

4-32

cific histochemical one for these groups. However, staining of alternate sections with AB at pH 1, AB pH 1–AY pH 2.5, and with PIC shows that NSM that possesses mainly weak acid groups also exhibits a secondary fluorescence. Yet, in a number of insect species, such NSM fluoresces weaker, and its fluorescence fades more rapidly in the ultraviolet light (Figs. 4-11 and 4-18). However, in *Calliphora vicina,* for example, the NSM of the A1 and A2 cells fluoresces equally intensely despite great differences in the ratio of their strong and weak acid groups (Figs. 4-21 and 4-22). Therefore, results of PIC staining are largely indefinite, like those obtained with PF technique.

Thus, staining with AB-AY or with PTh-PF permits a distinction of at least two different types of PF-positive NSM. But they do not reveal the so-called B cells, whose NSM remains acidophilic despite permanganate oxidation. In contrast to the data by Wendelaar Bonga (1970) on the staining of the same NSM with both AY pH 2.5 and phloxine in the snail *Lymnea stagnalis,* we have never observed this phenomenon in any insect studied (*Eurygaster integriceps, Calliphora vicina, Hyphantria cunea, Hyalophora cecropia*). The B cells of these insects were always as colorless as the background in sections stained with AB-AY technique (Figs. 4-7, 4-8, 4-13, 4-14). The B cells also cannot be seen in PIC preparations (Figs. 4-17–4-20, 4-23, 4-24).

The B cells differ sharply from various types of A cells in sections stained with ABPh or PThPh. These staining methods give basically similar results provided that AB stains at pH 2.5. Yet, the PThPh technique possesses two main advantages as compared with the ABPh stain. First, PTh stains the A-NSM a much more saturated deep blue in comparison to the bright blue of AB. This permits the distinction of very small granules of A-NSM after PTh staining. Second, AB at pH 2.5 sometimes stains the neuropil and all cell nuclei pale blue. On the other hand, PTh leaves

---

Figs. 4-25 through **4-32:** Demonstration of NSC in situ. P.F. *c.a.* = corpus allatum; *c.c.* = corpus cardiacum; *L1* = the 1st group of lateral NSC; *M1, M2, M3* = the 1st, 2nd, and 3rd groups of medial NSC.

Fig. 4-25. *Chironomus* sp., larva; the NSC of the group M3 and their processes (*arrows*) running into ipsilateral brain hemispheres are focused. × 144.

Fig. 4-26. The same; more deeply located NSC of the group M2 and their processes (*arrows*) running into contralateral brain hemispheres and then into subesophageal ganglion are focused. × 144.

Fig. 4-27. The same; a general view of the brain and retrocerebral endocrine gland complex. × 58.

Fig. 4-28. *Lyda nemoralis,* larva; an unusual position of NSC of the pars intercerebralis in the hind cellular cortex of the brain. × 96.

Figs. 4-29–4-32. *Hyphantria cunea,* mature larvae; staining of NSC *in situ* demonstrates distinctly an individual variability in position of the same NSC groups. × 96.

the background colorless if the fixation is satisfactory. This also increases the sensitivity of the PThPh staining method.

The B cells stain worse with light green (used separately or as a part of Halmi's mixture) as compared with phloxine. The contrast is weaker between acidophilic NSM and the cytoplasm in PF preparations counterstained with Halmi's mixture than in PThPh (or ABPh) preparations. This appears to be the reason why the B cells have not been found in, for example, the pars intercerebralis of many bugs (Dogra, 1967a).

For their relative simplicity, in situ methods of demonstration of neurosecretory elements in the whole ganglia or other organs have become widely used. Interest in these methods is also due to their permitting a comparison of neuronal distributions obtained by cobalt iontophoresis with the distribution of the NSC and their processes (Nijhout, 1975).

When using PF, PFAVB, PFAAB, PTh, and PFARF for in situ demonstration of the NSC, the possibility of revealing only some of several NSC types with these methods is valid. Staining with PFAVB or with PFAAB (at pH below 1.0) permits a clear detection of only those NSC and NSM that are rich in SS- and/or SH-groups. Other types of NSC are not visible or are only weakly visible in the whole ganglia. However, the advantage of these methods is their pale background coloration.

In contrast, by using PF or PFARF, we can, first, sharply reveal the NSC that were observed in PFAVB preparations. Second, the NSC are distinctly visible, NSM of which is poor in SS- and SH-groups but becomes basophilic after oxidation. As a result, the general number of the NSC observed in PF or PFARF preparations of the whole ganglia appears significantly larger than that seen in PFAVB or PFAAB preparations. This can explain the difference in NSC numbers revealed in specimens of the same species when using different staining techniques (Tandan and Dogra, 1966; Srivastava et al., 1974; Furtado, 1976).

PF and PFARF permit the observation of minute amounts of NSM, and they give sharper pictures than PFAAB or PFAVB. This is why they are particularly useful for following the course of neurosecretory fibers (Figs. 4-25–4-27). In this case, it is important to oxidize with performic acid and not with permanganate, the former oxidizing better the deeply located structures (Dogra, 1967b). However, as in sectioned material, the results of staining with PF and PFARF in toto are more indefinite, as far as NSC composition is concerned, than those obtained with PFAAB or PFAVB. Moreover, in the whole ganglia, it is more difficult to distinguish different cell types that can be incorrectly considered as asynchronously acting NSC of a single type.

In all cases, the staining in toto does not permit one to observe the B cells and other cell types that are poor in neurosecretion. Because of this, the set of brain NSC revealed by these methods appears to be very poor. For example, in the bug, *Dysdercus koenigii,* 18 NSC designated as the A cells were revealed in the whole pars intercerebralis after in toto staining (Awasthi, 1972). On the other hand, in sections of the brain of this bug, we found (Panov, 1972) that the pars intercerebralis includes 10 A, 8 a, about 45 B, 12 C, and 4 D cells; in other words, about 76 NSC of 5 different types or 4 times more than was found after in toto staining. Similar data can also be reported for other insects.

In 1963, Raabe described particular NSC, found in the tritocerebrum of many insects whose NSM stained well with azocarmine G after prolonged fixation in Helly's fluid but was GChH- and PF-negative. These NSC were further found to be numerous and widespread in the ganglia of the central nervous system. Those that are located in the thoracic or abdominal ganglia form a neurosecretory depot known as the perisympathetic organs (Raabe, 1971).

The azocarmine-positive NSC named by the authors "C cells" were then found to be histochemically heterogeneous (Raabe and Monjo, 1970; Baudry and Baehr, 1970). Some of them show an affinity to phloxine (Furtado, 1971) and therefore can be identified with the B cells revealed by classical neurosecretory stains.

Indeed, when staining the alternate sections of *Eurygaster integriceps* and *Eurydema spectrabilis* brains fixed in Helly's mixture with PThPh and with Heidenhain's azan, we found almost complete coincidence of structures with strong affinity to phloxine and to azocarmine. Attention should be paid to the fact that the sensitivity of staining with azocarmine for demonstrating acidophilic NSM is considerably higher than that with phloxine. Because of this, in sections stained with Heidenhain's azan, we could follow minute neurosecretory B-type fibers that were feebly visible after staining with PThPh.

Conclusions

The multiplicity of the neurohormones released from the insect nervous system appears to correspond to the diversity of the NSC types elaborating them. Tinctorial methods of NSC differentiation are one means to establish their diversity. Other approaches should also be used to estimate the cell status, i.e., particular cell types or different functional states of the same cell type. For instance, the subdivision of the A cells of the pars intercerebralis into A′ and A″ cells in several Lepidoptera as well as of the B cells into B_l and B_s cells in *Hyphantria cunea* could be mainly carried out taking into account the stable difference in size of their perikarya (Panov and Kind, 1963; Panov and Melnikova, 1974).

Present-day methods for demonstrating the NSM are unequal in their potentiality. This should be kept in mind when planning an investigation.   Staining of neurosecretory elements in situ can be recommended primarily for establishing the position of the NSC of certain types and for following the course of their processes.

In situ methods should not be used if the purpose of the study is the estimation of the complete set of NSC or the search for NSC involved in the control of a given function.

For staining in sections, PThPh (ABPh) and PTh-PF methods can be advised.   A combination of these techniques permits the resolution of the maximum diversity of the NSC types presently known in insects.   Taking into account the widespread occurrence of the azocarmine-positive NSC in insect nervous system, the complete investigation of the NSS should include staining with Heidenhain's azan method after prolonged fixation in Helly's fluid.

Appendix: List of Suppliers

1. Allied Chemical Co.
2. BDH Chemicals Ltd., Poole BH I2 4 NN, England
3. Chroma-Gesellschaft, Schmid & Co., Stuttgart-Untertürkheim, BRD
4. Edward Gurr Ltd., Michrome Laboratories, 42 Upper Richmond Road West, London S.W.I4, England
5. E. Merck A. G., Darmstadt, BRD
6. Fluka AG Chemische Fabrik, CH-9470 Buchs, Switzerland
7. Gee Lawson Chemicals Ltd., 6/8 Sackville Street, London, England
8. J. R. Geigy S. A., Bâle, Switzerland
9. Georg T. Gurr, Searle Scientific Service, High Wycombe, Buchs, England
10. Hopkins & Williams Ltd., Chadwell Heath, Essex, England
11. Kepec Chemische Fabrik GMBH, Siegburg, BRD
12. Schuchardt GMBH & Co., München, BRD
13. VEB Laborchemie, Apolda, Jena, DDR

Chapter 5

Histochemical Demonstration of Biogenic Monoamines (Falck-Hillarp Method) in the Insect Nervous System

Nikolai Klemm*

The Johns Hopkins University School of Medicine
Baltimore, Maryland

The biogenic monoamines (β-arylethylamines and β-arylethanolamines) dopamine (DA),[1] noradrenaline (NA), adrenaline (A), and 5-hydroxytryptamine (5-HT) are considered to be neuronal transmitter substances in vertebrates (Carlsson, 1974; Axelrod, 1975) and invertebrates (Sakharova, 1970; Murdock, 1971; Welsh, 1972; Gerschenfeld, 1973; Kerkut, 1973; Klemm, 1976). They were found to be stored mainly in vesicles (Adèn et al., 1969; Smith, 1972). Chemical and histochemical analyses for biogenic monoamine content have been performed on insect central nervous systems (CNS) (Frontali, 1968; Klemm, 1968a, 1971a, 1974, 1976; Björklund et al., 1970; Elofsson and Klemm, 1972; Klemm and Björklund, 1971; Klemm and Axelsson, 1973; Ramade and L'Hermite, 1971; Muskó et al., 1973; Robertson, 1976) and stomatogastric nervous systems (Klemm, 1968b, 1971b, 1978; Chanussot et al., 1969; Chanussot and Pentreath, 1973; Gersch et al., 1974), on nerve fibers innervating visceral muscles (Klemm, 1972, 1978) and salivary gland (Klemm, 1972; Bland et al., 1973; Fry et al., 1974; Robertson, 1975), and in the corpora

[1] Abbreviations used: A, adrenaline; α-MNA, α-methyl-noradrenaline; CA, catecholamine; CNS, central nervous system; COMT, catecholamine-O-methyltransferase; DA, dopamine; DOPA, 3,4-dihydroxphenylalanine; 5-HT, 5-hydroxytryptamine; 5-HTP, 5-hydroxytryptophan; MAO, monoamineoxidase; NA, noradrenaline; IA, indolylalkylamine.
* Present address: Universität Konstanz, Fakultät für Biologie, D-7750 Konstanz, F.R.G.

cardiaca (Klemm, 1971b, 1972, 1976; Klemm and Falck, 1978; Lafon-Cazal and Aurluison, 1976; Gersch *et al.*, 1974). Apart from in nervous systems, biogenic monoamines are also observed in the hemolymph (Wirtz and Hopkins, 1977) and in the epidermal system, where they are involved in the sclerotization process (Sekeris and Karlson, 1966; Andersen, 1974; Bodnaryk *et al.*, 1974; for review, see Murdock, 1971; Klemm, 1976). In addition, biogenic monoamines and/or their immediate precursor amino acids are involved in neurosecretion in insects and thus may be influenced by the physiologic state of the insect (Klemm, 1976; Klemm and Falck, 1978). Since biogenic monoamines subserve varied physiologic functions and have a wide anatomical distribution in insects, chemical biogenic-monoamine analyses of whole insects or whole tagmata are of limited value. Chemical biogenic-monoamine analyses should be accompanied by histochemical data to ensure that one is dealing exclusively with intraneuronal amines. Conversely, observations obtained by histochemical analyses should be confirmed chemically.

There is a paucity of information about synthesis, release, and inactivation of biogenic monoamines and their receptors in insects (for review, see Klemm, 1976). Obviously, biogenic monoamine metabolism in insects differs from that observed in vertebrates: monoamine oxidase (MAO) and catecholamine-O-methyltransferase (COMT), the two major degradative enzymes for monoamines in vertebrate nervous systems, are almost absent in insect nervous tissue (Blaschko, 1974; Nishimura *et al.*, 1975; Dyck and Robertson, 1976; Vaughan and Neuhoff, 1976; for review, see Murdock, 1971; Klemm, 1976). Instead, N'-acetyltransferase may be the main enzyme for monoamine metabolism in the insect nervous system (Dewhurst *et al.*, 1972; Evans and Fox, 1975; Nishimura *et al.*, 1975; Vaughan and Neuhoff, 1976) and N'-acetylated catecholamines are also known to take part in sclerotization processes (Sekeris and Karlson, 1966; Andersen, 1974; for review, see Klemm, 1976).

General Considerations

The histochemical gaseous formaldehyde method of Falck and Hillarp for the intraneuronal demonstration of biogenic monoamines is highly specific and sensitive. With this method as little as 5×10^{-4} pg (about 5×10^{-6} pmol) of DA and NA can be detected in a varicosity (Jonsson, 1971; Björklund *et al.*, 1975). A typical vertebrate varicosity was calculated to contain 6×10^{-6}–4×10^{-5} mol NA. Lower amounts were estimated for the cell body (6×10^{-8}–6×10^{-7} mol) and even lower levels for the preterminal area (Jonsson, 1971). The recently developed glyoxylic acid method is even more sensitive (especially after perfusion of the tissue with glyoxylic acid) for catecholamine (CA) (10^{-7} pmol could be detected) but not for 5-HT (Björklund *et al.*, 1975). Since the glyoxylic acid method has not yet been successfully applied to insect tissue, only the formaldehyde method will be treated in this article. Recently a method was described applying glyoxylic acid without perfusion of tissue in mammals (Watson and Barchas, 1977). In the future, this method may be applied also to central nervous tissue in insects.

A thorough understanding of the reactions taking place during the histochemical procedure of the Falck-Hillarp method is essential for its successful performance. The method is based on selective, sensitive, and controlled chemical reactions of the monoamine molecule, which is transferred into a strongly fluorescent compound within the neuron. The monoamines and their fluorophores are water soluble at temperatures above 2°C (Watson and Barchas, 1977), and the optimal treatment is in the gas-phase formaldehyde reaction, which prevents dislocation of the fluorescent compound from its original sites. Replacement of gaseous formaldehyde by formaldehyde solution has to date been unsuccessful (Sakharova and Sakharov, 1971).

The chemical reactions of fluorophore formation are now well understood (Corrodi and Jonsson, 1965; Caspersson *et al.*, 1966; Jonsson, 1971; Björklund *et al.*, 1975). In the initial step, the Pictet-Spengler reaction, the phenylethylamines or indolylethylamines react with formaldehyde to form 1,2,3,4-tetrahydroisoquinoline (Fig. 5-1, I,III) or 1,2,3,4-tetrahydro-β-carboline (Fig. 5-2, II), respectively. This reaction requires a high electron density at the point of ring closure of the amine, i.e., the 6-position of phenylethylamines and the 2-position of indolylethylamines (Figs. 5-1 and 5-2) (Whaley and Govindachari, 1951). Tyramine and octopamine lack substitutes at the 3-position of the ring and are thus nonfluorogenic (Björklund *et al.*, 1975). The initial reaction is catalyzed by H^+ ions (Whaley and Govindacari, 1951); therefore, small amounts of HCl gas, present during the formaldehyde treatment, can markely increase the fluorescent intensity (see HCl-Paraformaldehyde Treatment, p. 61).

The second step of the fluorophore formation is also promoted by formaldehyde and proceeds in two ways: (1) autocyclation (Fig. 5-1, II and V; Fig. 5-2, III) and (2) acid catalysis to 2-methyl derivatives (Fig. 5-1, II and V; Fig. 5-2, IV). Both substances have identical fluorescent spectra (Björklund *et al.*, 1975). At neutral pH isoquinolines appear in the highly fluorescent tautomeric quinoidal form (Fig. 5-1, III and VI) with fluorescence characteristics Ex_{max} 410/Em_{max} 470–480 nm. IAs, especially those substituted at the 5-position, react in a similar way yielding a highly fluorescent β-carboline compound (Fig. 5-2, III). In an acid environment, the nonquinoidal form persists, which is reflected by the spectrum (Fig. 5-1, II and V). Fluorophores with a hydroxy group at position 4 (derived from the β-hydroxy group of the side chain as NA) are trans-

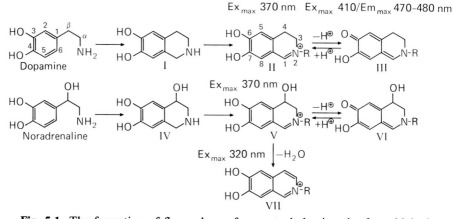

Fig. 5-1. The formation of fluorophores from catecholamines by formaldehyde. First step: Pictet-Spengler cyclization leads to the formation of 6,7-dihydroxy-1,2,3,4-tetrahydroisoquinoline (I) and 4,6,7-trihydroxy-1,2,3,4-tetraisoquinoline (IV). Second step: Two types of dehydrogenated isoquinolines are formed (II,V) —6,7-dihydroxy-3,4-dihydro derivatives (R = H) and 6,7-dihydroxy-3,4-dihydroiso-2-methyl derivatives (R = CH). At neutral pH the fluorophores are in their tautomeric quinoidal forms (III,VI). By short HCl treatment the fluorophores are converted to the nonquinoidal form (II,V). After prolonged HCl treatment the noradrenaline-fluorophore is converted into the fully aromatic 6,7-dihydroxyisoquinoline (VII). [Altered after Björklund *et al.*, 1975.]

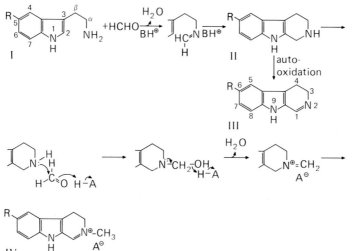

Fig. 5-2. The formation of fluorophores from indolylalkylamines (I) by formaldehyde. The reactions are analogous to that of Fig. 5-1, II, 1,2,3,4-tetrahydro-β-carboline; III, 3,4-dihydro-β-carboline; IV, 2-methyl-3,4-dihydro-β-carboline derivative. (Tryptamine, R = H; 5-hydroxytryptamine, R = OH.) [Altered after Björklund *et al.*, 1975.]

ferred into a fully aromatic isoquinoline compound, with an Ex_{max} of 320 mn after prolonged acid treatment (Fig. 5-1, VII) (Corrodi and Jonsson, 1966; Björklund et al., 1968). Differentiation of DA from NA is based on this reaction (see Microspectrofluorometry, p. 69).

The fluorophores generally appear in granules in the neuronal cytoplasm. Some diffusion from the granules can be seen even in good tissue specimens. Some of the extragranular fluorescence may result from loss from granular sites. The ultrastructural correlates of the fluorescent granules are not yet fully understood. Electron-microscopic studies have suggested the existence of clusters of vesicles (van Orden et al., 1970; Eränkö, 1972) as well as a nonvesicular amine pool (Tranzer, 1972) in vertebrate sympathetic ganglia (Eränkö, 1976). In insects, fluorescence intensity could be correlated with the quantity of vesicles (Schürmann and Klemm, 1973; Klemm and Schneider, 1975; Klemm, 1976).

The fluorescence intensity is directly proportional to the amount of amine in the neuron when present in low concentration. Despite the fact that higher concentrations of biogenic monoamines produce quenching of fluorescence, so that a 50% reduction in the pool in a neuron cannot be detected by a change of fluorescent intensity (Jonsson, 1971), and despite recent results suggesting that fluorophore formation is not quantitative (Lindvall et al., 1974), this method has been employed in semiquantitative estimations in certain areas of the vertebrate brain (Jonsson, 1971). However, semiquantitative analyses, based on fluorescence intensity, could not be performed in insect tissue since the fluorescence intensity varies from specimen to specimen even when treated the same way at the same time (Klemm and Schneider, 1975; Klemm, 1976).

Although the method is very sensitive, some structures might slip detection because of low concentrations of intraneuronal fluorogenic amines, e.g., DA, in certain areas of the vertebrate diencephalon (Brownstein et al., 1976). Serotoninergic neurons are usually very thin, and the 5-HT fluorophore fades fast (Elofsson and Klemm, 1972; Klemm, 1974, 1976; Klemm and Schneider, 1975).

Many compounds, chemically related to biogenic monoamines, can take part in the Pictet-Spengler reaction to form fluorophores. Alterations of the side chains of fluorogenic amines and amino acids do not alter their spectral characteristics (Björklund and Falck, 1973). The precursor amino acids 3,4-dihydroxyphenylalanine (DOPA) and 5-hydroxytryptophan (5-HTP) (but not tyrosine or tryptophan) will form fluorophores with the same spectral characteristics as their product amines. Although it is not yet possible to differentiate between precursor amino acids and their amines in tissue, their differentiation in model systems has been reported (Lindvall et al., 1975). Since the intraneuronal concentration of precursor amino acids is very low*, this drawback is less important. In contrast, alterations of the ring do alter spectral characteristics (Björklund and

* So far only small amounts ($\leqq 0.1$ μg/g.w.w.) of DOPA were detected in insect CNS (Klemm and Murdock, unpublished).

Falck, 1973). This is true also of peptide-bound fluorogenic monoamines or amino acids. Their initially low fluorescent yield can be increased by ozone treatment (Håkanson et al., 1971). Their spectral characteristics may differ somewhat from those of free monoamines but the spectral characteristics of NH_2-terminal DOPA and 5-HTP are similar to those of DOPA and 5-HTP, respectively (Håkanson et al., 1971; Partanen, 1975). Although mammalian neural tissue has provided no evidence for such compounds (Björklund et al., 1972), it is possible that they may appear in insect neurosecretory material (Klemm, 1971b, 1976; Klemm and Falck, 1978).

Procedure

Dissection

Decapitate the insect under light ether anesthesia. The animal under treatment must be fully alive. In moribund insects monoamine fluorescence is rarely achieved. There are two ways to handle the nervous tissue: (1) Treat whole tegmata, i.e., the nervous system inside the head or segments. Try to take off all heavily sclerotized parts such as mandibles, legs, wings. (2) Dissect out the nervous tissue and treat it separately. If the tissue to be investigated is small, place it on a little piece of vertebrate or mollusk tissue (heart, liver, pancreas) as a carrier. Better results have been obtained with method 1 (Klemm, 1968a).

Deep-freezing

For a detailed description of the theoretical background of the freeze-drying procedure see Björklund et al. (1972). Since the procedure has to be performed under nearly dry conditions, the tissue must be brought to the temperature of liquid nitrogen as rapidly as possible and then freeze-dried. To achieve this, the tissue is first dipped in propane:propylene (9:1, v/v) cooled in liquid nitrogen. Usually the commercially available propane is contaminated with an appropriate amount of propylene (pure propane solidifies at the temperature of liquid nitrogen and cannot be used). An immediate dipping of the tissue into liquid nitrogen would cause a layer of vaporized nitrogen around its surface and thus prevent the tissue from fast cooling. Fast cooling is essential because ice crystals can form and damage the tissue. After quenching in propane-propylene the tissue is transferred into liquid nitrogen.

Procedure for Deep-freezing

1. Condense the propane-propylene mixture in a metal beaker in liquid nitrogen and cover the container with a cloth to prevent oxygen and water condensation in the mixture.

2. Kill the animal and place the dissected sample on a paper strip with a code number on its back. Tagmata can be glued to the paper strip with hemolymph. Transfer the paper strip with forceps and dip it into the cooled propane-propylene mixture for 5–10 s and then transfer it into the metal holder covered by liquid nitrogen (Fig. 5-3). The tissue must be submerged in liquid nitrogen (for cryostat sectioning: Adhere the tissue, if necessary, with saline solution or hemolymph into the propane-propylene mixture. After 5–15 s transfer it into liquid nitrogen for storage).

Freeze-drying

This step is the most crucial in the entire procedure. Several available freeze-dryers were tried. The best results were achieved in the "cold finger" model (see Fig. 5-6) in which up to 100 preparations can be treated simultaneously. The freeze-dryer removes water from the rapidly frozen tissue by sublimation at a temperature below the freezing point of the tissue. The tissue must be placed in a vacuum. A trap ("cold finger" or phosphorous pentoxide) must be installed to remove water continuously. A two-stage mechanical pump in combination with an oil-diffusion pump (Fig. 5-4) is used to achieve a vacuum at about 10^{-6} Torr. A cooling mixture of solid carbon dioxide and acetone is used which gives a temperature below $-80°C$. The cup with the preparations (Fig. 5-5) is kept

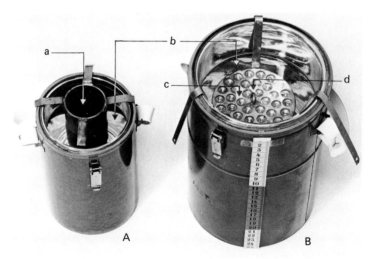

Fig. 5-3, A and B. Equipment for freezing tissue preparations. **A** The beaker with propane-propylene (*a*) is suspended in liquid nitrogen (*b*) in a Dewar vessel. **B** The metal holder with numbered holes (*c*) for frozen tissue specimens is hanging in a metal frame beneath the surface of liquid nitrogen in a Dewar vessel (metric scale). (*d*) Metal hook for the metal holder. [Reproduced with permission from Björklund *et al.*, 1972.]

in a cooling ethanol (96%) bath. The temperature has to be thermostatically regulated. The drying procedure lasts about 3–5 days. When the drying is completed the preparations must be kept above room temperature to avoid water condensation on the surface of the tissue, because the preparations are hygroscopic. When the preparations are exposed to air, dry conditions should be maintained and if necessary a cold trap (Fig. 5-4f) should be used.

Procedure for Freeze-drying

1. Switch on the compression pump 2–3 h before beginning to freeze-dry, and lower the temperature in the Dewar unit (Figs. 5-4,5-5) to −40°C. Immerse the glass cup (Fig. 5-5, d) into the alcohol bath and seal it with a plastic cover, using high-vacuum silicone grease, to prevent water condensation on the inner surface of the cup. Place 200 ml acetone into the cold finger.

2. Remove the cover from the cooled cup and move the metal block with the preparations from liquid nitrogen to the bottom of the cup (Fig. 5-5, d,f). After replacing the plastic cover, wait until the residual liquid nitrogen boils away (1–3 min). Grease the flange of the outer tube of the column with high-vacuum silicone grease.

3. Start the backstage pump. Take off the cover and fasten the cup to the column. Hold it in position until it is fixed by vacuum (Fig. 5-5).

4. Suspend in the alcohol bath. Set the thermostat (Fig. 5-5, c) at −20°C. This temperature should be maintained throughout the drying procedure.

5. After 20 min check the vacuum with a Tesler coil (high frequency test) and start the oil diffusion pump.

6. Fill the inner tube (Fig. 5-4, a, 6) with solid carbon dioxide (Fig. 5-6, e).

7. Freeze-dry 3–5 days, depending on size and type of tissue (for insect heads or body tegmata, 5 days). Check vacuum daily and measure the dry-ice column with a ruler.

8. Switch off the compressor one day before breaking the vacuum.

9. On the final day replace the alcohol bath and its cooling coil by a thermostatically controlled water bath (35°C) for 4 h. To facilitate removal of the cup, reduce the viscosity of the vacuum grease in the seal with warm air from a hair dryer.

10. Fifteen minutes before breaking the vacuum stop the oil diffusion pump.

11. Open the air inlet (Fig. 5-4, e) slowly. In case of high air humidity place a small Dewar vessel with liquid nitrogen around the cold trap (Fig. 5-4, f).

12. Remove the cup and close it with a plastic cover.

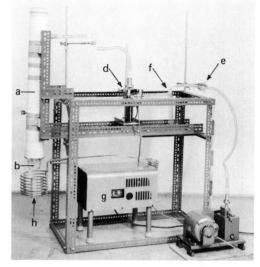

Fig. 5-4. The "cold finger" type freeze dryer. The freeze dryer column is enclosed in a fine-meshed metal net covered with gauze for safety (*a*); inner tube (cold finger) (*b*); two-stage mechanical vacuum pump (*c*); air-cooled oil diffusion pump (*d*); air inlet (*e*); cold trap (*f*); compression-expansion unit (*g*); cooling coil from the compressor (*h*). [Reproduced with permission from Björklund *et al.*]

Fig. 5-5. Detail of Fig. 5-4. Lower part of the freeze-dryer column (*a*); cold finger (*b*); thermometer to control the compressor thermostat (*c*). The glass cup (*d*) with the metal holder carrying the frozen tissue samples (*f*) is sealed to the flange of the outer tube of the column (*a*). Ethanol bath (*e*). [Reproduced with permission from Björklund *et al.*, 1972.]

13. The preparations in the metal block can be stored overnight in dry condition, i.e., in a desiccator with phosphorous pentoxide. However, from this point, delay in the procedure could be detrimental to the results. *Attention:* the freeze-dried specimens are very fragile. Use entomologic forceps.

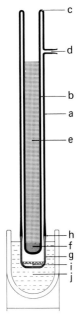

Fig. 5-6. The column of the cold finger type freezer dryer. Outer tube (*a*); inner tube (*b*); connection between inner and outer tube (*c*); the surfaces between both tubes are silvered. Connection to the vacuum line (*d*); solid carbon (*e*) and acetone (*f*) in the inner tube; glass cup, closing the outer tube (*g*) at a flange (*h*); metal holder (*i*) with holes for tissue samples; ethanol bath (*j*). [Reproduced with permission from Björklund *et al.*, 1972.]

Stretch Preparations

Thin tissue sheaths, visceral muscles, retrocerebral endocrine glandular complex, stomatogastric nervous system, salivary glands, and ventral nerve cord, especially of small animals, can be dried at room temperature or *in vacuo* as originally described by Falck (1962) for the rat iris. This method is easy to perform and was successfully applied to insects (Chanussot and Pentreath, 1973) and crustacea (Cooke and Goldstone, 1970). An additional method for freeze-drying very small specimens *in toto* was described by Sulston *et al.* (1975) for nematodes. A promising method for treating insect stretch preparations may be the glyoxylic acid method as published by Lindvall *et al.* (1974).

Procedure for Stretch Preparations

1. Dissect out the tissue, and stretch it out on a microscope slide or pin it on a planchet coated with a thin layer of silicone resin (Cooke and Goldstone, 1970).

2. (a) Freeze the tissue via propane-propylene and liquid nitrogen, and dry it in a freeze dryer (see p. 57); (b) air-dry the tissue at room temperature or dry it in a desiccator over phosphorous pentoxide for 1–4 h (in case of ganglia 4 h) and treat as for routine paraformaldehyde treatment.

Routine Paraformaldehyde Treatment

For histochemical treatment the water content of the paraformaldehyde is important. To adjust water content, heat the paraformaldehyde at 100°C for 1 h and store it in desiccators at constant relative humidity for 5–7 days before using. The relative humidity can be achieved over sulfuric acid of different concentrations in the desiccator (Hamberger *et al.*, 1965). High relative humidity (80%–95%) gives a strong fluorescence but shows diffusion. Distinct but weak fluorescence can be achieved under dry conditions (30%). For routine use, best results in most insect tissues were achieved at 70% relative humidity. The CNS of some insect species require different relative humidities.

Procedure for Routine Paraformaldehyde Treatment
Transfer the deep-frozen specimens carefully into a 1-liter glass vessel containing 5 g paraformaldehyde with an adjusted amount of humidity. Seal the vessel with a plastic cover and heat it in an oven at 80°C. During heating, paraformaldehyde depolymerizes to gaseous formaldehyde. With this treatment, the primary CAs develop maximal fluorescence intensity in 1–1.5 h. For adrenaline treatment, about 3 h are necessary.

After paraformaldehyde treatment, the specimens remain hygroscopic and should be embedded immediately. If the specimens must be stored overnight, keep them in a desiccator over phosphorous pentoxide and in darkness. The quality of fluorescence is reduced after storage.

HCl-Paraformaldehyde Treatment

5-HT reacts less with paraformaldehyde than primary CA (Fuxe and Jonsson, 1967). The reaction can be enhanced by the presence of a minute amount of HCl gas during treatment with paraformaldehyde originally used to demonstrate tryptamine and 3-methylated phenylethylamines (Björklund and Stenevi, 1970).

Procedure for HCl-Paraformaldehyde Treatments
1. The 1-liter vessel with 5 g paraformaldehyde at a relative humidity of about 50% is provided by a multisocket cover connected to a manometer (Björklund *et al.*, 1972). The humidity of the specimens should be standardized. Place the preparations into the vessel. This procedure can also be performed on deparaffinized sections.
2. Evacuate the vessel.
3. Suck HCl-saturated air from a 1-liter vessel containing 200 ml 6 N HCl. Control the amount of HCl gas with a manometer. HCl gas at 100–300 Torr is optimal for insect tissues.
4. Evacuate the vessel.

5. Repeat step 3 (when treating deparaffinized sections, step 5 can be omitted).

6. Let normal air at atmospheric pressure into the reaction vessel and heat to 80°C (for another method, see Procedure for Routine Paraformaldehyde Treatment, p. 61).

Paraffin Embedding and Sectioning

Proceed with embedding as soon as possible. Embedding should be performed *in vacuo*. A suitable container for vacuum embedding is shown in Fig. 5-7. It consists of two parts (a, b) connected by a standard ground-glass fitting, and a side channel (Fig. 5-7, c) connected to a vacuum pump, interrupted by a stopcock.

Procedure for Paraffin Embedding and Sectioning

1. Put degassed paraffin wax of 52–54°C melting point into the chamber (a) and immerse it in a water bath (60°C).

2. Place specimens in chamber b and connect both parts of the container.

3. Evacuate the container and tip the specimens into the melted paraffin by tilting the whole vessel. The infiltration time is about 10 min.

4. Remove the specimens and reincubate in paraffin wax with a higher melting point (60°C). (Not all types of paraffin wax give equally good results. Best results were achieved with WIV Paraffin für Acetoneinbettung, Heidelberg, West Germany.) Be careful not to overheat the tissue. Hot paraffin may extract the fluorophore (see Björklund *et al.*, 1972) and increase background fluorescence. If whole tegmata were freeze-dried, scale off the paraffin wax together with the peripheral sclerotized tissue and re-embed.

Fig. 5-7. Glass container for vacuum embedding in paraffin. Vessel (*a*) contains paraffin and is immersed in a water bath. The freeze-dried specimens are in vessel (*b*). After evacuation through the outlet (*c*) the tissue sample is tilted into the vessel (*a*). [Reproduced with permission from Björklund *et al.*, 1972.]

5. Prepare a paraffin block (blocks can be stored for several weeks).

6. Prepare slides: Clean in dichromate solution overnight and wash in running water for 12 h; rinse in alcohol and air dry.

7. Section preferably at 8–14 µm. For routine examination, put the sections on a slide on a hot plate (60–65°C). Allow the sections to spread slowly. Press the parts not touching the slide with a smooth, dust-free brush. Do not allow the paraffin to melt. As soon as the paraffin starts to get transparent, take the slide off and cool it before the sections become opaque. If the sections are treated according to this protocol, it is unnecessary to remove the paraffin with xylene (Fig. 5-8) which may damage the section (Klemm, 1968a).

For post-treatment (see p. 62) and microspectrofluorometric analyses (see Procedure for Paraffin Embedding and Sectioning, step 7), cover the slides with albumin-glycerol. Heat the slides to 70–80°C to evaporate water, and let them cool. Proceed as for routine treatment. Remove the slide with the slightly melted preparations from the hot part of the plate and leave it on a cooler part of the plate for a while.

Fig. 5-8. Oblique frontal section through the brain of the adult fly *Calliphora vomitoria* after treatment with formaldehyde. The fluorescent structures demonstrate CA. *c* cell body; *cb* central body complex; *nl* neurilemma; *g* fore gut. Scale 50 µm.

8. Now the preparations can be examined with the fluorescent microscope. Light scattering can be reduced by mounting in nonfluorescent immersion oil and adding a coverslip.

9. Examine the preparations as soon as possible. Covered sections should not be stored longer than 24 h at room temperature. (Noncovered sections should be examined within a few hours of sectioning.) They can be stored at −20°C for up to 2 weeks.

Plastic Embedding

Very thin sections (less than 0.5 μm) can be embedded in epoxy resins instead of paraffin (Hökfelt, 1965). The drawbacks are low fluorescence intensity and a higher background fluorescence.

Procedure for Plastic Embedding

Instead of paraffin, the specimens are infiltrated in vacuo in a Durcupan ACM (Durcupan ACM-Fluka, Switzerland)—epoxy resin:964 hardener:dibutylphthalate:964 accelerator (8:12:0.10:0.1). Use the same mixture for further infiltration for 2 h at 50°C and for final embedding (Hökfelt, 1965). Section on ultramicrotome under dry conditions.

Post-treatment

Preparations for histochemical visualization of biogenic monoamines can be post-treated for histology. Silver impregnation of histochemical sections has so far been unsuccessful (Elofsson and Klemm, 1972). However, post-treatment for "neurosecretory staining" has been achieved (Klemm and Falck, 1978).

Procedure for Post-treatment of "Neurosecretory Staining"

1. Cover microscope slides with albumin-glycerol and allow water to evaporate (see Procedure for Paraffin Embedding and Sectioning, step 7).

2. Proceed as for fluorescence histochemistry.

3. Make photographs of the sections, and mark their positions.

4. With a scalpel remove as much of the surrounding paraffin as possible.

5. If necessary, postfix in alcoholic Bouin's solution for 2–5 h; otherwise proceed to 6.

6. Dip the preparations into xylene carefully, to dissolve paraffin.

7. Dip the slide into a mixture of 100% alcohol:ether (1:1).

8. Dip the slides into a solution of 1% celloidin in ether for a few seconds. Remove and allow excessive celloidin solution to run onto filter paper.

9. Treat the sections further as for routine chrome hematoxylin-phloxine or aldehyde fuchsin staining (see Klemm and Falck, 1978).

10. Before mounting remove the celloidin with a 100% alcohol:ether mixture prior to the xylene step.

11. Mount as usual.

In Vitro Studies

Neuronal transmitter substances are released from presynaptic sites and act on pre- and postsynaptic receptors. Their action is terminated by inactivation of the active compound. This is accomplished by reuptake into the releasing sites (uptake 1). Since the uptake process is directed against the concentration gradient, it requires energy and is thus temperature dependent. It also requires the presence of certain cations. Nontemperature-dependent accumulation of externally applied drugs is considered nonspecific, although some specificity concerning the sites of accumulation is possible (Klemm and Schneider, 1975; Klemm, 1976).

Because the uptake process is not very selective, higher concentrations of applied substances are taken up by other neurons, e.g., NA by 5-HT neurons (Snyder and Coyle, 1969), or even other tissue (Ascher *et al.*, 1968). Requirements for uptake are also fulfilled by false transmitters, such as α-methyl-noradrenaline (α-MNA) and 6-HT, which after treatment with paraformaldehyde give a stronger fluorescent compound than the natural amines (Fig. 5-9).

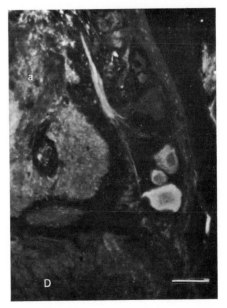

Fig. 5-9. The brain of the adult desert locust *Schistocerca gregaria* Forsk after incubation for 10 min in $5 \times 10^{-6}\,M$ α-MNA at 25°C. The periplasma of three CA-containing cell bodies and their processes, which have taken up α-MNA, can be followed. *a* α-lobe; *D* deutocerebrum. Scale 50 μm.

Incubation studies should be performed at concentrations below the K_m values for the drug uptake into sliced or nonhomogenized tissue (Ross, 1976; Ross and Renyi, 1975).   Unfortunately no data are available on transmitter uptake in insect systems; thus one must extrapolate from data from vertebrate tissue, bearing in mind that the insect nervous system does not resemble the vertebrate nervous system in all respects.   High-affinity uptake of 5-HT into cockroach crude synaptosomal fractions was found to have a K_m of $5 \times 10^{-8} M$ (N. N. Osborne and N. Klemm, unpublished).

Uptake studies allow the detection of neurons containing nonfluorogenic transmitter substances, which are probably closely related to fluorogenic ones (e.g., octopamine) and which share the uptake specificity with other aminergic neurons (Klemm and Schneider, 1975; Klemm, 1976) (Fig. 5-10).

The uptake abilities can be used (1) for neuroanatomical studies (Fig. 5-10) and (2) for pharmacologic studies.   The latter studies are beyond the scope of this article, and the interested reader is referred to Klemm and Schneider (1975), Klemm (1976), and Klemm and Falck (1978).

Procedure for Anatomical Mapping

1. Dissect out the tissue to be examined.

2. Preincubate the tissue in a tube for 5 min in adequate saline solution at 20–25°C in a water bath under constant shaking.   Preincubate control tissue at 0°C.

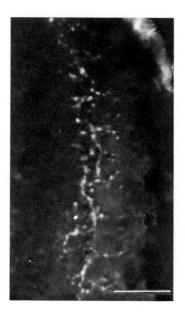

Fig. 5-10. The medulla of the optical lobe of *Schistocerca gregaria,* treated as in Fig. 5-9. Longitudinal arrangement of aminergic fibers, which do not contain fluorogenic amines, but selectively take up α-MNA and 6-HT.   Lateral is to the *right*.   Scale 50 μm.

3. Incubate both control tissue (0°C) and samples (20–25°C) in 10^{-5} M 6-HT, 10^{-6} M α-MNA, or 5×10^{-5} M L-DOPA[2] (you can also use the natural compounds 5-HT, DA, and NA for intensifying intraneuronal fluorescence) with constant shaking for 5–10 min.

4. Stop the incubation by placing the tubes in ice water or by replacing the saline solution with new 0°C saline solution.

5. Treat further as usual (see Routine Paraformaldehyde Treatment, p. 61).

Drugs such as biogenic monoamines, their immediate precursor amino acids, or false transmitters can also be injected into hemocoel. However, since these drugs are also accumulated in other tissues, such as excretory organs, fat body, and neurilemma, these studies should be preferentially performed as autoradiographic detection. After injection of ^{3}H-DA, label was shown to accumulate in α-lobes in cockroaches (Rutschke *et al.*, 1976). Combined fluorescence and autoradiographic studies are possible using Durcupan for embedding and further treatment for autoradiography as originally described by Lam and Steinman (1971).

Fluorescence Microscopy

Several types of fluorescence microscopes are available. Ordinary light microscopes equipped for fluorescence can be used. The source of excitation light is a high-pressure mercury lamp (Osram HBO 200 or HBO 100). Sufficient light intensity is achieved by a light beam preferably bent only once by a metalized front-surfaced mirror and focused into a dry dark-field condensor. Oil dark-field condensors give the highest contrast, but cover only a small part of the section. For lower magnification use a dry dark-field condensor. In some microscopes an additional insertion of a collector lens at the base of the tube often improves the light intensity. The author prefers the *Grosse Fluoreszenzanlage* (C. Zeiss, Oberkochen, West Germany) for transmitted light and the *Grosses Fluoreszenzmikroskop* (C. Zeiss, Oberkochen, West Germany) for incident light observations. Incident light is advantageous in only a few cases, e.g. when using objectives with high numerical aperture such as Apo 40.100 öl (C. Zeiss, Oberkochen) (Klemm, 1968a).

Filters: A heat absorption filter should be inserted first, close to the lamp. The primary filter(s) should be selected to achieve excitation as close as possible to the activation maximum (absorption maximum) of the fluorophores. For CA and 5-HT fluorescence Schott BG 12 or BG 3 filter(s) should be used. The background fluorescence can be lowered at

---

[2] These concentrations are given as general guidelines. Optimal concentrations should be determined for each insect species.

the expense of light intensity by using thicker or several filters of the same type.

For secondary filters (barrier filters) use filters absorbing below 500 nm (Schott OG 4). ' The combination of filters absorbing below 500 and 470 nm is optimal. This filter combination excludes the blue component of the emitted light and thus appears green (Em_{max} about 475 nm) for CA and yellow (Em_{max} 520 nm) for 5-HT and gives a good contrast within the spectral range of the optimal sensitivity of the human eye. For strong fluorescence, selection filters Schott SAL 480 or SAL 525 can be used as secondary filters. For originally blue CA fluorescence, the primary filters Schott AG 1 or BG 3, the interference filter Leitz AL 405 in combination with BG 3, or a monochromator (Ploem, 1969) can be used.

Characterization of the Fluorophores

While CAs exhibit green or blue fluorescence, 5-HT produces yellow fluorescence and can be easily differentiated from CAs. At high CA fluorophore concentrations, however, the CA fluorophore appears yellow to the eye, giving the impression of the presence of 5-HT fluorophore, but still has an Em_{max} of 475 nm [Betzold-Brücke effect (see Ritzèn, 1967)]. To avoid misinterpretation in case no microspectrometer is at hand, use more dry paraformaldehyde and/or reduce the exposure time to paraformaldehyde. 5-HT fluorophore fades faster than CA fluorophores in excitation light, which might help to differentiate between CA and IA fluorophores. Primary CAs can be differentiated from secondary CA (A). Secondary CA needs prolonged paraformaldehyde treatment to show optimal fluorescence. Adrenaline has not yet been found in insect nervous tissue (see Klemm, 1976). The primary CAs, DA and NA, can be differentiated by spectral analysis (see Microspectrofluorometry, p. 69).

Specificity Test

To avoid misinterpretations, specificity tests are necessary. Several simple methods can be used to test specificity:

1. Compare paraformaldehyde-treated and untreated specimens. Be aware that slight contamination of the room air by paraformaldehyde may cause specific fluorescence.

2. CA fluorophores and especially 5-HT fluorophore show photodecomposition when exposed to blue-violet light. Autofluorescence is almost stable to blue-violet light. However, trachea and tracheols often show green or yellow fluorescence, which fades in excitation light (Klemm, 1974).

3. Specific fluorescence, but often also that of the treacheal system, can be quenched when the sections are exposed to water.

4. Diffusion of the fluorophore occurs, especially in tissue not optimally treated for specificity: No diffusion of autofluorescence has yet been observed in insects.

5. Autofluorescent structures have different spectral characteristics. Thus, different primary filters drastically reduce the specific fluorescence intensity but often not the nonspecific ones. Spectral analysis is the safest way to identify specific fluorescence (the autofluorescence of tracheols of the CNS of some flies may expose the same Em_{max} as CAs but differs in Ex_{max}).

6. Treatment with alcoholic sodium borohydride ($NaBH_4$) reduces the fluorescent dihydroisoquinoline and dihydroxy-β-carboline to their corresponding nonfluorescent tetrahydro derivatives (Corrodi *et al.*, 1964). The following treatment with formaldehyde gas can regain the fluorescence. This method is somewhat difficult to perform and thus methods 1–5 are recommended.

Microphotography

Because the histochemical preparations cannot be stored, they must be photographed. Black and white pictures can be used for routine protocol pictures. For routine pictures Kodak Panatomic-X and Kodak Plus-X Pan have been used. The exposure time is dependent on the fluorescence intensity and was found to be between 30 s and 3 min. Structures with fast-fading fluorophores (5-HT) can be photographed with Kodak Tri-X Pan (ASA 400, 27 DIN) (exposure time, 10–45 s). To discern green from yellow fluorescent structures, color films must be used. Not all color films distinguish well between both colors. Several types of films were tested by the author. For CA fluorescence, Kodakchrome KX-135 (ASA 64, 19 DIN) gave best results (Fig. 5-11, Plate 3). Ektachrome High Speed (ASA 160 and ASA 400, 23 DIN or 27 DIN) was found to differentiate best between green and yellow fluorophore when optimal exposure time was used (Fig. 5-12, Plate 3). With excessive exposure time green fluorescence appears yellow. Ektachrome EX-135 (ASA 64, 19 DIN) can be used for both green and yellow fluorescence.

Microspectrofluorometry

Routine Microspectrofluorometry

The correct interpretation of the fluorophores can only be achieved by microspectrofluorometric analysis. The instrument must enable both excitation and emission spectra to be recorded. The excitation spectrum is obtained by measuring the varying wavelengths of the excitation light

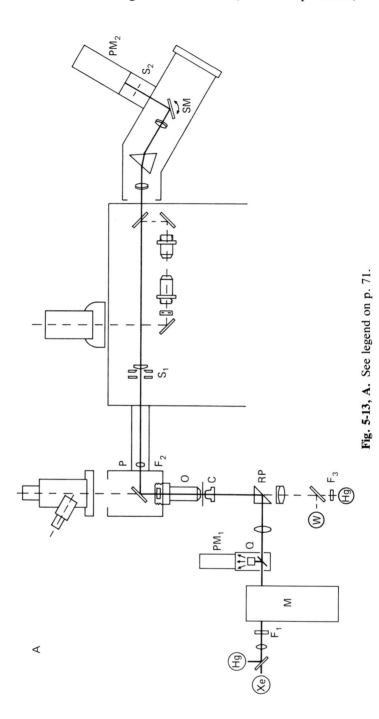

Fig. 5-13, A. See legend on p. 71.

Fig. 5-13, A and B. The modified Leitz microspectrofluorometer: **A**, schematic drawing. **B**, anterior view. Two illumination systems are installed. One pair of lamps (below in the figure) consists of an Osram HBO 200 mercury high-pressure lamp (Hg) and one tungsten lamp, which are used for scanning and searching purposes. F_3 is the filter of Hg (Schott BG3 or UG1). The other illumination unit (to the left) consists of an Osram HBO 200W/2d.c mercury high-pressure lamp (Hg) used for recording emission spectra, and an Osram XBO 75 high-pressure xenon lamp (Xe) for recording the excitation spectra. A mirror allows the selection of either of these light sources. F_1 Schott KG heat absorption filter; M Zeiss M 20 grating monochromator; Q quantum counter, with photo multiplier (PM_1) according to Ritzèn, 1967 (a small part of the light from the monochromator is reflected by a quartz plate onto a rhodamine β-containing cuvette in the quantum counter). RP rotating prism to switch between monochromator light and scanning light. Fluorescent light passes a cardioid dark field immersion glass or quarz condensor (C). O objective; F_2 barrier filter; P projective; S_1 slit placed in the focal plane of the projective (P) for adjustment of size and shape of the measuring area. This slit entrance is also the entrance slit of the analyzing prism monochromator. SM swinging mirror that receives the light from the analyzing prism monochromator and from which it goes to the exit slit of the prism monochromator (S_2). PM_2 measuring photo multiplier. The signals from the photo multiplier as the wavelength are recorded on an x-y recorder. [Reproduced with permission from Björklund *et al.*, 1972.]

while the intensity of the emitted light is registered at a fixed wavelength. The emission spectrum is recorded at fixed excitation light, close to the Ex_{max}. The excitation spectrum is recorded between 300 and 460 nm and the emission spectrum from 400 nm throughout the visual part of the spectrum.

Several instruments were described (see Björklund *et al.*, 1972). The instrument used by the author is the Leitz microspectrofluorometer as seen in Fig. 5-13.

To allow comparison between spectra achieved in different laboratories and different instruments, the spectra should be corrected and expressed in terms of quanta per unit of wavelength. For the emission spectrum, an instrument-calibrated curve is required, which can be obtained from measurements with a calibrated tungsten lamp or by the fluorescence intensity of a fluorescent reference solution with known emission characteristics. For correction of the excitation spectrum the intensity of the excitation light must be measured. This varies among different lamps and throughout the lifetime of a lamp. For details the interested reader is referred to Ritzèn (1967).

Procedure for Routine Microspectrofluorometry
For routine recording of the monoamine spectra, mount freeze-dried and formaldehyde-treated sections on microscope slides, mount in nonfluorescent immersion oil, and cover with a coverslip. Excitation spectra from 360 nm onward can be recorded using an ordinary optical glass system.

Differentiation between DA and NA

DA and NA fluorophores cannot be distinguished under standard conditions in the fluorescence microscope. After acidification, both the fluorophores will be converted from their quinoidal form into their nonquinoidal form, which is accompanied by a shift of the Ex_{max} from 410–415 nm to 370–380 nm. Upon further acidification, the Ex_{max} of the DA fluorophores will remain at 370–375 nm, whereas the NA fluorophore will be converted into a fully aromatic isoquinoline (see Fig. 5-1) with an Ex_{max} at 320 nm (Björklund et al., 1968).

Procedure for Differentiation between DA and NA
For spectral recordings below 360 nm the optical quartz system and quartz slides or, better, coverslips should be used (Björklund et al., 1970).

1. Glue several successive paraffin sections on a microscope coverslip pretreated with albumin-glycerol (see Procedure for Paraffin Embedding and Sectioning, step 7).
2. Remove paraffin from all sections by gently heating on a hot plate and using blotting paper. From this point, one of two different procedures may be followed: (a) expose the deparaffinized sections for some seconds to HCl gas; (b) replace paraffin by liquid paraffin saturated with HCl (prepared by mixing liquid paraffin with concentrated HCl and waiting until liquid HCl settles out). Pipette HCl paraffin close to sections and let all sections be floated with HCl paraffin. Be aware that all sections are equally exposed to HCl paraffin. After 2–5 min replace HCl paraffin with ordinary liquid paraffin. The second procedure seems to yield better results.
3. Invert the coverslip down on the stage with the sections facing the

quartz dark-field condensor.   Use liquid paraffin as the immersion liquid, and measure the excitation and emission spectra of the structure in the first section.

4. Remove the coverslip, remove as much of the liquid paraffin as possible with a blotting paper, and expose the sections to liquid HCl paraffin again for about 2–5 min.   Replace HCl paraffin with ordinary liquid paraffin.

5. Same as 3, but the emission spectrum need not be measured.   Measure only well-preserved tissue and fluorescence.   In case of tissue damage and/or photodecomposition, shift to the successive section.

6. Same as 4.

After 5–10 min of HCl treatment, the excitation peak shifts from 415–420 to 370–380 nm.   After prolonged treatment in the case of the DA fluorophore, the Ex_{max} persists at 370–380 nm, and a smaller peak arises at about 320 nm that might rise slightly during the procedure.   In the case of a NA fluorophore, prolonged treatment causes diminution of the peak at 370–380 nm and gives rise to a major peak at about 320 nm.

When the time course of shifting peaks in a certain tissue is established, the tissue can be exposed to several sequences of 2- to 5-min exposures to HCl paraffin until the time a shift is expected, and then measured cytofluorometrically.   This has the advantage of dealing with good tissue and fluorophore quality.

Acknowledgments

I wish to express my gratitude to Professor B. Falck (Histological Institute, Lund, Sweden), who made his laboratory facilities available to me during my frequent stays between 1967 and 1976.   The description of the equipment is that used by Professor Falck and his co-workers.   These studies were supported by the Swedish Medical Council Nos. B70-14x-56-06 and B70-14x-05 and since 1970 by grants from the Deutsche Forschungsgemeinschaft.   I wish to express my gratitude to Professor B. Falck and Dr. A. Björklund (Histological Institute, Lund) for permission to use certain figures.   For technical assistance I thank Mr. R. N. DeHaven, Dr. E. D. London, and Dr. M. J. Kuhar (Johns Hopkins University, Department of Pharmacology, Baltimore, Maryland).

Chapter 6

The Bodian Protargol Technique

G. E. Gregory

Rothamsted Experimental Station
Harpenden, England

The methods currently used to impregnate nerve tissue with reduced silver fall broadly into two classes: methods of block impregnation and the so-called silver-on-the-slide techniques. In the earlier block impregnation procedures, such as those of Golgi, Cajal, and Bielschowsky, and their modifications, whole pieces of tissue are treated with silver solutions and only afterward sectioned and mounted. In the silver-on-the-slide methods the material is embedded and sectioned first, and then the sections are stained after they have been mounted on slides. Block-impregnation methods are discussed elsewhere in this volume (see Chap. 9) and need not be considered further here. Procedures for impregnating mounted sections, usually of paraffin wax-embedded material, are numerous, but they can be subdivided roughly according to the type of silver compound used as the impregnator. This is generally either an inorganic silver salt, most often silver nitrate, or a silver-protein complex, such as protargol.

Protargol was first used to stain mounted paraffin sections of fish (bullhead) brain by Bartelmez and Hoerr (1933), who substituted it in a modified Bielschowsky method—a useful historical survey of earlier reduced silver techniques is given by Silver (1942a). Bodian (1936), working in the same laboratory on the brain of the opossum, found that staining could be speeded up and made more reliable by adding a small quantity of metallic copper or mercury, or nitric acid, to the protargol impregnating bath, and afterward reducing the silver with a hydroquinone/sodium sulfite solution. This method was subsequently much used on vertebrate tissue, and many writers reported the effects of variations in the procedure. Bodian himself (1937) studied the influence of different fixatives, and Davenport and his co-workers, in an extended series of investigations (Davenport and Kline, 1938; Davenport et al., 1939, 1947; Bank and Davenport, 1940; MacFarland and Davenport, 1941), tested numerous other methods of fixation. They also examined the effects of changes in most of the stages in the procedure. Romanes (1950), too, analyzed some of the variables in his investigation of the mechanism of silver staining. However, Holmes (1942, 1943) criticized the Bodian method on the grounds that protargol was "a substance of uncertain and variable composition," samples of which, from different sources, varied greatly in their effectiveness. He proposed instead the use of an ammoniacal solution of silver nitrate for impregnation, later (Holmes, 1947) modifying this to a silver nitrate solution buffered with boric acid/borax. This technique was the forerunner of a variety of silver nitrate methods, together with concomitant studies of variations in procedure and of the mechanism of silver staining in general, such as those by Palmgren (1948), Romanes (1950), Samuel (1953a–d), Peters (1955a–e), Wolman (1955a,b), Blest (1961), and Rowell (1963). Chapter 7 in this volume deals with the use of silver nitrate in more detail.

Other silver-protein impregnators have been tried. Glassner et al. (1954) reported the effects of varying the steps in the technique using a German product (from Serumvertrieb, Marburg, West Germany), and Polley (1956) tested another, of French origin (Laboratoire Roques, 36 Rue St. Croix de la Bretonnaire, Paris, France), and found it to have many similarities to protargol.

A combination of protargol and silver nitrate was used by Davenport et al. (1939) in their much-shortened (2-h) staining method, and FitzGerald (1964) recommended a similar double impregnation in his method. Various modified procedures have been devised for impregnating frozen or celloidin sections (Silver, 1942b; Foley, 1943; FitzGerald, 1963; Moskowitz, 1967).

Bodian's original method was adapted for staining insect nerve tissue by Power (1943), who used it first to show the general neuroanatomy of the brain of the fruit fly Drosophila and later more detailed aspects of the brain (Power, 1946) and the neuroanatomy of the thoracicoabdominal ganglia (Power, 1948). He employed double impregnation in two baths of protargol and copper, each followed by reduction in a hydroquinone/sodium sulfite solution. Subsequent workers (Hess, 1958a,b; Pipa et al., 1959; Chen and Chen, 1969) followed this procedure with only minor modification, though Rogoff (1946) used only a single impregnation to stain ganglia of mosquitoes. A summary of protargol and other methods for the insect nervous system has been given by Strausfeld (1976).

In a study of the neuroanatomy of the cockroach central nervous system, the present author (Gregory, 1974a) found that Power's modification of the Bodian method yielded better results than other silver-on-the-slide methods tried—in some instances strikingly so—but not consistently. A long series of tests revealed at least some of the factors responsible for this variability and showed the critical importance of fixation in obtaining consistent results (Gregory, 1970). The following account is based on some ten years' experience with the method, used to stain the nervous systems of, in addition to the cockroach *Periplaneta americana* (L.), the desert locust, *Schistocerca gregaria* (Forskål), house fly, *Musca domestica* L., and honey bee, *Apis mellifera* L. It covers probably most of the more important variables that affect the results, but new factors can intervene unexpectedly from time to time (Gregory, 1974b), and continuous vigilance is recommended. The practical procedure is described first, followed by a detailed consideration of the various steps, and finally some guidance is offered on tracing faults.

Practical Procedure

Preliminary Treatment

1. Fix for 16–24 h (overnight) at room temperature (20–25°C) in alcoholic Bouin (Duboscq-Brasil) (Gray, 1954) "aged" for 40 days at 60°C (Gregory, 1970), or its synthetic substitute (see below).

2. Wash out in 70% ethanol, dehydrate in ascending grades of ethanol, clear, infiltrate with paraffin wax, and embed in usual way.

3. Cut serial sections at 10–30 μm and mount on chemically clean slides.

4. Dewax sections and take down through descending grades of ethanol to distilled water.

Live animals are narcotized, if necessary, with carbon dioxide and either whole portions of the body (e.g., fly heads or thoraces) removed or selected parts of the nervous system dissected away under an appropriate saline solution. Material should be fixed without delay, as stressed by Power (1943), but preservation does not seem to be improved either by injecting fixative into the body cavity before dissection, as recommended by Pipa *et al.* (1959), or by dissection under the fixative. However, it is sometimes helpful to remove some of the air from the tracheal system of whole portions of the body after placing them in fixative by *gently* reducing pressure briefly, then transferring them to fresh fixative. Strausfeld (1976), though, warns that under no circumstances should house-fly brains be evacuated. Fixation should not be prolonged much beyond the times given. The yellow color caused by picric acid can be removed rapidly during washing out by means of a little solid lithium carbonate or 10% ammonium acetate in the 70% ethanol baths (Gray, 1954), but this gives no better staining, and lithium carbonate crystals tend to adhere to the specimens and are not easily removed. Washing can be conveniently speeded up by increasing the temperature. Times needed for dehydration, clearing, and infiltration will naturally depend also upon the size of the material, but for pieces the size of cockroach ventral nerve cord ganglia 20 min in each bath has been found sufficient, and for house-fly thoraces, 30 min. Most workers will have their own favorite processing regime, but the following has proved reliable and is used routinely by the author:

1. Transfer to 70% ethanol, 3 baths ⎱   at 37°C
2. 90% ethanol, 2 baths ⎰
3. Absolute ethanol, 2 baths ⎫
4. Absolute ethanol/xylene, 1:1, 1 bath ⎬ at room temperature
5. Xylene, 2 baths ⎭
6. Xylene/paraffin wax, 1:1, 1 bath, on top of oven
7. Paraffin wax, 4 baths

Small specimens, such as individual ganglia, can be stained briefly with eosin (saturated in absolute ethanol) (Pantin, 1946) after dehydration, to aid later orientation in the block for sectioning, then rinsed with ethanol before transfer to the ethanol/xylene bath. Slides used for mounting sections should be scrupulously clean, to prevent loss of sections during the long staining process. Overnight soaking in chromic acid cleaning mixture, followed by thorough washing in running tap water and drying with a clean, soft cloth, is an efficient method. Sections should be attached to slides by floating on albumen solution rather than by smearing the adhesive on the slide with the finger, because this leaves epidermal cells that stain with silver.

Staining Method

1. Impregnate for 18–24 h at 37°C in 2% protargol solution adjusted to about pH 8.4 with a little 10% ammonium hydroxide and containing clean metallic copper.
2. Rinse briefly in distilled water.
3. Develop 10–15 min in hydroquinone/sodium sulfite solution.
4. Wash in distilled water, three changes, 3 min each.
5. Repeat steps 1–4, using fresh protargol and copper.

6. Intensify in 1% gold chloride ($NaAuCl_4 \cdot 2H_2O$) solution, 15 min.
7. Rinse briefly in distilled water.
8. Reduce in 2% oxalic acid solution, 10 min.
9. Rinse briefly in distilled water.
10. Remove residual silver salts in 5% sodium thiosulfate solution, 10–15 min.
11. Wash in distilled water, three changes, 3 min each.
12. Dehydrate in ascending grades of ethanol, clear, and mount in neutral Canada balsam or similar mountant.

The protargol used in all the work described here has been Protargol-S (Winthrop Laboratories, 90 Park Avenue, New York, N.Y. 10016) but other kinds, manufactured by Merck, Germany (*Albumoses Silber*), and Roques, France (*Proteinate d'Argent*), are also said to work excellently (Strausfeld, 1976). Many batches of Protargol-S have been used but staining has shown no noticeable variation between different batches. The protargol solution is made just before use by sprinkling the powder gently onto the surface of the water, preferably already warmed to 37°C, and leaving it to dissolve (10–15 min at 37°C). The ammonium hydroxide solution is then added, with stirring, while pH is monitored with a glass electrode. The copper can be in various forms (foil, turnings, etc.), but small-gauge wire is convenient and readily obtainable. The weights given later refer to 20 gauge (British Standard) or 19 gauge (American) (0.91-mm diameter) wire. The quantity needed depends on various factors, as will be explained below. It is cleaned most simply by being rubbed briefly with household scouring powder on a damp cloth, rinsed and dried, but other forms of copper may need to be cleaned in nitric acid, then rinsed in water. The wire is then cut into 5- to 10-mm lengths, which are placed at the bottom of Coplin staining jars before adding the slides and protargol solution. No more than five slides are used per jar. Greenish crystalline strands may grow from the copper during impregnation, increasingly so at lower pH, but can be prevented from obscuring sections by mounting these toward the upper ends of the slides. A short length of glass rod placed at the bottom of the jar to raise the slides higher in the solution also helps; 65 ml of solution are required to cover slides completely. Impregnation is routinely done in the dark in a closed incubator but whether darkness is essential is not known. After impregnation excess protargol solution can be wiped quickly from the backs and ends of slides before they are rinsed and placed in the developer; labeling the upper side with a writing-diamond before staining helps to avoid accidents! The developer (reducer) should be made up from the solids shortly before use, because both hydroquinone and sodium sulfite oxidize in solution. Concentrations required depend on a number of factors, as will be discussed later. The developer solution can often be kept and used again the second day, but it is usually better to make a fresh solution for the second development. The gold chloride solution seems to keep indefinitely, and both the oxalic acid and thiosulfate last for some time, depending on the numbers of slides processed. Gold toning has always been done under normal laboratory lighting, but whether "bright light" (Strausfeld, 1976) is necessary has not been investigated. Impregnation and development are the most important stages, and times in the subsequent baths are far less critical (see later). Sections change in appearance during staining as follows. After development (Fig. 6-1, Plate 4) the nervous tissues are stained in shades of yellow or brown, darker after the second impregnation, and nuclei stand out clearly, golden yellow or brown; surrounding tissue, such as fat body, should generally be colorless, though fat body of female cockroaches tends to stain somewhat, more than that of males; any muscle stains heavily, dark brown. During gold toning colors change to shades of gray, with black nuclei; light purple tints should appear in darker-stained nerve fibers and larger cell bodies and are a good indication that staining has been successful. In oxalic acid the final colors appear: shades of red, purple or blue, according to the conditions of staining, with nuclei purplish- or bluish-black (Figs. 6-2 and 6-3, Plates 4 and 5).

Factors Affecting Staining

The final result produced is greatly influenced by both fixation and the conditions of staining, particularly of impregnation and development, and the following account considers in some detail the factors so far found to be important. It is based largely on experiments carried out by the author during attempts to make the method more reliable (Gregory, 1970).

Silver is deposited in the tissue during impregnation in two forms (Samuel, 1953b): (1) minute particles ("nuclei") of reduced silver and (2) unreduced silver combined with the tissue components. This unreduced silver is redistributed during development. Some of it is reduced and is thought to be deposited around the silver nuclei, which act as catalytic centers for the reduction; some is reduced in situ in the tissues; and some is removed into solution in the developer. The proportions that suffer these different fates depend upon the composition of the developer (see later). Methods derived from those of Samuel (1953b) were used to determine how the various factors that affect staining influence the two silver fractions deposited during impregnation. The distribution of the silver nuclei was studied by first removing the unreduced silver with a 2.5% solution of sodium sulfite for 5 min and then developing in the physical developer (that is, one containing silver in solution) of Rowell (1963). This deposits silver on the silver nuclei, so making them visible. Only a single impregnation in protargol was used because the usual double impregnation gave too dense a stain. The physical developer of Pearson and O'Neill (1946) and the glycine mixture of Peters (1955b) were also tried but were less successful. Unreduced silver was revealed by developing it in situ with 1% hydroquinone solution, which without sodium sulfite does not remove silver from the tissues. For these tests double impregnation was used. In all the experiments, protargol concentration was 2% and incubation temperature 37°C (± 1°C); some sections were intensified with gold, others were not.

Fixation

It was found during initial tests of the Bodian method that fixation in alcoholic Bouin (Duboscq-Brasil) (picric-formaldehyde-acetic-ethanol) gave far better preservation and staining than any other method tried, but even so, results were not consistent. Eventually it became apparent that only old samples of the fixative, made up a year or so before, gave really good results. With fresh solutions fine fibrous neuropil had a characteristically matted appearance, larger fibers were shrunken and poorly and unselectively stained, and the glial cells stained heavily, further reducing contrast and obscuring detail. Precipitation of chromatin at the inner side of the nuclei, clearly visible in all the illustrations (Figs. 6-1–6-4, Plates 4 and 5) is a characteristic artifact produced even by old solutions, and is apparently caused by the diffusion inward through the tissues of the alcohol in the fixative (Baker, 1950). Various attempts were made to age solutions more quickly, and the method that emerged as the most convenient was simply to put the bottle of fixative in the wax-embedding oven at 60°C for 40 days or so. The solution then fixed as well as an old solution. During the aging process, fixation improved rapidly during the first 7–10 days but then the rate slowed down. Forty days was chosen as a safe, convenient period, but fixation is adequate some time before this and prolonging the aging process somewhat beyond 40 days makes little difference. After aging, the solution was found to keep well for a year or more on the shelf at room temperature (longer if refrigerated), though the quality of preservation eventually began to decline, fiber shrinkage increasing and staining becoming gradually darker and more "clogged" in appearance.

Recently the opportunity occurred to analyze, mainly by gas-liquid chromatography and mass spectrometry, the volatile constituents of the Bouin solutions and to examine the effects of aging (Gregory et al., in preparation). The following is a summary of the initial results.

The percentages of the volatile constituents found in freshly mixed and 40-day aged solutions are shown in Table 6-1. They clearly change con-

Table 6.1. Percentages of Principal Volatile Constituents in Alcoholic Bouin Fixative: Original Formula, Freshly Mixed Solution, and Solution Aged for 40 days at 60°C

	Formaldehyde	Ethanol	Methyl acetate	Acetic acid	Ethyl acetate	Diethoxymethane
Original formula	11	55	—	6.88	—	—
Freshly mixed	10.2	54	0.1	6.1	1.3	0.4
Aged solution	5.7	35	0.5	3.4	5.0	14.7

siderably during the aging process. However, the concentration of picric acid, shown by other methods, such as thin-layer chromatography, seems to remain fairly constant. The amounts of the volatile primary fixatives (formaldehyde, ethanol, and acetic acid) are all nearly halved, while the reaction products, especially diethoxymethane, increase correspondingly. It is this last that gives the matured solution its characteristic, pleasant fruity odor. Also present in the aged solution were traces of methanol, dimethoxymethane, and ethoxymethoxymethane. Although methanol-free ethanol was used to make up the test solutions a small amount of methanol was present as a stabilizer in the formaldehyde solutions used, which accounts for the presence of these compounds. Interactions evidently begin as soon as the constituents of the fixative are mixed, for even the freshly mixed solution had a slightly different composition from the proportions of the reagents used to make it. As expected, the aging process continued rapidly during the first few days but then slowed down as a more stable composition was approached. "Overaged" solutions, which had been aged for 40 days and then kept on the shelf for a year or two, and were giving less good fixation, contained still less of the volatile primary fixatives and more of the reaction products.

Solutions of "aged" alcoholic Bouin made up from the component reagents according to the figures in Table 6-1 were found to give fixation indistinguishable from that given by the normally aged mixture. However, the concentrations of the component reagents do not seem to be excessively critical, and the following simplified mixture has worked well in all trials so far.

Synthetic "aged" alcoholic Bouin:

Formaldehyde (ca 40% aqueous solution)	15 ml (=ca. 6% HCHO)
Ethanol	35 ml
Acetic acid	3.5 ml
Ethyl acetate	5 ml
Diethoxymethane	15 ml
Picric acid	0.46 g
Distilled water to	100 ml

The picric acid is dissolved in the alcohol, the other reagents added, and the volume made up with water, with periodic shaking to allow the volume to contract as the ethanol and water mix. Ordinary industrial methylated ("denatured") ethanol seems quite adequate in place of the methanol-free ethanol used in the analytical work, and no differences in fixation have been detected using it. Diethoxymethane, $CH_2(OC_2H_5)_2$, can be obtained from the Eastman Kodak Company. The synthetic mixture has not yet been in use for very long, but there seems no reason why it should not keep as well as normally aged Bouin, particularly if refrigerated. Its great advantage, in any event, is that it can be made up immediately when required, without the 40-day wait usually necessary. It also provides a known starting point for modifications, and further experiments are in progress to see whether the mixture can be improved upon and to study the effects of the various components on staining. If no diethoxymethane is available, initial tests indicate than an increase in the amount of ethyl acetate to 25 ml is a satisfactory substitute, though the keeping properties of this mixture have yet to be tested.

The pH of Impregnation

Freshly prepared 2% protargol solution has a pH, measured with a glass electrode, of around or just above 8.0, but this can be varied between 7.0 and 9.0 by means of a buffer such as 0.25 M borate (Rowell, 1963). A series of experiments, using normal hydroquinone/sodium sulfite development, showed the effects of such changes in pH on staining to be as follows. Raising the pH gave a darker, redder stain but with reduced selectivity, as staining of the fine fiber background of the nerve tissue increased. Also, above about pH 8.5 sections mounted with albumen adhesive tended to become detached and lost during subsequent treatments. Lowering the pH gave a progressively weaker, bluer stain with increased selectivity down to about pH 7.5, but below this selectivity diminished rapidly. Other silver methods respond similarly (Peters, 1955a; Polley, 1956; Rowell, 1963; Samuel, 1953a). The color changes are thought to be caused by the Tyndall effect of the colloidal gold particles (Rowell, 1963): Smaller particles scatter shorter wavelengths and appear redder by transmitted light, and larger particles scatter longer wavelengths and appear bluer. The paler, bluer color produced with lower pH would therefore suggest that fewer but larger gold particles are present after toning and reduction and hence fewer, larger particles of silver after impregnation and development. Tests using physical development confirmed this, showing progressively fewer reduced silver nuclei as pH was lowered. Also, reduced silver decreased faster in the fine fiber background than in the larger nerve fibers, so that these became more selectively stained as pH fell.   At the same time, as staining became weaker with lowered pH, the amount of unreduced silver, shown by development in hydroquinone without sulfite, also decreased slightly.   Below about pH 7.5 the larger nerve fibers seemed to possess too little reduced silver to stain selectively and the overall stain was then weak and had little contrast.   In unbuffered protargol solutions containing copper, the pH falls gradually during impregnation (Holmes, 1943), slightly increasing selectivity and reducing staining intensity.   In all experiments done subsequent to these on the effects of pH, the initial pH of the protargol was standardized at 8.5, by the addition of a little 10% ammonium hydroxide solution.   Buffer was not used because it would have masked any effect on pH of other factors. However, for routine staining an initial pH of 8.4 was eventually settled upon, as it was found safer for avoiding the detachment of albumen-mounted sections.

The Copper Content of the Impregnation Solution

If no copper is used in the protargol solution all tissues stain dark red after gold toning and reduction, with nuclei and medium-sized nerve fibers almost black.   The addition of copper produces a lighter stain, particularly in the peripheral layers of the ganglia (neural lamella, perineurium, and glial cell layer) and in the fat body.   It also results in more selective staining of the fibers of the ganglion core (neuropil *sensu lato*).   With increase in copper content from 0.5 to 5 g per 65 ml of protargol solution both the large and the very fine nerve fibers stain progressively less, until finally they are almost colorless.   Medium-sized fibers retain their deeper color longer and so at first stain selectively, but fewer of them stain as the copper concentration increases further, so that finally the tissue is almost entirely unstained apart from the nuclei.   Increase in the copper content also changes the final color of the stain from red, through shades of purple, to blue.   This effect is greater than the similar one caused by change in the pH of the impregnation solution but its mechanism seems the same.   Before gold toning, sections appear dark brown in color when no copper is used, but with increasing amounts of copper they are a progressively paler yellow.   Development in hydroquinone without sulfite shows that this change is associated with a marked decrease in the amount of unreduced and hence developable silver in all tissues.   The nuclei and some nerve fibers at first stain more than other tissue components, but with more copper the contrast falls and nuclei and fibers do not stain selectively until after gold toning and reduction.   Physical development shows that when only a small quantity of copper is present the larger fibers contain considerable amounts of reduced silver and so stain strongly; as the copper concentration is increased the reduced silver in the larger fibers decreases, faster than in the intermediate-sized fibers and nuclei, so that these then stain selectively.   The fine fiber background contains little reduced silver when copper is absent, but more when it is used, though the amount of reduced silver increases only slightly with increasing copper content.   Because less unreduced silver is present at higher copper concentrations, the background stains little even though it contains some silver nuclei, and the intermediate-sized fibers and cell nuclei, which contain most silver nuclei, stain selectively, apparently at the expense of the other tissues.

Copper produces these effects in at least two ways.   It causes most of the fall in the pH of the protargol solution during incubation, and it decreases the concentration of silver in the solution by electrochemical displacement (Holmes, 1943).   The amount of copper added determines the change in pH: Without copper, but with sections present, the pH measured with a glass electrode was found to decrease from 8.5 to 8.4 during 24-h incubation at 37°C; with increasing amounts of copper, ranging from 0.5 to 5 g per 65 ml, the final pH changed from 8.35 to 8.1.   The fall in pH must play a part in the action of the copper, as thought by Holmes (1943), because the effects of both copper and lowered pH are similar. However, the copper produces a far greater effect than could result from the change in pH alone: preventing the fall in pH by buffering the solution with borate only partly compensates for the amount of copper present, and staining is always weaker and more selective as copper content is increased.   The decrease in silver concentration in the solution by the copper satisfactorily explains the decrease in unreduced silver in the tissues and hence the weakening of the stain as copper concentration rises, for the unreduced silver is in equilibrium with the silver in the solution (Peters, 1955a).   The decreased silver content of the solution is also probably the main cause of the decrease in reduced silver in the tissues, for in silver nitrate impregnations the rate of formation of reduced silver nuclei increases with silver concentration (Peters, 1955a).   It remains unknown, however, whether the passage of copper into the solution and its deposition in the tissues (Bodian, 1936) modifies impregnation, as Romanes (1950) suggested; also, whether it is the quantity of copper added or its total surface area which is most important in determining the character of staining.   Acceleration of the rate of impregnation by copper, observed by Davenport and Kline (1938) in Hofker-fixed vertebrate material, has not been apparent after alcoholic Bouin fixation.   The optimum quantity of copper needed for each incubation thus depends very much on the type of staining required.   It is also affected by the thickness of the sections to be stained, as explained later.

The Duration of Impregnation

Power's modification of the Bodian method specified two impregnations, each lasting 24–48 h.   To see whether variation in the duration of impregnation had any effect, sections were incubated for two equal periods of between 16 and 48 h. The temperature (37°C) was not varied, though Rowell (1963) found with silver nitrate that raising the temperature increased the rate of impregnation. The present experiments revealed little change between 16 and 24 h. After 48 h, however, staining was paler and less selective: more fibers were impregnated but all were stained similarly. Development in hydroquinone without sulfite showed much less unreduced silver in all tissues, which would account for the paler staining. It was probably due to a continued fall in the silver content of the impregnation solution during the additional incubation time. The pH also decreased gradually and would have a slight contributory effect. Physical development showed that the amount of reduced silver increased with longer incubation, as happens in silver nitrate methods (Peters, 1955a): The fibrous core of the ganglion was impregnated more completely and, in particular, more was present in the fine fiber background. This would account for the changed pattern and decreased selectivity of the stain, and also the paler and more uniform color, because less unreduced silver is available for reduction at many more sites. From these experiments it was evident that the duration of impregnation was not especially critical, and at 37°C a time of between 16 and 24 h for each of the two incubations was most effective. Other experiments have shown that for test purposes and for thick sections of tissue that stains too densely with two impregnations a single impregnation may be sufficient, but it gives too weak a stain for general use. However, the effects of variation in impregnation time are still the same.

Rinses

Wiping unwanted protargol solution from the slide and rinsing it briefly in distilled water after each impregnation gives cleaner preparations and, as pointed out by Glassner *et al.* (1954), prevents "clogging" of the tissue by the indiscriminate precipitation of silver. Longer washes, of up to several minutes, offer no advantages and merely remove unreduced silver from the tissues, making the stain paler. The rinse should thus be kept short, say three dips for 10-μm-thick sections or six dips for 20-μm ones. An additional effect of this rinse is that, even in thinner (10-μm) sections, it tends to remove silver more from the exposed upper surface of the section than from the deeper regions closer to the slide. This is of great assistance when neurons are being traced through series of sections, as it enables the upper and lower surfaces of each section to be readily distinguished by their slightly different staining, the upper surface having a clearer appearance than the lower. The similar short rinse after gold toning also gives cleaner preparations, as noted by Davenport *et al.* (1939).

The washes after development need to be longer to remove unused reducer but, within limits, their duration is not critical. Extended washing times (30–40 min total) give a slightly bluer final stain with rather less color in the fine fiber background, but are not recommended for routine use.

The brief rinse after oxalic acid does not stop reduction (Glassner *et al.*, 1954), which ceases only in the thiosulfate bath (Davenport *et al.*, 1947). The lengths of the washes to remove thiosulfate and unwanted silver salts are again not critical and the times given are safe minima.

For all washes the author routinely uses distilled water for consistency, but for the longer washes running tap water could probably be used, if followed by a distilled water rinse. Various samples of distilled or deionized water have been tried but no effects on staining have so far been observed.

The Composition of the Developer

The main variables in development are temperature, time, pH, and developer composition, but by far the most important of these in practice is the composition of the developer. Temperature is unimportant as long as it is high enough for the developer to act at reasonable speed, and normal laboratory temperatures are usually quite adequate. Similarly, the duration of development matters little as long as it is sufficient for reduction to be completed, which occurs fairly rapidly at room temperatures of 20–25°C. The times recommended in this account are probably well in excess of the minimum required. The effect of pH is discussed below.

A range of developers of various activities was tested with vertebrate material by Davenport and Kline (1938), but in the present work Bodian's original combination of hydroquinone and sodium sulfite has been used exclusively. The sodium sulfite (or other alkali) differentiates the stain (Davenport and Kline, 1938) by competing with the developing agent (reducer) for the unreduced silver in the tissues (Samuel, 1953a). This silver is removed from areas with fewest reduced silver nuclei, where development is slowest, and part of it is reduced and redeposited in areas where silver nuclei are most abundant and development is therefore most vigorous (Samuel, 1953b). The degree of differentiation thus depends on the ratio of the concentration of sulfite to that of the reducer, hydroquinone. More sulfite gives greater differentiation, more hydroquinone gives less. Without sulfite· hydroquinone produces no differentiation additional to that due to impregnation: Sections stain uniformly, except for the usually much darker cell nuclei. The amount of differentiation influences the final color of the stain, as do the pH and copper content of the impregnation solution. When only a small amount of copper is used the final stain is always red, but with more copper present the color becomes more blu-

ish as differentiation is increased. The degree of differentiation necessary is related to the quantity of unreduced silver present in the tissues,
which is determined chiefly by the conditions of impregnation. The pattern of differentiation is governed mainly by the distribution of the reduced silver nuclei formed during impregnation. The outermost glial
cells, smaller nerve cell bodies, and other peripheral tissues of ganglia
usually lose color first, followed by the fine fiber background of the ganglion core. Contrast between the background and the larger nerve fibers
thus increases, until a maximum is reached, beyond which the larger fibers
themselves lose color. The developers used in the present study
contained 0.25%–1% hydroquinone with 1%–12% sulfite. The most appropriate developer composition depends on several factors, as will be
explained under Choice of Staining Conditions (p. 90). All percentages
of sulfite quoted here refer to the hydrate, $Na_2SO_3 \cdot 7H_2O$; if the anhydrous
salt is used the amounts given must be halved. The pH of these mixtures
ranged from about 8.2 with 1% sulfite to 8.7 with 12%. Raising it was
found to give slightly darker, less differentiated staining and lowering it
had the opposite effect, but little is to be gained by adjusting it because the
pH range within which hydroquinone acts is small. Chen and Chen
(1969), in their modified procedure, recommended substituting 5% sodium
sulfate for sulfite in the developer used after the second impregnation.
However, sodium sulfate seems to have no effect on staining, and the
same result would presumably be obtained by omitting it altogether.

The Purity of the Sodium Sulfite

An unexpected variable came to light when Bodian staining was used
again after an interval and failed to yield the strong reds, purples, and
blues it had produced consistently for some years previously. Instead it
gave a dull, grayish-mauve or lilac color (Fig. 6-4, Plate 5), which showed
little of the variation in tone between different structures formerly found
so valuable in neuroanatomical studies. A series of tests finally traced
the cause of the change to the use of a different sample of sodium sulfite in
the developer (Gregory, 1974b).

When a range of samples of sulfite, mostly of analytical purity and from
various manufacturers, was screened, only about half were found to give
the desired red and blue colors (termed for brevity ''red'' sulfites), while
the rest produced the mauve stain (''mauve'' sulfites). The color given
was constant among samples from the same batch but was unrelated to
whether the sulfite was hydrated or anhydrous. Mixtures of red and
mauve sulfites gave stains of intermediate colors, tending toward the
shade of whichever type of sulfite predominated.

The two kinds of sulfites differed not only in the final color they produced but also, to a lesser extent, in the way they differentiated the

stain. Mauve sulfites characteristically left color in the outermost layer (neural lamella) of ganglia and in the fat body, whereas red sulfites at the same concentration normally destained both regions completely. The effects could be seen well before the final colors appeared. Development in a mauve sulfite after the first impregnation yielded a dull, brown stain, with darker brown nuclei, instead of the clear yellow or golden brown shade given by red sulfite. Differences were less obvious after the second impregnation and development because the stain was then darker and tended to obscure them. During subsequent gold toning, the light purple tints normally seen after development with red sulfites never appeared when mauve sulfites had been used.

How the mauve color arises is still uncertain. The differences in color after development suggest that the two types of sulfites give silver deposits that differ in the sizes of their particles. If, as already noted, the final color of the stain is determined by the size of the colloidal gold particles ultimately present in the tissues, the mauve shade may indicate either a mixture of different-sized silver particles deposited during development, and hence gold particles later, or a more homogeneous deposit of intermediate-sized particles. The darker, brown color after development and the grayish tint of the final stain suggest that some larger particles are present, so perhaps the former explanation is the more probable.

Purification of samples of red and mauve sulfites (see below) revealed that it was the mauve sulfites that were contaminated. Emission spectrographic analysis failed to show any characteristic metallic impurities so it appeared that an anionic one was to blame. None of the main commercial impurities of sodium sulfite (sodium sulfate, carbonate, and chloride) when added to red sulfite gave a mauve stain, but the color was changed to mauve if a more complex sodium sulfoxy salt such as thiosulfate or dithionite ($Na_2S_2O_4 \cdot H_2O$) was added. In addition, the dithionite decreased differentiation, closely simulating the effect seen with mauve sulfites. Whether the mauve color given by these sulfites is indeed caused by traces of one or more of the more complex sodium sulfoxy compounds is still not certain, for proof of this would require more detailed chemical analysis, but at present it remains the most likely explanation.

Fortunately the impurity was found to be easily removable by recrystallizing the sulfite from water at about 50°C. This had no effect on red sulfites, but every mauve sulfite purified in this way afterward yielded the red stain. The method was effective even with a sample of nonanalytical purity.

Therefore, if Bodian staining does not yield the clear red, purple, or blue colors expected, it may be well to suspect the type of sulfite being used. Workers employing the method for the first time, and unfamiliar with the colors it normally produces, might be well advised to test a recrystallized sample of their sulfite anyway, to see if this gives any improvement.

Gold Toning and Reduction

Samuel (1953d) examined the processes of gold toning and reduction after silver nitrate impregnation and concluded that variations in these processes have far less effect on the final result than have impregnation and development. With protargol impregnation, the same appears to be true.

Gold is thought to be deposited on the surface of the silver particles in the tissues during toning as a result of the following reaction (Davenport and Kline, 1938; Samuel, 1953d):

$$3Ag + AuCl_3 \rightarrow Au + 3AgCl$$

Oxalic acid reduction then deposits further gold from the gold chloride still present in the section and probably also causes flocculation of the gold-plated silver particles (Samuel, 1953d).

The concentration of the gold chloride solution used for toning does not seem to matter greatly; stronger solutions merely act faster (Davenport and Kline, 1938; Glassner *et al.,* 1954). The duration of toning also seems to be unimportant, provided that it is long enough for toning to be completed. Extending the toning time from 15 to 40 min at room temperature was found to produce no perceptible effect. The addition of acetic acid to the gold solution is sometimes advocated but seems to make no obvious difference, as was also found by Davenport and Kline (1938) and Glassner *et al.* (1954).

The concentration of the oxalic acid reducer also seems not to be critical. The duration of reduction may, however, be important. To display particular structures it may be desirable to observe the progress of reduction under the microscope, as recommended in some procedures, and to stop it before it is complete. If so, it should be borne in mind, as already mentioned, that reduction apparently continues in the subsequent distilled water rinse and is stopped only by the thiosulfate bath that follows it. For routine staining of general neuroanatomy, however, reduction can normally be left to go to completion, and its duration is then of no great consequence, though it is probably best not to continue it unnecessarily. Extending it from 10 to 50 min at room temperature was found to give a slight bluish staining of the albumen used to mount the sections, though the sections themselves appeared unaffected.

Choice of Staining Conditions

The Bodian procedure offers a wide scope for variation in the conditions of staining, and the choice of the most suitable ones for a given piece of material can at first sight present problems. The optimum conditions must depend to some extent on the characteristics of the tissue and will no doubt vary somewhat from one species to another. However, although few species have yet been tested, among those so far examined by the author less variation has been found than expected. Strausfeld (1976) mentions that Hymenoptera are particularly resistant to the Bodian technique and are better stained by silver nitrate methods, but the present author has found honey-bee thoracic ganglia to stain readily using the procedure described here. Fixation also influences the staining properties of the tissue, and the importance of good fixation in obtaining satisfactory and consistent results must again be emphasized.

Selection of the most appropriate staining conditions for a particular batch of sections is made easier than might at first appear because, as already indicated, many of the variables seem to be relatively unimportant or can easily be standardized, and only impregnation and development seem to be critical. Even in impregnation and development some variables, such as time, temperature, and pH, can be standardized, and only two need really be considered: the copper content of the impregnation bath, and the relative concentrations of hydroquinone and sodium sulfite in the developer. Choice of the best combination of these is determined chiefly by the type of result desired and the thickness of the sections to be stained.

Increases in both copper concentration and sulfite concentration (with the amount of hydroquinone kept constant) have fairly similar effects, giving lighter, more selective, and at higher copper concentrations, bluer staining. Within limits, therefore, an increase in one can be balanced by a decrease in the other. Thus various combinations of copper and sulfite concentrations can give approximately similar results (Table 6-2). Section thickness mainly determines the amount of differentiation required: Thick sections need more differentiation than thinner ones to give the same sort of result. Thus if sections of different thicknesses are impregnated using similar amounts of copper, the thicker ones will require a higher concen-

Table 6-2. Combinations of Staining Conditions Giving Approximately Similar Results with *Periplaneta* Ganglia

Section thickness (μm)	Copper content of impregnation bath (g/65 ml 2% protargol)	Sodium sulfite content of developer (%$Na_2SO_3 \cdot 7H_2O$ in 1% hydroquinone)
10	1.3	4
	2.6	3
	3.9	1.5
20	1.3	10
	2.6	5
	3.9	3

tration of sulfite in the developer than the thinner ones. Similarly, more copper must be used to impregnate thick sections than thinner ones if both are to be developed in the same concentration of sulfite. However, if the effective limits of copper or sulfite concentration are exceeded they can no longer compensate for one another. Results then diverge increasingly. The lower limit for copper content in the impregnation bath is reached sooner with thicker sections, which then stain too heavily and cannot be differentiated sufficiently by the developer. Likewise, the effective maximun amount of copper is arrived at sooner with thinner sections, which then become too weakly stained even with minimum differentiation. It is these effects that enable the Bodian method to be used to give a variety of different types of results, according to the choice of the user. Low copper concentration always gives an almost complete, dark red stain, whereas high copper concentration gives a severely selective blue stain. For general neuroanatomy of insect central nervous systems, where most nerve fibers are to be stained (see Fig. 6-2, Plate 4), impregnation with 1.3 g copper (= 2% w/v protargol solution) followed by development in 1% hydroquinone with 4% or 5% sulfite has proved consistently good for 10-μm-thick sections, and 2.6 g copper and 1% hydroquinone with 5% sulfite for 20-μm sections. For more selective staining of individual nerve fibers, blue against a clearer background (see Fig. 6-3, Plate 5), 3.9 g copper and 1% hydroquinone with 5% sulfite have worked well with 10-μm sections, but for 20-μm sections 5 g or more of copper and 10% sulfite in 1% or less of hydroquinone may be needed. However, this more selective type of staining has now largely been superseded for tracing neuron pathways by the various methods of filling individual neurons or groups of neurons with dye, which are discussed in other chapters of this book. Other types of results, between the extremes described, can be obtained by employing suitable intermediate conditions. The combinations recommended have also given good results with insect peripheral nerves.

Fault Tracing

In a lengthy method such as this, having many different steps, many different errors can arise, though the procedure given should help to reduce them to a minimum. However, the causes of most faults are usually fairly readily identifiable with a little experience. It is usually worthwhile and instructive to check the state of the sections after, particularly, the first development, to ensure that staining is progressing as intended. The presence of light purple tints after gold toning and before oxalic acid reduction is also a good confirmatory sign, as noted earlier. The most common defects and their usual causes are summarized in Table 6-3, but it should be borne in mind that more than one fault can occur at a time.

Table 6-3. Defects in Bodian Preparations and Their Likely Causes

Defect	Cause
Nerve fiber shrinkage; staining light or uneven, with glial cells darkly stained	Poor fixation: alcoholic Bouin insufficiently aged
Nerve fiber shrinkage; staining darker than expected, with "clogged" appearance; glial cells paler stained	Poor fixation: alcoholic Bouin too old
Shrinkage of all tissues; staining unexpectedly dark	Fixation unduly prolonged; faulty dehydration and/or clearing procedure
Lighter, bluer-stained patches in otherwise satisfactory sections	Sections not flattened properly on slide—silver washed from lifted areas
Sections become detached from slide	Slide not sufficiently clean; albumen adhesive faulty; pH of impregnation baths too high
Sections unstained or with only faint overall granular deposits of silver	Ineffective sample of protargol; grossly excessive amounts of copper and/or sulfite relative to section thickness
Staining too dark, red	Insufficient copper in impregnation baths
Staining darker than expected, with "clogged" appearance	Rinses after impregnations too short
Staining too light, pink	Only single impregnation used; rinses after impregnations too long; sections too thin; insufficient oxalic acid reduction
Staining too light, purplish or bluish	Only single impregnation used; too much copper in impregnation baths; pH of impregnation baths too low; too much washing after development; too high sulfite/hydroquinone ratio in developer
Staining dark enough but insufficiently differentiated	Too low sulfite/hydroquinone ratio in developer; sulfite oxidized or not sufficiently pure
Staining light, even, and insufficiently differentiated	Impregnations unduly prolonged
Stain overdifferentiated, red or pink	Too high sulfite/hydroquinone ratio in developer
Stain overdifferentiated, purplish or bluish	Too much copper in impregnation baths; too high sulfite/hydroquinone ratio in developer
Staining grayish-mauve, not red, purple, or blue	Impure sample of sulfite used in developer—test recrystallized sample
Dark purple- or black-stained band across section	Growth of crystalline strand across section from copper during impregnation
Albumen adhesive stained bluish	Too long in oxalic acid bath

Addendum

Fixation

Further experiments (Gregory, in preparation, II) have shown that the chief factor in the improvement of alcoholic Bouin by aging is the reduction in concentrations of formaldehyde and ethanol; also that formaldehyde mainly increases staining of glial cytoplasm, ethanol increases tissue shrinkage and lightens overall staining, acetic acid improves preservation, and picric acid decreases glial staining but has few other effects within a wide range of concentrations, though omitting it seriously impairs results. Decrease in ethanol and increase in acetic acid content both improve results. Omitting formaldehyde virtually eliminates glial staining, without apparently affecting overall preservation, and seems a promising modification to combat the problem of staining the neurons but not the glia around them. Its only disadvantage apparent so far is the formation of fine black granules, mainly in the fibrous core region of ganglia, but these are noticeable only at higher magnifications and do not interfere appreciably with observation. Adding just a small amount of formaldehyde to the fixative (see below) greatly reduces the formation of these granules, but glia then stains somewhat. From this work a new fixative mixture has been developed.

Improved synthetic alcoholic Bouin:

Formaldehyde	0–15 ml
(ca. 40% aqueous solution)	(=ca. 0–6% HCHO)
Ethanol	25 ml
Acetic acid	5 ml
Ethyl acetate	5 ml
Diethoxymethane	15 ml
Picric acid	0.5 g
Distilled water to	100 ml

The formaldehyde content is adjusted to yield the amount of glial staining desired and to minimize granule formation; usually about 2% (5 ml) formaldehyde is a good compromise. As in synthetic "aged" alcoholic Bouin, diethoxymethane can be replaced by increasing the ethyl acetate concentration to 25 ml, with again some increase in staining, especially of the neural lamella. These "25:5" (ethanol:acetic acid) solutions give better preservation than any of the previous mixtures, and also increase the intensity of impregnation. They have produced excellent results with both cockroach and locust material, and also with honey bee.

Single Impregnation Method

Overstaining of thicker (20 μm) sections can often result from the usual double impregnation of material fixed in the "25:5" mixtures. This has led to the devising of a time-saving single impregnation Bodian procedure (Gregory, in preparation, III), which differs from the usual method as follows.

Ganglia are fixed in either improved synthetic alcoholic Bouin or, usually better, its ethyl acetate-substituted version, which gives darker staining. Formaldehyde content is selected as above. Sections are impregnated as usual but once only, using 1.3 g copper per 65 ml. The subsequent distilled water rinse is either omitted (10-μm cockroach sections) or reduced to 1–3 dips (20-μm cockroach sections; 10-μm and 20-μm locust sections), and developer sulfite concentration is reduced to 2.5%–4% for 10-μm sections, but is kept normal (10%) for 20-μm sections. Subsequent treatment is as for double impregnation. Results are a little lighter than with double impregnation but seem entirely satisfactory for examination of neuroanatomy.

Chapter 7

Reduced Silver Impregnations Derived from the Holmes Technique

A. D. Blest

Department of Neurobiology
Research School of Biological Sciences
The Australian National University
Canberra, Australia

P. S. Davie

University of Canterbury
Christchurch, New Zealand

Techniques for the reduced silver impregnation of vertebrate central nervous tissues have a long history, but until recently none of them could be made to give consistent results with arthropod materials, and the best results obtained were usually far inferior to those from vertebrate preparations. This chapter describes a number of modifications of Holmes' method for vertebrate peripheral nerve (Holmes, 1947). The original method was first applied to honey-bee brains by Vowles (1955), but although the staining obtained was promising and allowed a certain amount of gross anatomy to be resolved, it still fell far short of the standard routinely required in vertebrate neuroanatomical work. Blest (1961) introduced modifications that employed alkylated pyridines as vehicles for the silver, and they were further modified by Strausfeld and Blest (1970 cf. Blest, 1976). Weiss (1972) held, with justification, that the published results of these techniques still failed to meet modern requirements, and he made a number of adjustments to the fixation, embedding, and impregnating routines that yielded, in his hands, far greater resolution and differentiation from background. Meanwhile, Rowell (1963) substituted a "physical" developer for the hydroquinone-sulfite developer used in the other variants.

A major disadvantage of all these derivatives of the Holmes technique is that they are capricious: Even with careful handling, staining is often inconsistent within batches, and particular batches may fail altogether for no apparent reason.   Two causes for this behavior are probable: First, the impregnating baths, even in the more recently devised procedures, use very low concentrations of silver and are thus vulnerable to trace contaminants.   Second, the performance of alkylated pyridines varies between batches from different manufacturers and appears, as well, to relate to the freshness of the samples used.   In general, workers intending a sustained anatomical investigation are probably best advised to employ Bodian or Ungewitter techniques, but the former requires much practice, and the latter some experimentation with fixation and mordanting.   For more modest purposes, Holmes' methods offer greater simplicity, and should always be tried first.

Basic Strategies of Silver Impregnation

General Considerations

All reduced silver methods follow similar sequences of treatments.   Sections of fixed material, usually on slides, are first of all *impregnated* in a bath containing a silver compound, with or without some form of pretreatment; the silver taken up by them is reduced, in part, by a *developer;* the silver particles are replaced by gold in a two-stage process that consists of *toning* in a bath of either chlorauric acid or sodium chloraurate, and a further reduction in a bath of oxalic acid.   Bound but unreduced silver still present in the sections that would darken on exposure to light is removed by a so-called *fixation* in a solution of sodium thiosulfate.

There has been much discussion in the literature as to the precise chemistry of these processes (Davenport and Kline, 1938; Holmes, 1947; Palmgren, 1948; Peters, 1955a,b; Wolman, 1955a,b; Rowell, 1963; Gregory, 1970), but it must be admitted that a full understanding of the factors that influence their fine control has not yet been achieved.   Much depends on the materials being treated, and attempts to investigate a given routine are always lengthy and fraught with difficulties.   Comparisons between methods should, ideally, be made on sections prepared from one piece of tissue, so that variations in prior physiologic state and differences arising from "accidental" inconsistencies in fixation and handling are obviated. A large vertebrate brain or piece of spinal cord can provide material for numerous comparative exercises, each replicated many times.   A small insect brain suffices for no more than a simple comparison between two routines performed once each, and even here a critically informative region may be represented on only one section.   Invertebrate results, too, seem to be more variable, even under the most favorable circumstances, so that confidence in the outcome of a comparison requires abundant repetition. In a context where each step in a multistage process requires that not only the subsequent procedures be adjusted to it, but also the preceding routines, it is clear that the complete experimental rationalization of any one staining method is a formidable task. It has only been attempted in a few instances, of which Gregory's (1970) examination of the double Bodian method and its modification for the notoriously intractable cockroach central nervous system (CNS) is the outstanding example.

An obvious consequence of this situation is that, while the technology of silver staining has developed rapidly during the last two decades, it has done so piecemeal and by the efforts of several different workers: Each histologist modifies an existing technique until it suits his own material, at which point he sees no point in pursuing matters further. The shifts in methodology are largely empirical, and when they are justified by their authors on general chemical grounds, the arguments tend to have a perceptibly post hoc flavor.

Impregnation

All of the successful reduced silver methods for insects achieve more-or-less similar end results: After gold toning, axons appear in various shades of pink, blue, or black, the former color being more likely to be assumed by tracts and by axons of large diameter, and the latter by the smaller axons, and especially by those that lie within neuropil. Rowell (1963) and Gregory (1970) have discussed the origin of these colors, and attribute them to the sizes of the particles of gold laid down in toning. Particles that are small compared to the wavelengths of light produce red colors by the Tyndall effect, short wavelengths being scattered more than long ones, which become the major fraction of the light transmitted. As particle size increases, so the hues of the transmitted light shift through pink, purple, blue, and finally to black. The sizes of the gold particles must, certainly, relate to those of the silver particles that they displaced; Rowell (1963) considers that silver particle size is controlled by the conditions of impregnation and that it is not much affected by the speed of development, as was sometimes claimed by earlier authors. Our general experience bears this out, but in some of the variants described here and in the previous chapter, it is clear that particle size may be almost wholly controlled by a mordanting stage prior to silver impregnation and that it may be further modulated by the introduction of pyridine derivatives into the developing solutions. A weakness of all discussions of these phenomena is that no direct study of particle sizes in relation to conditions of processing has yet been made.

The final outcome of a successful method, then, must achieve a rather delicate balance: It should be blue or black enough to photograph well, but should it be perceptibly granular it will fail in resolution, for such granular aggregates of gold may be coarser than the finer processes that they are intended to resolve. It is also helpful to the observer if there is some degree of chromatic differentiation, so that tracts appear in colors that are different from those assumed by solitary axons in neuropil, for example.

Of equal importance is the quality of *differentiation*. It has sometimes been supposed that these methods are specific for nervous tissues because silver attaches to nothing else; both Peter's (1955a,b) and Gregory's (1970) analyses have demonstrated that this is not the case. The specificity of impregnation depends on two factors: the uneven affinity between silver and various tissue components, which causes nuclei of reduced silver to be differentially distributed by the end of impregnation, and, second, the chemical events during development. The latter are complex, for not only is some bound silver redistributed during the process, but silver carried over from the impregnating bath is also reduced and deposited in association with existing silver nuclei (Peters, 1955b; Gregory, 1970). The developer, then, in a sense, "amplifies" the tissue specificities of a relatively feeble kind, which the impregnating bath can only inadequately exploit. Gregory (1970) prefers to call this stage "differentiation."

It is unfortunate that so little is certain about the chemistry that causes silver to be bound to neural components and to be reduced to silver nuclei by some of them. Peters (1955a) claimed to have eliminated a number of possible reactive groups, including sulfhydryl, by the systematic use of blocking agents, and proposed that silver is reduced by the imidazole component of histidine in a scheme derived from photographic chemistry. He supported this conclusion by showing that the intensity of staining of protein smears was a function of their known histidine content. Wolman (1955a,b), on the other hand, using different histochemical strategies, concluded that carbonyl and sulfhydryl groups are responsible for the greater part of reduction, but failed to identify some residual reducing groups without excluding histidine. It is likely, in practice, that silver is reduced by both components; certainly, many photoreceptor membranes are intensely stained by reduced silver methods, and under normal conditions of fixation the sulfhydryl groups of their rhodopsin complement will have been exposed by bleaching. Insect muscle stains intensely, and here it is likely that its high content of methyl-histidine is concerned. Electron-microscopic studies by Blackstad (1970) indicate that silver attaches to microtubules (cf. Strausfeld, 1976), whose tubulin contains sulfhydryl groups that may be available after fixation.

Palmgren (1948) suggested that the charges on both proteins and silver nuclei might be important in determining the course of both impregnation and development, but Peters (1955a) considered that the distribution of such charges during impregnation had not been established, and noted that they would become negative when the sections were immersed in the developer.

The role of the materials with which silver is combined to make the impregnator is equally obscure. Holmes' methods can use pyridine or any of its alkylated derivatives that are sufficiently miscible with water. Broadly speaking, the addition of side chains to the pyridine molecule increases the amount of differentiation that is ultimately obtained, and the choice of derivatives thus far employed (picolines, lutidines, collidines, and 2-methyl, 5-ethyl-pyridine) has been determined by price and commercial availability. Vinylpyridines polymerize when mixed with the buffered silver solution and cannot be used. The addition of alkyl groups to pyridine reduces miscibility with water, and it is likely that a physical interaction between the vehicles and the lipid constituents of neuronal membranes is involved, but the problem cannot be usefully discussed without a knowledge of the effects of the fixatives employed, which is thus far lacking.

The following sections will describe techniques of fixation in some detail, outline a basic procedure for impregnation, and show that for difficult materials, superior differentiation and resolution can be obtained if the simple schedules are preceded by the cobalt-mercury mordanting procedure (Chap. 8).

Fixation

The development of reduced silver impregnations for arthropod central nervous tissues has not been accompanied by a parallel attack on the problems of fixation, and only a few fixative solutions have been employed. Carnoy's and Duboscq-Brasil solutions were developed in the nineteenth century for other purposes; Bodian's fixatives were originally formulated for vertebrate tissues and the Bodian No. 2 has been used successfully by Weiss (1972) on a variety of insect brains; the formol-acetic acid-ethanol fixative (FAA) recommended by Strausfeld (1976) closely resembles it. In general, attempts to use other solutions have not been successful, although Ribi (1975) hints at some interesting exceptions. In particular, aqueous or seawater Bouin's solutions are incapable of yielding material that can be differentiated after silver impregnation.

Fixatives that are consistently successful contain ethanol, but the experience that ethanol alone is ineffective is common to all workers in this field. Some solutions (e.g., FAA, Duboscq-Brasil, Bodian No. 2) contain formaldehyde, but it was the failure of formaldehyde alone to provide adequate fixation of arthropod nervous tissues in any way comparable to the results obtained with vertebrate tissues that impeded arthropod neuroanatomy for so long. Many workers have experimented with other aldehydes (e.g., glutaraldehyde, crotonaldehyde, acrolein, etc.) but with no greater success (N. J. Strausfeld, personal communication). It is apparent that the problems posed by arthropod nervous tissues are different from those offered by vertebrate material.

If the amount of differentiation given by the various classical fixatives is compared, it is found that maximum differentiation and resolution are given by Carnoy's solution, and the least by Duboscq-Brasil, although in Gregory's (1970) modification the evenness, clarity, and lack of distortion of the preparations to some extent compensate for this, provided that photography is not required. It is sometimes supposed that the role of ethanol in these solutions is to ensure rapid penetration; Carnoy's contains the greatest proportion of ethanol and Duboscq-Brasil, the least. But the effects observed can be just as plausibly interpreted if it is supposed that the main virtue of ethanol consists in its not being water and that the effective component of the solutions is acetic acid.

Davenport and Kline (1938) substituted mixtures of n-butanol and n-propanol for ethanol in Hofker's fixative (ethanol, trichloroacetic acid, and acetic acids), and formic acid for the acetic acid. They showed that fixation of spinal cord was superior to that obtained with the original fluid and that there was less shrinkage of the tissues. Trichloroacetic acid is not a useful component of fixatives for arthropod CNS, but without it Hofker's solution merely approximates a fixative of the Carnoy type. We have substituted mixtures of n-propanol (propan-1-ol), n-butanol (butan-1-ol), and, optionally, n-octanol (octan-1-ol) for ethanol, and formic acid for acetic acid in non-chloroform-containing solutions of the Carnoy's type, and they are not wholly empirical. Formic acid is a stronger acid than acetic, with the ambiguity, interesting in the present context, of a structure that can be regarded as either a carboxylic acid or a hydroxyaldehyde. One of the disadvantages of fixation with Carnoy's solution is that, although bulk shrinkage of insect brain tissue is not severe (Strausfeld, 1976), individual axons tend to acquire a somewhat distorted appearance and must undoubtedly shrink; Carnoy's and all the other fixatives listed above clearly make no attempt to match their osmolarity to that of the tissues on which they are employed. Davenport and Kline (1938) noted that the shrinkage of spinal cord fixed in their solution was minimal and was less when the two alcohols were combined than when either was used alone, or than when ethanol was employed as the vehicle. Paraffin alcohols become less hydrophilic as their chain length increases, and we have assumed that it is possible to balance the mixture of butanol, propanol, and formic acid so that entry of the latter into the tissue is matched by the amount of water permitted to diffuse out into the fixative. The formula given later (FBP) produces remarkably little shrinkage or distortion of the axons of fly or of bee brains, and has the advantage of giving a very high degree of differentiation. The effect of altering the proportions of the constituents to any great extent or of substituting ethanol for one or other alcohol is to induce shrinkage and distortion, often most dramatically seen in the perpendicular fibers of the lamina; if ganglia are transferred to the fixative with adherent "shells" of saline solution or hemolymph, distortion of the peripheral regions of neuropil is also observed.

Davenport and Kline noted that they were unable to substitute tertiary butyl alcohol for *n*-butanol in their formula; it is likely that the molecular configurations of the vehicles used in both fixatives and impregnating baths affect the nature of the results obtained.

Composition of Standard Fixatives for the Holmes and Ungewitter Techniques

The fixatives are listed in the approximate order of differentiation which they yield, starting with those that give *least*.

   1. *Duboscq-Brasil solution, Gregory's (1970) modification:*

Absolute ethanol	120 ml
Distilled water	30 ml
Picric acid	1 g
40% formaldehyde	60 ml
Glacial acetic acid	15 ml

Either the solution should have been stored for at least 1 year before use *or* it must be incubated at 60°C for 40 days before use at room temperature. Duboscq-Brasil solution that has not been either aged or incubated produces distortion of neuropil. The successful fixation of moth and fly brains obtained by Strausfeld and Blest (1970) and Strausfeld (1970) resulted because a stock solution that had been stored for several years was fortuitously used throughout, although the effect had not been described at that time.

For neuroanatomical purposes, the fixation must be allowed to proceed for *at least* 48 h, following which the tissues should be washed in 80% alcohol, until all picric acid has been removed. This normally takes many days, but for most purposes washing may be accelerated by using 0.5% nitric acid in 80% ethanol, or 10% ammonium acetate in 80% ethanol (Gregory, 1970), followed by a further thorough wash in 80% ethanol alone.

   2. *Carnoy's solutions:*

With chloroform		*Without chloroform*	
Absolute ethanol	60 ml	Absolute ethanol	75 ml
Chloroform	30 ml	Glacial acetic acid	25 ml
Glacial acetic acid	10 ml		

Fixation should be allowed to proceed for no longer than 2 h at room temperature, after which the tissues should be passed through 90% ethanol and then washed thoroughly overnight in an ample volume of 70% ethanol.

3. *Formol-acetic acid-ethanol fixatives:*

Bodian No. 2 (Weiss, 1972)		*FAA* (Strausfeld, 1976)	
80% ethanol	90 ml	Absolute ethanol	85 ml
40% formalin	5 ml	40% formalin	10 ml
Glacial acetic acid	5 ml	Glacial acetic acid	5 ml

Weiss recommends that fixation should not proceed for longer than 1–3 h, although Strausfeld (1976) recommends 8–10 h. Our own experience and that of Dr W. Ribi with Carnoy's solutions suggest that tests with new material should always include the shorter periods of fixation.

4. *Formic acid–n-butanol–n-propanol fixative* (*FBP*) (*Blest and Davie, 1977*):

n-Butanol	46 ml
n-Propanol	46 ml
Pure formic acid	8 ml

Mixtures of formic acid and paraffin alcohols esterify extremely rapidly, and small volumes of fixative should be made up *immediately* before use. We have found that the inhalation of vapor from the solution tends to produce headaches, and it should be used in a fume cupboard. The duration of fixation is critical: It should be for 1 h at room temperature, following which the tissues are washed for 30 min in 90% ethanol, followed by an overnight wash in 70% ethanol. Ganglia should be transferred to fixative with the minimum amount of adherent saline solution or hemolymph; if necessary, they should be very gently blotted with filter paper.

Practical Considerations

All reduced silver methods require that nervous tissue should be fixed as rapidly as possible following the death of the animal concerned, and speedy penetration of the fixative is essential. Because insects are small, the penetration of nervous tissue poses few problems provided that it has been exposed by dissection, but it must be emphasized that even in the case of quite small and fragile species immersion of the intact insect is inadequate, even though it may suffice for conventional histologic purposes. Workers should practice whatever dissection is needed so that it may be performed briskly, and here it is desirable that the instruments should be equal to the task; we regard the use of fine forceps and iridectomy scissors as essential, an old pair of the latter being used for cutting through cuticle. The insects may be killed by injection of fixative; in some cases (e.g., Diptera and Lepidoptera), removal of the mouth parts and injection of fixative via a fine hypodermic needle inserted into the thorax allows rapid perfusion with a large volume of solution. Alternatively, the insects may be anesthetized with carbon dioxide. If dissection under a Ringer's solution is necessary, the saline solution used must be physiologic for the particular species of insect. It is advisable always to employ generous volumes of fixative.

As soon as fixation is complete, the tissues should be transferred to the appropriate grade of ethanol, and dissected free from any remaining cuticle before commencing the wash. This is desirable because not only is washing more efficient, but the nervous tissue will become more brittle during its course, and thus more easily damaged.

Embedding and Sectioning

Insect ganglia that have been fixed for neuroanatomical purposes are always hard, and require careful processing if they are to be sectioned successfully. Material fixed in Duboscq-Brasil or Carnoy's solutions is the least difficult to handle, and ganglia fixed in FBP the most. A successful solution to the problem of embedding, however, has to recognize that nervous tissue needs to be infiltrated with paraffin wax for the shortest time and at the lowest temperature possible (Rowell, 1963; Weiss, 1972; Strausfeld, 1976). However, vacuum embedding causes any minute quantities of air left in the tracheal system to expand and damage the neuropil, so it should never be employed. Thus, if circumstances allow, waxes with melting points of 54°C or 56°C are to be preferred over those that melt at 60°C or over. Dehydration through an alcohol series to absolute ethanol is followed by clearing in cedarwood oil or methyl benzoate, and each stage should occupy the minimum effective time. The cedarwood oil is removed in benzene before transferring the tissues first to a 1:1 mixture of wax and benzene, and then to two to three successive baths of pure wax. Alternatively, tissues can be transferred from methyl benzoate to a benzoate-wax mixture, and then to two changes of pure wax, held just above its melting point.

In the case of fixatives that harden severely, and especially in the case of FBP, it may be best to employ the embedding procedure of Weiss (1972), which takes the tissues through an alcohol-terpineol series before reaching benzene. Orientation of these small structures should be accomplished before quenching the wax by manipulating them under a stereomicroscope with a heated needle. Quenching should not be violent, and the wax may be allowed almost to solidify before being immersed in water at room temperature.

Sectioning requires that the microtome knife be sharp and that the mechanics of the microtome itself be in good order. This does not necessarily mean that the equipment should be elaborate; many of the preparations illustrated in this chapter and in Chapter 8 were made with an old Beck rotary microtome of dilapidated appearance, which gave good ribbons at 3 μm thickness, and which tended to suffer from slight but troublesome backlash only after it had been cleaned and oiled. The optimal rake angle of the knife for trouble-free sectioning can depend on the nature of the tissue, the hardness of the wax (and therefore the ambient temperature at which the sections are cut), and the degree of bevel at the knife edge and is best determined by trial and error. Slow knife strokes of uniform speed give the best results.

The hardness of insect nervous tissue blocks causes static electricity to be a severe problem, especially in dry climates; its effects may be overcome by positioning a source of low-penetrance α-radiation near the cutting region of the knife.

Adhesion of expanded ribbons to the slides must be managed with some care, because of the prolonged and violent processing stages that are employed. We stir 10–15 drops of a 1:1 mixture of egg-albumen and glycerine into 50 ml distilled water. The stock mixture will keep for several months in the light without added preservative, and we have not found it necessary to filter it after preparation. The diluted solution is used to cover the surface of the slides, and the ribbons are placed on the liquid film before expanding them on a hot plate maintained just below the melting point of the wax. Good expansions are rapid, and the adhesive should not be allowed to evaporate. It is carefully drained from beneath the sections after cooling and the ribbons shunted into final place with a needle. The slide is warmed again for a few seconds on the hot plate, and propped narrow end down to drain. Drying of the slides may be accomplished in an incubator or overnight in a desiccator. Drying for 2 h on a hot plate is faster but less reliable.

A Basic Schedule for Holmes-Blest Impregnations

Sections (7–10 μm) on slides are dewaxed and brought to distilled water.

1. 20% aqueous silver nitrate   2–4 h
2. Distilled water   5 min
3. Incubate at 37°C in:
   M/5 boric acid   27.5 ml
   M/20 borax   22.5 ml
   1% silver nitrate   5–20 ml
   Pyridine or pyridine derivative   2–10 ml   16–48 h
   Distilled water   to 250 ml
   pH = 8.4

(a) The amounts of the pyridine derivatives that may be added are:

Pyridine	10 ml
2:4 or 2:5 lutidine (dimethyl-pyridines)	6 ml
2:6 lutidine (dimethyl-pyridine)	10 ml
2,4,6-collidine (trimethyl-pyridine)	8 ml
2-methyl-5-ethyl-pyridine	2 ml
1:1 2,4,6-collidine/2,6-lutidine (recommended)	10 ml

Pyridine is readily miscible; the other derivatives should be added with vigorous shaking.

(b) Rowell (1963) provides a table of the relative proportions of $M/5$ boric acid and $M/20$ borax to give a range of pH of 7.0–9.0. Lower pHs produce a highly selective impregnation of larger axons; at pH 8.4 (recommended here as optimal), small fibers are impregnated, and there is more background staining. We prefer to control background staining by appropriate fixation and premordanting rather than by reducing the pH of the impregnating bath.

(c) The temperature of the impregnating bath can be varied, in principle, between 30°C and 70°C. However, temperatures higher than 60°C tend to result in detachment of the sections. Increase in temperature merely increases the rate of uptake of silver, so that more small fibers are impregnated, and there is more background staining. Similar effects follow increase in the duration of impregnation; 24 h is suggested for an initial series.

(d) Because the final outcome of the development stage depends upon the exchange of silver between binding sites, it is much affected by the thickness of the sections. For Holmes' methods, optimal results are obtained with sections that are 7–10 μm thick.

4. *Develop in:*

Hydroquinone	3 g	
Sodium sulfite (anhydrous)	15 g	
Distilled water	300 ml	5 min

After reduction, sections are usually a very pale yellowish-brown, and individual components cannot be distinguished at this stage. Gregory's comments on the effects of varying the hydroquinone-sulfite ratio in the context of the Bodian method also apply here (Chap. 6).

5. Wash in running tap water, or in several changes.   5 min
6. Rinse in distilled water (optional).
7. Tone in 0.2% chlorauric acid or sodium gold chloride.   variable: 1–10 min
8. Rinse briefly in distilled water.   30 s
9. Reduce in 2% oxalic acid.   5 min

Stages 7–9 are repeated until a satisfactory result is obtained. The duration of toning in the gold solution that is necessary can vary greatly, and it is recommended that an initial 2- to 3-min treatment followed by reduction be used to assess the amount of toning that will be needed for a given preparation. Gregory (1970) has pointed out that there is in fact no need to use distilled water in place of tap water at this stage, provided that the supply is not grossly impure and that there is also no advantage to be gained from acidifying the gold solution with acetic or citric acids.

10. Wash in tap water.
11. Fix in 5% sodium thiosulfate.   3 min
12. Wash thoroughly in tap water.
13. Dehydrate through an ethanol series, clear in xylene, and mount in a synthetic medium.

The Holmes-Rowell Method

Rowell (1963) reviewed a number of reduced silver methods and proposed that those of the Holmes type could be improved by the use of a physical developer in place of the simple hydroquinone-sulfite solution customarily employed. The Holmes-Rowell method uses the same procedure as given above until the completion of impregnation (3), which is carried out in a bath containing the high concentration of silver nitrate specified. Thereafter, the sections are washed thoroughly in distilled water, immersed in 1% anhydrous sodium sulfite for 2 min, washed thoroughly in distilled water, and developed for 5–10 min in the following solution:

5% silver nitrate	7.5 ml
4.5% anhydrous sodium sulfite	250 ml
0.5% hydroquinone	16.5 ml

The sections are then washed in distilled water followed by a 5-min wash in tap water. They are gold-toned and reduced in oxalic acid in the usual way.

Holmes-Rowell methods tend to favor large fibers, with respect to whose populations they can be somewhat selective. They also tend to give a somewhat granular result. They can be admirable for tracts, but are rarely satisfactory for neuropil.

Weiss' Method

Weiss (1972) published a method that gives successful results with a range of difficult materials. Fixation in Bodian's No. 2 solution for 3–18 h is followed by a prolonged (1–2 days) wash in 70% ethanol, and a careful embedding procedure as follows:

1. 82% ethanol	2 min
2. 95% ethanol	3 min
3. 3:1 95% ethanol-terpineol (v/v)	4 min
4. 1: 95% ethanol-terpineol (v/v)	5 min
5. 1:3 95% ethanol-terpineol (v/v)	6 min
6. Terpineol—2 baths totaling	1½ h
7. Benzene—2 baths of 30 s each with *constant* agitation	
8. Benzene—wax (1:1, fully liquefied)	2 min
9. Wax in oven, 2 baths, 5 min each	
10. Embed in fresh wax	

Waxes of either 56–58°C or 50–52°C melting points were used, with the oven held just above their melting points. The duration of the various stages appears to be nicely adjusted to the degree of tissue hardness given by the fixative and probably requires modification should the embedding routine be used for tissues fixed in other solutions.

Ten-micrometer sections on slides are processed by the collidine Holmes-Blest method, following by toning in 1% gold chloride for 7 min and reduction in oxalic acid for 10 min.

Weiss' method gives good resolution of fine fibers, with little background staining. Examples kindly provided by Dr. Weiss are shown in Figs. 7-1 and 7-2.

Pretreatment with Dilute Nitric Acid

Pretreating sections with 2% nitric acid before the 20% silver nitrate bath increases the uptake of silver; the results (Fig. 7-3) tend to be highly selective for particular axon species.

Cobalt-Mercury Mordanting

In common with many other workers, we have experienced difficulty in obtaining satisfactory preparations from such materials as the brains of Hymenoptera, Orthoptera, and Crustacea by means of standard Holmes' procedures. The principle of mordanting with mercury devised for the modified Ungewitter method (Blest, 1976, and Chap. 8) can be applied to the present schedules (Blest and Davie, 1977). Sections are pretreated in the mordanting solution described in Chapter 8, followed by the same sequence of iodine and thiosulfate. After thorough washing and a passage through distilled water, they are then impregnated by the full Holmes-Blest routine described above.

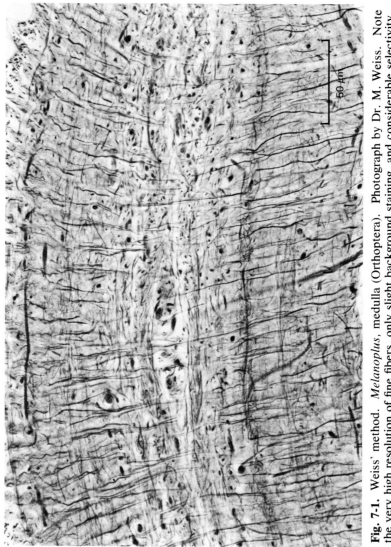

Fig. 7-1. Weiss' method. *Melanoplus*, medulla (Orthoptera). Photograph by Dr. M. Weiss. Note the very high resolution of fine fibers, only slight background staining, and considerable selectivity.

Fig. 7-2. Weiss' method. *Melanoplus*. Central body. Photograph by Dr. M. Weiss.

20 μm

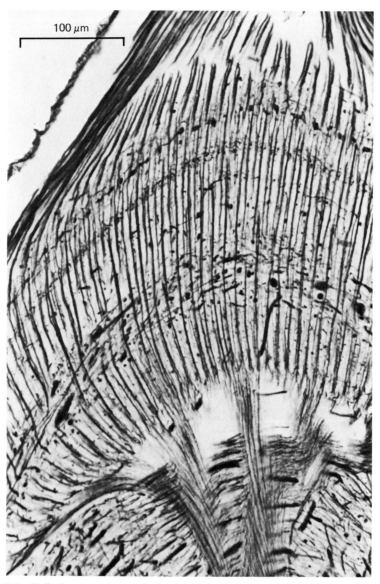

Fig. 7-3. *Calliphora erythrocephala.* Medulla. Holmes-Blest impregnation, preceded by treatment of the sections for 1 h with 2% nitric acid. 2,6-Lutidine/2,4,6-collidine variant. Note the high selectivity for perpendicular components and the poor resolution of fine fibers. Fixation in FBP for 1 h at 20°C.

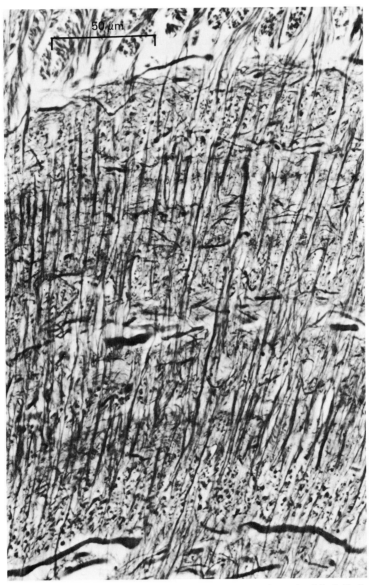

Fig. 7-4. *Scylla* (mud crab) (Crustacea, Decapoda, Brachyura). Medulla interna. Holmes-Blest, 2,6-lutidine/2,4,6-collidine variant, preceded by cobalt-mercury mordanting. Fixation in FBP for 1 h at 20°C. Note the equal impregnation of all components and the relative lack of selectivity.

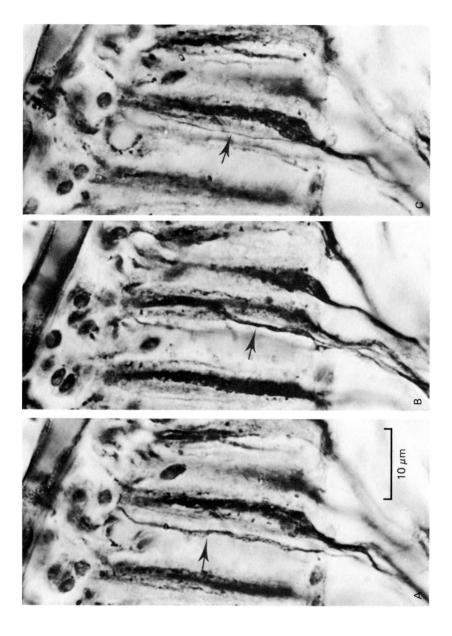

The best results from this strategy are obtained if the amount of mercuric chloride in the mordanting bath is reduced to 0.035 g/100 ml and treatment is continued for 60 min. The subsequent immersion in 20% silver nitrate should be for 4 h, and incubation in the impregnating bath, which contains the higher concentration of silver nitrate given, should be for 48 h. Results obtained from this method are shown in Fig. 7-4 and in Fig. 7-10 (Plate 6).

Varying the Processing Strategies

The foregoing account suggests that Holmes' routines are flexible and can be modified in a number of different ways—so many, in fact, that the inexperienced histologist may well find them confusing. The following summary provides some simple guidelines that can be used in the selection of a particular routine for new material.

1. *Choice of fixative*. Gregory's Duboscq-Brasil and Carnoy's should be tried first. Higher-resolution fixatives (which produce handling difficulties) can then be employed later if these fail to give good results.

2. *Mordanting*. Although mordanting with the cobalt-mercury procedure gives the highest degree of differentiation and is probably necessary for difficult materials, it requires some experimentation with times and concentrations if it is to succeed with a given tissue; it is best to omit it in a first attempt. The effects of varying the relative proportions of cobalt and mercury in the mordanting solution are the same as those described for the Ungewitter methods in Chapter 8.

3. *Pretreatment with nitric acid alone*. We have performed relatively few experiments in which nitric acid pretreatment is given independently of the cobalt-mercury mordant, but they do not suggest that it yields any advantage that would not be improved by the full mordanting procedure.

4. *20% Silver nitrate pretreatment*. Most materials benefit from the longer periods of treatment, and 4 h is suggested as an appropriate duration for tests.

5. *Impregnating bath*. The longer the duration of incubation, the greater the uptake of silver. Easy materials, such as Diptera, can give

---

Fig. 7-5, A–C. *Musca domestica* (Diptera). A lamina cartridge, photographed with the microscope focused at three different depths. *Arrowed* structures are: **A**, the apposed axons of retinula cells R_7, R_8; **B**, laminar monopolar cells L_3, L_4; and **C**, L_5. Ethanol-acetic acid-formaldehyde fixation, followed by Holmes-Blest impregnation using pyridine. Despite the relative lack of contrast, this is the only method to reveal all components of a laminar cartridge, including L_5, which is usually refractory to impregnation. Preparation courtesy of Dr. N. J. Strausfeld.

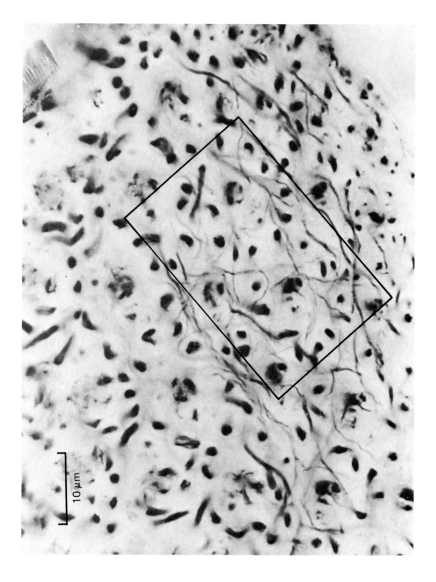

satisfactory preparations after no more than 24 h; difficult materials require longer. The most satisfactory combination of pyridine derivatives employs a 1:1 mixture of 2,6-lutidine and 2,4,6-collidine, and yields a purplish picture. 2,4,6-Collidine alone gives a reddish or brownish result, and 2,6-lutidine alone tends to produce a bluish and often rather granular impregnation that can be excellent for photography, but whose toning is more difficult to control.

General Comments on Handling Sections

Reduced silver techniques are often attempted by physiologists with little experience of conventional histology and who are frequently disappointed by the results. Although many of the beliefs that have evolved around silver histology are merely superstitious in the context of the present techniques (e.g., the use of water double- or triple-distilled in glass for key stages), the methods do require a more fastidious approach than most, and observation of the performance of apprentices in the field has suggested that a number of points are worth mentioning.

1. Stainless steel instruments should be used for dissections under saline or fixative solutions. Metallic impregnations are vulnerable to trace contamination by irrelevant metals, and the effective fixatives are to various degrees mildly corrosive.

2. For similar reasons, it is advisable to start operations with new glassware; some materials are difficult to remove from glass surfaces, and others, especially mercury salts, may be absorbed by glass.

3. Staining dishes should be arranged in a logical sequence on ample bench space. Racks of slides should be moved briskly from one solution to another; they should not be waved about in air while the worker ponders his next move or discovers that he has omitted to fill the appropriate dish with solution. Workers are sometimes prone to waste time on finicky maneuvers intended to avoid spillage of solutions on the bench or on themselves, so that "a rapid dip" becomes an unintentional 15-s soak, and so forth. It is better to be quick and dirty.

4. Although the processing stages do not have precisely specified durations, an appreciation of their various degrees of latitude and of the ways

---

Fig. 7-6. *Musca domestica* (Diptera). Tangential section of the lamina. Thick and thin (within rectangle) lamina tangential fibers are resolved, the latter arising from amacrines. Fixation in aqueous Bouin's solution, followed by Holmes-Blest impregnation using 2,6-lutidine. This is the only combination of fixative and impregnation that reveals the amacrine tangentials. Preparation courtesy of Dr. N. J. Strausfeld.

in which they may be adjusted for particular situations and materials only comes with experience; until the worker is confident about such matters, he is well-advised to obey the schedules given.

5. Nervous tissues usually need to be examined as serial sections. Practice in and mastery of the simple technique of arranging sections in orderly rows when the slides are drained after expansion can save a great deal of time later on, when the results are being analyzed.

6. Slides of reduced silver preparations improve slowly over periods of days or even weeks as the mounting media penetrate the sections, and this effect is notable even with synthetic media of low viscosity. Critical observations and photography should, ideally, be delayed until this process is complete.

Results

The results of various Holmes-Blest routines are shown in Figs. 7-1–7-6 and in Figs. 7-7–7-14 (Plates 6–9). Black-and-white photographs were made with a Leitz Orthoplan microscope through yellow or orange filters, using Ilford Pan F 35-mm film developed in ID11 for 8.5 min at 20°C. Color photographs were made on Agfachrome Professional 35-mm 50 L film.

Chapter 8

Reduced Silver Impregnations of the Ungewitter Type

A. D. Blest

Department of Neurobiology
Research School of Biological Sciences
The Australian National University
Canberra, Australia

Ungewitter (1951) described a reduced silver method in which silver nitrate was combined with an excess of urea in a bath that yielded good impregnations of vertebrate peripheral nervous tissue after incubation at 60°C for only 20 min. Silver incorporated into the sections was developed in a conventional hydroquinone-sulfite solution that was also saturated with urea. Gold toning was not employed, and the results resembled Bielschowsky preparations in that uptake of silver was sufficient to stain fine axons in various shades of brown and yellow. Although it is rarely used for critical neuroanatomical work, the Ungewitter method and its variants are still in use in the clinical context for the rapid processing of biopsy material.

The speed and simplicity of the technique make it attractive as a starting point for the development of methods for arthropod central nervous systems, but in its original form it has failed to give useful or even adequate results either with invertebrate materials (Blest, 1976) or with mammalian brain tissues (Professor R. W. Guillery, personal communication). In the former case, these failures relate to three problems: (1) The high temperature of incubation coupled with the high concentration of urea tends to detach sections of small area from their slides. (2) In my hands, precipitates form very readily on both slide and sections during the development stage. (3) The uptake of silver by sections of arthropod brains is relatively feeble, as it is with all reduced silver methods. This means that development must be followed by a gold-toning stage, but the outcome of toning is coarsely granular and the overall differentiation poor.

Attempts to devise modifications of Ungewitter's method were a consequence of the difficulties experienced in obtaining fresh samples of pyridine derivatives in New Zealand, thus making variants of the Holmes technique (Chap. 7) precarious as routine methods. Some provisional solutions to the various difficulties (Blest, 1976) have also suggested a number of techniques that can be employed to improve the resolution and differentiation of the more widely employed Holmes techniques in the context of difficult materials, such as the brains of Hymenoptera. This chapter describes some modified Ungewitter routines and the general strategies by which they may be varied for particular materials (see Chap. 7, which deals with the more difficult and capricious Holmes methods). The reader should bear in mind that all the strictures on methods of embedding tissues and handling sections, as well as the relative virtues of different fixative solutions contained in that chapter, apply equally to the methods detailed here.

Mordanting: The Use of Mercury Pretreatment

The original Ungewitter technique added very small traces of mercuric cyanide and picric acid to the urea-silver nitrate bath (1 drop of each per 100 ml). The choice of mercuric cyanide was presumably dictated by its stability and by the fact that it does not form an insoluble product with silver salts; it was said to have a "catalytic" effect on impregnation, and the picric acid was said to improve clarity and resolution, but no attempt was made to explain the role of either component. The effect of mercury is interesting in several respects: (1) Wolman (1955a) showed that the treatment of sections of mammalian central nervous tissue with solutions of various heavy metals (Hg, Bi, Cu, Fe, Pb) followed by prolonged gold toning resulted in the staining of axons only in the case of mercury, where it was intense; myelin was weakly stained by mercury and also by bismuth, but the remaining metals failed to stain axons at all. (2) Nakata *et al.* (1971) combined a thiolacetic acid–lead nitrate technique for nonspecific vertebrate cholinesterases with a Bielschowsky method in order simultaneously to demonstrate motor end-plates and peripheral nerve. Although the Bielschowsky method alone did not reveal end-plates, and the deposits of lead sulfide in the latter were faint in the absence of silver, end-plates were strongly stained by the combination, suggesting that prior impregnation by a heavy metal augments the subsequent accretion of silver. Finally (3) Herdman and Taylor (1975) preceded the original Holmes technique with a simple mercury treatment and found that fine peripheral axons in frozen sections of mouse kidney showed enhanced differentiation as a result.

These diverse observations suggested that it might be appropriate to mordant arthropod tissues with mercury solutions *before* reduced silver impregnation. However, the Herdman and Taylor routine, which treats sections with a solution of 0.5% mercuric chloride, removes the bulk of the bound mercury with an alcoholic solution of iodine, and then bleaches residual iodine with sodium thiosulfate before starting the silver impregnation, did not give satisfactory results with blowfly brains when either the Holmes or Ungewitter method was employed; the staining tended to be coarsely granular. Promising results were obtained when the sections were pretreated with very dilute mercury solutions without subsequent exposure to iodine, but the effective concentrations of mercury were of the order of 0.0001%, and such solutions are vulnerable to trace contaminants and to absorption of the mercury onto the walls of the glass vessels in which processing is carried out. Very dilute solutions of mercuric chloride, mercuric and mercurous nitrates acidified with nitric acid, mercuric cyanide, and mercury acetamide all gave similar pictures.

A possible solution might employ a higher concentration of mercury in the presence of a considerable excess of some other metal; the two metals might be supposed to compete for sites of attachment, and if the second metal were loosely bound it might be possible to remove it in its entirety at a subsequent processing stage. Cobalt or nickel seemed plausible choices, because of the readiness with which they form organic complexes. Empirical tests showed that the anticipated effects do, indeed, occur. It was also found that pretreatment of sections (or of whole tissue subsequent to fixation) with dilute nitric acid greatly increases the uptake of silver and that both procedures can be combined in one.

Basic Schedule for a Urea-Silver Nitrate Impregnation

Fixation

Fixation should be in either Gregory's Duboscq-Brasil (Gregory, 1970), Carnoy's solution, with or without chloroform (see Chap. 7), or formic acid-butanol-propanol (FBP) (Blest and Davie, 1977). Paraffin sections, prepared by standard embedding routines, can be 2–15 μm, but the method is best for thin (2–5 μm) sections.

Mordanting

Sections are dewaxed in xylol and brought to 70% ethanol. They are treated as follows:

1. Mercuric chloride	0.05 g/100 ml	⎫
Cobalt nitrate	10.0 g/100 ml	⎬ 30–60 min
Nitric acid	2.0% by volume	⎭
2. Wash in 70% ethanol		1–3 min
3. 0.5% Iodine in 70% ethanol		3 min
4. Wash in fresh 70% ethanol		1–3 min
5. Wash in tap water		1–3 min
6. 5% Aqueous sodium thiosulfate		1 min
7. Wash in several changes of tap water and bring to distilled water		3–5 min

The duration of stage 1 can be varied in relation to particular materials; good results are obtained with *Calliphora* using 30 min, but more difficult materials such as Orthoptera and Hymenoptera require longer periods. The relative proportions of mercury and cobalt determine the color and granularity of the final result. The lower the proportion of mercury, the redder and finer the impregnation, but in practice it is desirable to achieve some degree of chromatic differentiation, and in the case of dense neuropils, preparations that are restricted to different shades of red and pink are confusing to the eye and do not photograph well. When the method is explored for a new material, tests should be run by holding the cobalt nitrate at the given concentration and varying the mercuric chloride between 0.01 and 0.05 g/100 ml. All solutions can be used repeatedly without exhaustion. The cobalt nitrate should be of the highest purity available. Merck analytical grade is to be preferred, and the same quantity and purity of nickel nitrate may be substituted.

Impregnation

8. Incubate in the following solution, preheated at 55°C in the dark:

Silver nitrate	3.0 g/100 ml	⎫
Urea	3.0 g/100 ml	⎬ 1–4 h
Saturated picric acid	1 drop/100 ml	⎭

This solution greatly reduces the concentration of urea employed by Ungewitter, and it omits the trace of mercuric cyanide. The virtues of the picric acid appear to be no more than marginal, but it is retained. The solution, which often yields poor results when it is first employed, improves rapidly with storage and use, and like the mordant can be re-used many times; crystalline precipitates can be removed by filtering through cotton wool, and the bath is stable to light. Excellent preparations have been obtained with solutions that have been stored for 6–8 months under room daylight.

Development

9. Quickly remove the slides from (8) and dip them rapidly in and out of distilled water once only.
10. Plunge into the following solution:

Hydroquinone	0.5 g/100 ml
Anhydrous sodium sulfite	5.0 g/100 ml
Pyridine	1.0% by volume

or

2,6-lutidine, 2,4-lutidine, or
2,4,6-collidine to saturation   5 min

In my laboratory, processing is carried out in rectangular staining troughs each of which contains 250 ml of solution, and ten slides are carried in a glass rack. This arrangement creates a high degree of turbulence when slides are transferred briskly from one solution to another, and there is some evidence that this is critical as the sections pass from the impregnating bath, through the rapid wash and into the developing solution, because the use of Conklin jars and the transfer of slides individually is not attended by success, and turbulence is minimized by the latter procedure.

The purpose of including pyridine in the developer is to obviate the formation of precipitates on the sections without the need to incorporate a high concentration of urea. Low concentrations of the urea-silver nitrate complex carried over into the developing bath are reduced very rapidly, before they have diffused away from the sections. Silver pyridinium nitrate, however, is only reduced slowly, and precipitates do not form within the 5-min period of development, so that the role of pyridine is to trap excess silver compounds in a relatively stable form. Although no differences other than cleanliness can be seen between sections processed with and without pyridine, the substitution of lutidines and collidines enhances differentiation, just as it does in the case of the impregnating bath of the Holmes method, implying that events during development are of considerable complexity. Lutidines and collidines are poorly miscible with solutions of high salt concentration, and it is best to saturate the developer by mixing in the derivative with vigorous stirring *immediately* before immersing the sections. Droplets of separated derivatives and crusts that may form on the surface of the bath can be ignored, and as the temperature of the solution is raised by the immersion of the still-warm slides and rack, further separation usually occurs during development;

unless the separated materials cling to the surfaces of the sections, which rarely happens, no untoward effects result.

Gold Toning

11. The slides are washed in several changes of tap water. Following step (10), the bath has a surface layer of separated pyridine derivative sometimes mixed with a crust of separated sulfite. So that these materials do not contaminate the sections, the tap water should be introduced from the supply by a rubber tube whose orifice is inserted to the bottom of the bath, thus removing the surface materials by flotation as water is supplied.
12. Transfer to distilled water.   3 min
13. Tone in 0.2% sodium chloraurate.   1–2 min
14. Pass through distilled water.
15. Reduce in 2% oxalic acid.   5 min

Although it is perhaps safer to use distilled water for stages 12 and 14, Gregory (1970) suggests that there is little reason why most tap waters should not be employed.

After development, successful impregnations usually appear brown, and when examined under a ×40 objective, the larger axonal components may be clearly seen. The duration of toning in sodium chloraurate relates to the amount of silver deposited in the sections, and it is sound practice to commence with a toning period of one minute followed by reduction in oxalic acid, at which point the result can be inspected under the microscope. Usually, a number of short, successive repetitions of the toning sequence are necessary before a satisfactory intensity is obtained. Recognition of the stage at which to stop depends on experience, and criteria cannot be given, because the appearance of the preparations changes radically after mounting. However, it should be noted that it is possible to fix the sections, take them to xylol and inspect them, and then, if necessary, return them to distilled water and continue toning until a satisfactory picture is obtained.

Fixation

16. Wash in distilled water.   3 min
17. 5% sodium thiosulphate   3 min
18. Wash in several changes of tap water.   5 min
19. Dehydrate through an alcohol series and
    xylol, and mount in a synthetic medium.

The purpose of the thiosulfate treatment is to remove bound but unreduced silver that would darken on exposure to light. Of the synthetic media available, Fisher's Permount is preferable because of its low viscosity and rapid penetration of the sections. As with all media, *full* penetration is a lengthy process, and for critical observations of small structures, or for photography, sections continue to improve for a period of several weeks or months after preparation.

Choice of Variants

The method may be adjusted for specific materials at a number of stages:

1. *Different fixative solutions produce distinct and characteristic results.* *Gregory's Duboscq-Brasil* combined with the standard routine given above, using pyridine in the developer, yields a red-pink picture for most materials (see Figs. 7-11–7-12, Plate 8). There is usually some diffuse, pinkish background staining that prejudices the resolution of the smallest fibers, and the impregnation of medial protocerebral structures is usually unsatisfactory. Neuronal somata and their nuclei are adequately stained; in contrast to other techniques, the passage of axons can often be traced between neuropil and somata, and this can be an advantage in the analysis of the grosser anatomical relationships. In *Calliphora,* this technique yields consistently good results with the lamina, where L_4 collateral fibers (Strausfeld, 1971) are often well resolved (Blest, 1976), but it is capricious in the context of the medulla, lobula, and lobula plate. It is a valuable method for preliminary surveys, because tissues can be stored in the fixative idefinitely, and any period of fixation longer than 24–48 h gives good results; thus there is no need, initially, to arrive at a critical duration of fixation by trial and error, as is necessary in the case of the fixatives that give higher degrees of differentiation. *Calliphora* lamina is shown in Fig. 8-1.

Carnoy's solutions give excellent differentiation and only trivial background staining. Although they evoke little overall shrinkage of nervous tissues (Strausfeld, 1976), they appear to produce some distortion by shrinkage of individual axons. The color within neuropil, using the standard routine, tends to be more purplish or bluish than after Gregory's Duboscq-Brasil, and although excellent results may be obtained for the optic lobes, the medial protocerebrum does not respond well to these fixatives. *Calliphora* optic lobe after Carnoy's fixation is shown in Blest (1976), Fig. 6.

Formic acid-butanol-propanol (FBP) (Blest and Davie, 1977) gives the highest degree of differentiation, and the minimal background staining. In *Calliphora,* it yields the optimal staining of the medial protocerebrum (see Figs. 7-13–7-14, Plate 9). Correctly employed (see Chap. 7), distortion of individual components is minimal. In *Calliphora,* axons in neuropil

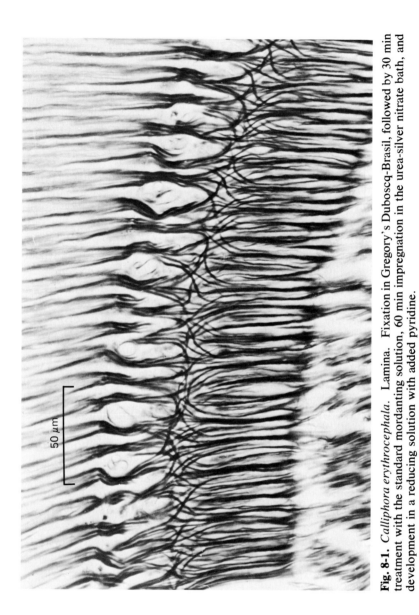

Fig. 8-1. *Calliphora erythrocephala.* Lamina. Fixation in Gregory's Duboscq-Brasil, followed by 30 min treatment with the standard mordanting solution, 60 min impregnation in the urea-silver nitrate bath, and development in a reducing solution with added pyridine.

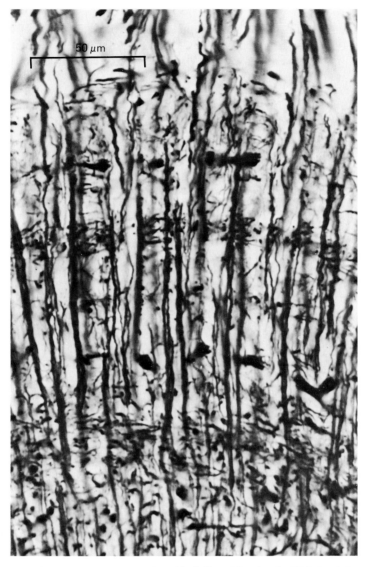

Fig. 8-2. *Calliphora erythrocephala.* Medulla. Fixation in FBP for 1 h at 20°C, followed by treatment with the standard mordanting solution for 60 min, and impregnation in the urea-silver nitrate bath for 60 min. Development was in a reducing bath containing added 2,6-lutidine. Horizontal and perpendicular elements are stained with equal intensity, and the majority of components are represented.

appear black against a clear, white background (Fig. 8-2), and tracts between neuropil are in various shades of pink to red.   This fixative has the disadvantage that neither the neuronal somata, their nuclei, nor the elements linking them to neuropil stain distinctly; it offers the advantages that very small axons (ca. 0.25 μm diameter) may stain intensely and that in the present context it is the least selective of all methods available, so that the majority of components within neuropil are likely to be resolved, including tangential elements in the medulla, which are often poorly impregnated after other fixatives.   This relative lack of selectivity, however, can be a disadvantage when low-power survey pictures are required.   The columnar organization of fly medulla, for example, is less evident than it is in Holmes' or Bodian preparations, although at high magnifications sections prepared by the present method are often more informative.   An example of a low-power survey picture is shown in Fig. 8-3, and may be compared with those given in Chapters 6 (Bodian) and 7 (Holmes).

   2. *The duration of mordanting, and the composition of the mordanting solution.*   In addition to the points noted above regarding the effects of changing the relative proportions of mercury and cobalt, it should be remembered that the nitric acid component of the bath also progressively increases the uptake of silver; the greater the uptake of silver, the shorter is the subsequent period of gold toning required, and the smaller the particle size of the gold finally deposited (see Chap. 7 for a fuller discussion). The metallic salts and the nitric acid in the mordanting bath, therefore, interact in a complex manner, and for particular materials it may be helpful to experiment by varying proportions and durations in the way that these comments imply.   It is possible to separate the nitric acid treatment from the other components by soaking the brains in 0.5%–2.0% nitric acid after fixation and then thoroughly washing them in 70% ethanol before embedding.   After fixation in Gregory's Duboscq-Brasil, this is also a rapid and efficient method of removing picric acid from the tissues.

   3. *The impregnating bath.*   The longer the duration of impregnation, the greater the uptake of silver, which also relates directly to the temperature of incubation.   Although incubation at 55°C for 1–4 h gives good results, adequate preparations may result if sections are left in the bath overnight at room temperature.   The concentrations of silver nitrate and of urea given are optimal in terms of the behavior of the sections during development.

   4. *The developer.*   Unlike the phenomena noted by Gregory (1970) in the context of the Bodian technique, where the ratio of hydroquinone to sulfate in the developer is critical in controlling the extent of differentiation, no such striking effect has been found for the present techniques, although the point has not been systematically explored.   Comment has already been made on the effects of substituting alkylated pyridines for pyridine.

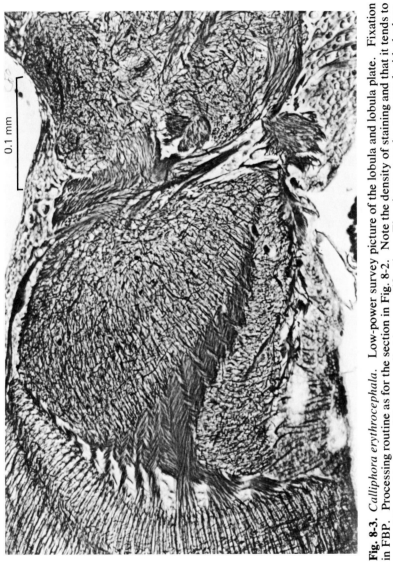

0.1 mm

Fig. 8-3. *Calliphora erythrocephala.* Low-power survey picture of the lobula and lobula plate. Fixation in FBP. Processing routine as for the section in Fig. 8-2. Note the density of staining and that it tends to conceal the architecture of the neuropil at low magnification. The picture may be compared with the low-power illustration of a section processed with a Holmes variant in Chapter 7.

Comparative Conclusions

For "easy" tissues, such as the brains of Diptera and Lepidoptera, it is evident that good preparations may be obtained by combining fixation in Gregory's Duboscq-Brasil with 30 min mordanting, followed by 1 h of incubation in the impregnating bath, and treatment in a developing solution that contains pyridine. This simple routine has given satisfactory results, too, with cockroach brains, the metathoracic ganglia of Orthoptera, the freshwater crayfish *Paranephrops* (Crustacea), and jumping spiders (Salticidae). The use of lutidines or collidines in the developer has given adequate results with the optic glomeruli of Dinopid spiders (Blest and Land, 1977), but for all these difficult materials it is necessary to increase the duration of impregnation to at least 4 h. Superlative results are obtained for Diptera and Lepidoptera by preceding the standard routine with careful fixation in FBP (see Figs. 7-11–7-14, Plates 8 and 9).

Acknowledgments

The methods described here and in Chapter 7 developed from studies which were initiated at University College London in 1960, and completed at the University of Canterbury, Christchurch, New Zealand. I am deeply indebted to Professor B. B. Boycott F.R.S. for much early encouragement and advice, and to Dr. N. J. Strausfeld, who collaborated with me from 1965 to 1968. Various aspects of the work have been supported by grants from the United States Public Health Service, the Science Research Council (U.K.), and the University of Canterbury Research Fund. Parts of these chapters were prepared during a visit to the School of Biological Sciences, Sydney University, and while occupying a Visiting Fellowship in the Department of Neurobiology, Research School of Biological Sciences, the Australian National University; I thank Professor D. T. Anderson F.R.S. and Professor G. A. Horridge F.R.S. for the hospitality of their respective Departments.

Chapter 9

The Golgi Method: Its Application to the Insect Nervous System and the Phenomenon of Stochastic Impregnation

N. J. Strausfeld

European Molecular Biology Laboratory
Heidelberg, F.R.G.

This chapter discusses aspects of methods that impregnate single neurons in a more or less unpredictable fashion. The basic methods are discussed with respect to strategies of fixation, chromation, and metal impregnation. This is followed by discussions of a possible mechanism of impregnation based on direct observations of neurons as they are impregnated and on model substrates. It is argued that silver (or mercury) chromate precipitates are not nucleated at a single point in the cell but usually begin to form throughout the cell. Some suggestions are also made about why cells rarely stain adjacently. Factors such as diffusion rates of silver ions into the tissue, permeability barriers, and reducing properties of chromated tissue are considered, and a model of stochastic impregnation is proposed (Appendix I).

The chapter also outlines various modifications of the original method and describes various combinations of chromation and impregnation that can be used on the insect central nervous system. Appendix II lists the compositions of useful fixatives and chromating solutions and describes three methods that are reliable for a range of insects.

Historical Background

The most spectacular procedures for revealing the shapes of single nerve cells are derived from a set of fortuitous experiments performed by Camillo Golgi in the 1870s. These techniques, which are now collectively known as the "Golgi methods," constitute one of the most powerful means for structural analysis of the brain.[1] They are also the least understood in terms of their chemistry.

Probably the original procedure was discovered by chance during investigations of the different affinities of fixed tissue to silver nitrate. In about 1872 or 1873, Golgi observed black anastomizations in brains that had been fixed in ammonium dichromate and sodium sulfate before immersion in a solution of silver nitrate. Similar observations were obtained from brains that had been treated with dichromates and mercuric chloride (Golgi, 1873). The use of Golgi's discovery provided the historical foundation for all contemporary knowledge about the cellular components of the central nervous system, if not all that is known about the structural relationship between neurons. It is ironic that Golgi's interpretations of his material made but little contribution to this knowledge.

The discovery of the *reazione nero*[2] within nerve cells came to the attention of only a few neurologists. However, in 1886, Simarro showed his friend Ramón y Cajal preparations made according to Golgi's original formula and brought to Cajal's attention Golgi's classic monograph describing the method and its interpretations. This account described what appeared to be supporting evidence for Gerlach and Meynert's theory of neuronal connections via a protoplasmic reticulum: Golgi's support for this doctrine was based upon observations of nerve terminals in the griseum, which he believed to be a diffuse and continuous network. He also proposed that lateral arborizations, which we now know as dendrites, were nutritive components that played no direct role in integrative processes.

According to Cajal's autobiography, Simarro was distrustful of Golgi's method. This was not because he disagreed with the reticulum theory, as did such notable neurologists as Forel and His, but because he felt that the method produced inconsistent results. These inconsistencies still hallmark the technique in that no two preparations are identical. However, Simarro was probably referring to the fragmentary nature of impregnated elements because only partial impregnation of neurons is achieved by the original method of immersing tissue in dichromate alone. Despite such reservations, Cajal immediately recognized the po-

---

[1] For insects, another is the resolution of neuronal assemblies by cobalt injection (Strausfeld and Obermayer, 1976; Strausfeld and Hausen, 1977).

[2] Black reaction.

tential analytic power of the method and, like Golgi, he modified it by combining osmic acid to the dichromate solution. Unlike Golgi, he introduced a simple but fundamental novelty, which was to recycle the procedure after silver impregnation in order to reveal an abundance of nerve cells.

Cajal publicly renounced the reticulum theory in 1888, by demonstrating that axon terminals in the griseum did not anastomize but arborized freely: Nerve cells were seen to be discontinuous elements. Cajal's observations lent support to theorists, such as His, who opposed the reticular doctrine and argued that neuronal processes would be expected to have discontinuous arrangements by analogy with the free endings of motor axons onto muscle. Cajal's observations demonstrated that axon terminals appeared to contact but not fuse with cell bodies and dendrites, and he proposed that conduction of nervous impulses progressed via a contiguous arrangement of nerve cells rather than via a continuum. Except for a brief "reticularist" revival in the early 1900s, and Golgi's continued insistence on his original proposals, the picture we now have of neuronal arrangements is the same as Cajal's.

Use of the Method and Its Selectivity

Few place absolute faith in the Golgi method and many workers harbor the suspicion that it consistently fails to impregnate parts of nerve cells or that it fails to impregnate some types of nerve cell at all. Even though black deposits have been demonstrated in neurons (Blackstadt, 1970), the entire spread of a neuron may not be revealed. Some neurons whose forms have been stained by other techniques, such as the Falck-Hillarp procedure for biogenic amines, are rarely revealed by the Golgi methods, and then only in part. An example is the interplexiform neuron of the mammalian retina (cf. Boycott *et al.,* 1975; Ehinger and Falck, 1969). In insects, large neurons are often partially impregnated, particularly when several occur together. Possible reasons for this are discussed later (pp. 167, 170) with respect to the reactants involved.

Quantitative studies on single nerve cells have provided basic information about general structural features of neurons in, for example, the mammalian cortex (Sholl, 1956) and have been used to describe neuronal morphologies in experimentally modified or developing brain (see, e.g., Altman and Anderson, 1972, 1973; Altman, 1973, 1976). However, it is often recommended that the method not be used as a quantitative technique. The method cannot be used alone to define relative populations of nerve cells nor can it show up precise spatial relationships or functional connections. Golgi-impregnated neurons can be related to un-

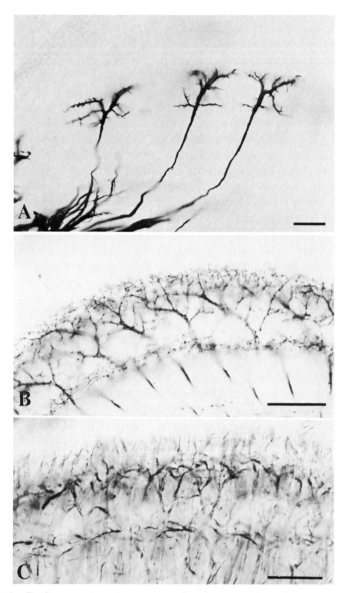

Fig. 9-1, A–C. A comparison between random impregnation of nerve cells by the Golgi method (single cycle, Colonnier technique) and nonrandom resolution of a single population of identical neurons after uptake of cobalt and the subsequent silver intensification of CoS profiles (**A** and **B**, respectively). Both are compared with the results of a block Cajal reduced silver preparation (**C**) of the the same region (lobula, *Calliphora erythrocephala*) which also reveals cytoskeletal Gestalten of this type of neuron as well as other elements. The scale on this and other figures represents 20 μm unless otherwise stated.

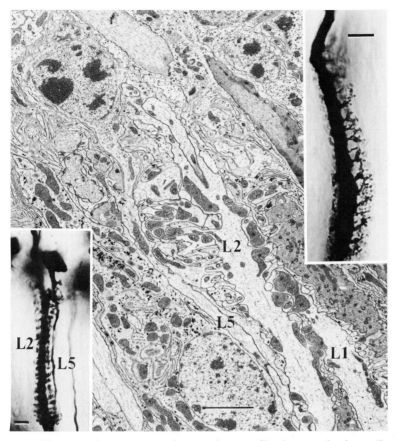

Fig. 9-2. Electron microscopy rarely reveals a profile that can be immediately identified as belonging to one or another species of neuron. However, in this section the profile of an L5 monopolar cell can be identified and related to the forms of L5 monopolars in Golgi preparations. One of these is shown in the *lower left inset,* displayed mirror image, with an L2 monopolar cell of the same retinotopic subunit, the optic cartridge. Even if electron microscopy reveals a profile of an identifiable nerve cell, parts of neurons are shown only as two-dimensional slices of a complex structure. To match these with intact cells, it is necessary to employ either serial reconstruction or combined Golgi electron microscopy (Fig. 9-4). For example, the spines of monopolar cells *(inset, top right)* are revealed in their entirety by selective impregnation: In so-called ultrathin sections they are resolved as closed profiles, usually separated from their parent fiber. Scale = 2.5 μm. Electron micrograph by J. A. Campos-Ortega.

stained or counterstained elements only when these can be identified as belonging to a distinct category of nerve cell.

One example may be drawn from studies of the retina where horizontal cell arborizations have been mapped against unstained receptor terminals and other horizontal cells because these were easily recognized by inter-

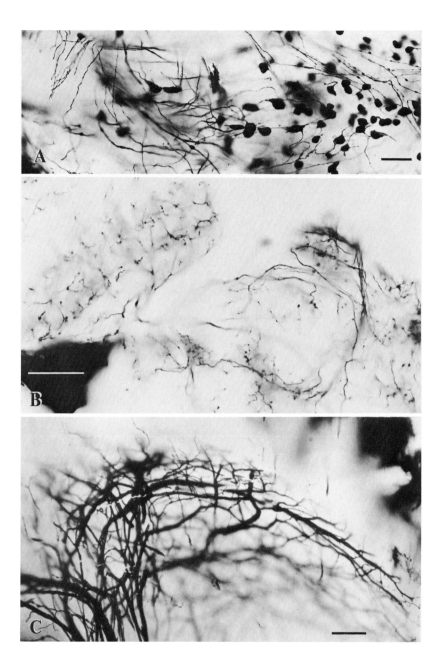

ference phase contrast microscopy or by counterstaining (Wässle and Riemann, 1978). Analysis of synaptic relationships between retinal neurons was achieved by combining Golgi impregnation and electron microscopy (see Boycott and Dowling, 1969; Kolb, 1970). Likewise, in the external plexiform layer of Dipterous insects, elements have such strictly defined geometric relationships that quantitative analysis is possible with reference to interneurons and receptors that are resolvable as total populations by reduced silver techniques (Strausfeld, 1971).

It should always be borne in mind that the Golgi method unpredictably selects single neurons (Fig. 9-1,A–C). Interrelationships that are revealed by selective impregnation thus bear no relevance whatsoever to real functional connections. The Golgi method basically serves to demonstrate neurons that deserve scrutiny at another mode of resolution, such as electron microscopy (Fig. 9-2).

Critics of the method have claimed that it selects moribund neurons, or conversly, neurons that were active at the time of initial fixation. Again, in insects, this seems unlikely because it is possible to guide the impregnation to specific brain regions after chromation, as is described later. The gravest misgiving, which is still relevant, is that some classes of nerve cells are refractive to many modifications of the method. It is known that some regions of the central nervous system are more susceptible to impregnation than others and that juvenile tissues are refractive. These apparent drawbacks, and a complete ignorance of the mechanism of impregnation, have done much to stimulate the many published modifications. Cajal saw these difficulties and sought to overcome them by intuitive variations of the technique (Fig. 9-3,A–C).

Summary of the Basic Procedures

The Golgi procedure is one of the simplest histologic techniques. The original method was to immerse pieces of tissue in Müller's fluid (a mixture of ammonium or potassium dichromate and sodium sulfate) for sev-

---

Fig. 9-3, A–C. The selectivity of the Golgi method can be altered by initial fixation and chromation. All three preparations shown here have been fixed in various ways and were chromated for 18 h in glutaraldehyde-sodium dichromate. **A** shows impregnation en masse of cell bodies and their neurites after fixation in PIPES-buffered glutaraldehyde-paraformaldehyde. **B** was fixed in phosphate-buffered glutaraldehyde and postfixed in osmium tetroxide-sodium dichromate. **C** was fixed in phosphate-buffered glutaraldehyde-paraformaldehyde and then postfixed in veronal-buffered osmium tetroxide: This aids selective impregnation of large-diameter neurons. The methods are outlined in Appendix II. **A** and **B** are from *C. erythrocephala;* **C** is from *Drosophila melanogaster's* lobula and lateral protocerebrum.

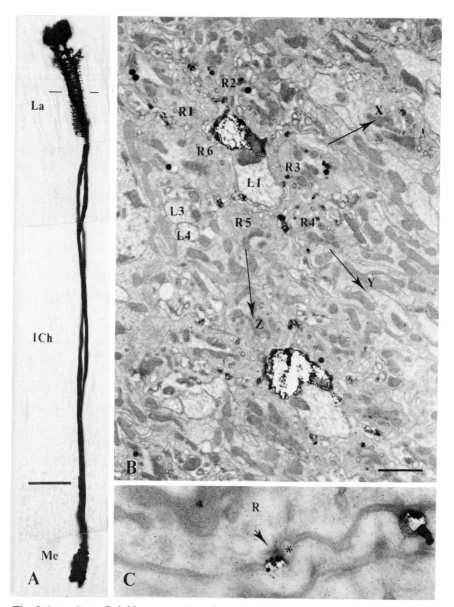

Fig. 9-4, A–C. A Golgi impregnation of a pair of L2 monopolar cells in the lamina (*La*) of *C. erythrocephala,* which pass via the first optic chiasma (*1 Ch*) to the medulla (*Me*). The material was fixed in Karnovsky's phosphate-buffered glutaraldehyde-paraformaldehyde, postfixed in potassium dichromate-osmium tetroxide, and impregnated by a single cycle of the Golgi-Colonnier technique. The electron micrograph (**B**) is of a section cut at the level indicated in the lamina, at right angles to the two neurons. The pair of L2 axis fibers are filled with silver chromate deposits and are seen to be located at identical positions in two adjacent optic

Table 9-1. The Original Golgi Methods[a]

Stage	Composition	Time
Fixative (F) and chromation fluid (C)	Ammonium or potassium dichromate (2.5–3.0 g)	
	Sodium sulfate (1.0 g)	1–7 days
	Distilled water (100 ml)	
Metal impregnation (M)	Silver nitrate (0.5%–1.0%)	Several days
	or	
	Mercury II chloride (5.0%)	Several weeks
Alkali treatment after HgCl₂ impregnation	Ammonium hydroxide	1–8 h

[a] See Golgi, 1873, 1878, 1879, 1891.

eral days followed by a solution of silver nitrate. This sequence will sometimes resolve cell bodies, sometimes parts of dendritic trees or axonal arborizations, or combinations of these.

Its initial modification was the addition of osmic acid to the dichromate, and the double impregnation procedure is usually referred to as the Cajal "Golgi rapid." Golgi first substituted silver nitrate for mercuric chloride, retaining the original fixation: Cox simplified this method by combining the mercuric chloride with the chromating bath, subsequently treating the tissue with a weak alkali to resolve, as oxides of mercury, a translucent primary precipitate whose main composition is now known to be mercurous chloride (Fregerslev *et al.*, 1971a). Cajal's rapid procedure and Cox's modification of Golgi's sublimate procedure were the first significant improvements of the original method. Later, Kopsch (1896) substituted formalin instead of osmium, though proper credit for this should be accorded to Hoyer (1894).

A third generation of Golgi methods includes those modifications that were devised to overcome specific technical problems, such as background crystallization, or that were devised to resolve hitherto refractive neurons and glia. These changes involve both primary and secondary adjustments to the chromation and metallic impregnation such as the use of zinc dichromate (Fox *et al.*, 1951), the addition of soporifics such as chloralhydrate to the dichromate, or the use of mercuric or mercurous nitrate in lieu of silver nitrate (Bertram and Ihrig, 1959; Blackstadt *et al.*, 1973). The importance of an initial fixation prior to chromation and of

cartridges. Unimpregnated profiles of other monopolar cells are indicated, as are "axes" natural to the neuropil mosaic (x,y,z). Receptor terminals are numbered R1–R6. Silver chromate-filled dendritic spines extend between receptor terminals, against which they terminate. A single dendritic spine is shown in C. It forms a dyadic arrangement postsynaptic to the receptor terminal with an unimpregnated spine derived from the L1 cell. The seagull-shaped presynaptic ribbon is reasonably well preserved despite Golgi impregnation. Scale = 2.5 μm.

Table 9-2. Derivatives of Golgi's Original Methods[a]

1	F and C	Potassium dichromate	6%–7%	10 parts	1–7 days
		Osmium tetroxide	1%	3–3.5 parts	
	M	Silver nitrate	0.5%–1%		24 h
2	F	Potassium dichromate	2.5%–3%		3–4 days
	C	Potassium dichromate	3.0%	20 parts	3–4 days
		Osmium tetroxide	1.0%	6 parts	
	M	Silver nitrate	0.75%		24–36 h
3	F and C	Potassium dichromate	2.5%	20 parts	8–14 days
		Osmium tetroxide	1.0%	6 parts	
	Post-C treatment	Potassium dichromate	3.0%	1 part	
		Copper sulphate or copper acetate	5.0%	1 part	1–14 days
	M	Silver nitrate	0.75%		24–36 h or longer
4	F and C	Potassium dichromate	2.5%	40 parts	1–2 days
		Osmium tetroxide	1.0%	10 parts	
	M	Silver nitrate	0.75%		24–36 h
	2nd C	Potassium dichromate	2.5%	40 parts	1–2 days
		Osmium tetroxide	1.0%	10 parts	
	2nd M	Silver nitrate	0.75%		24–36 h

5	F and C	Potassium dichromate	3.5%	40 parts	Two changes; 24 h
		Formaldehyde	(about 30.0%)	10 parts	
	3rd C	Potassium dichromate	3.5%		2–6 days
	M	Silver nitrate	0.75%		24 h
	F and M	Potassium dichromate	5.0%	20 parts	2–6 months
		Mercury II chloride	5.0%	20 parts	
		Potassium chromate	5.0%	16 parts	
		Distilled water		40 parts	
6	Alkali on celloidin sections	Sodium carbonate (or potassium bisulfate)	5.0%	1 part	1–several minutes
		Sodium, potassium, or ammonium hydroxide	3.0%	1 part	

[a] (1) Golgi, 1875. (2) Cajal's "mixed rapid" method. (3) Cajal's "indirect rapid" method. (4) Cajal's "double (or triple) rapid" method. (5) The "Golgi-Kopsch" method (see also Strong, 1895; Lachi, 1895). (6) The "Golgi-Cox" dichromate-sublimate method of Cox (1891).

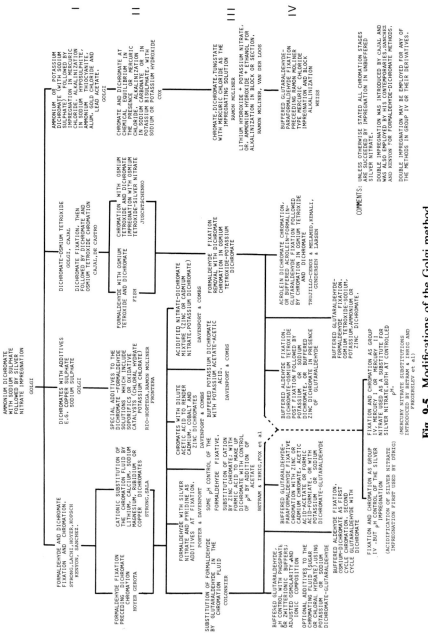

Fig. 9-5. Modifications of the Golgi method.

appropriate adjustments of the pH was realized by Bertram and Ihrig (1957), and their observations must count as a significant improvement for imparting some control over the procedure. The subsequent development of aldehyde-based fixatives that use physiologic osmolarities and ionic concentrations have lead to recent refinements that allow combined light and electron-microscopic examination of impregnated neurons in the context of their immediate surroundings of unimpregnated neighbors (Stell, 1964, 1965; Blackstadt, 1965; and for insect material, Strausfeld and Campos-Ortega, 1972a,b, 1973; Campos-Ortega and Strausfeld, 1973; Ribi and Berg, 1980) (Fig. 9-4,A–C).

The present account will generally avoid a description of the plethora of published modifications. Instead the introduction of some methods are summarized in Tables 9-1 and 9-2 and in Fig. 9-5 in the form of a genealogic table. The family tree is, of course, incomplete and shows only those modifications that the present author considers significant developments in terms of displaying refractive neurons or resolving juvenile and embryonic nerve cells, or which can be applied to both insects and crustacea.

Golgi Procedures for Insect Central Nervous System

Introduction

Insect neuropil is wholly refractive to Golgi's original procedure and is partially refractive to most classic modifications such as the Golgi-Rapid, Golgi-Kopsch, and Golgi-Cox methods. Kenyon (1896a,b) was the first to impregnate neurons of the insect brain using formalin instead of osmium in the chromating solution. However, his illustrations of Hymenopterous neurons are now recognized to represent partial impregnations. Cajal's double impregnation procedure was used by Pflugfelder (1937) and Goll (1967) on the brains of Hemiptera and Hymenoptera, and the Golgi-Kopsch procedure, also with double impregnation, was successfully employed by Cajal and Sánchez (1915) and by Sánchez (1916, 1925, 1933, 1935, 1937) on many species of insects. Their accounts constitute a basis for modern structural research on the insect central nervous system (CNS). Colonnier's (1964) important modification of the Golgi-Kopsch procedure, which substitutes formaldehyde by glutaraldehyde, was first applied to the insect brain by Boycott in 1965 (see Strausfeld, 1976) and has since been found to be generally applicable to a variety of arthropods. Nevertheless even the original Colonnier procedure is capricious and favors some species more than others. By trial and error, the original formula has been much modified by employing various initial fixatives, as well as by introducing combinations that employ osmium tetroxide as a secondary fixative. Substitution of potassium dichromate by other dichromates or substituting silver nitrate by mercuric or mercurous nitrates has enabled further control over the method. The rationale for the design of these modifications is discussed later and in Appendix II: Methods.

Applications of the Golgi Method

Students of vertebrate neuroanatomy nowadays seldom choose to investigate a brain or spinal cord that represents a tabula rasa in terms of its cellular constituents. Modern structural analysis of the nervous system extends classic anatomical studies and is employed in conjunction with high-resolution microscopy. Golgi studies are used in conjunction with experimental analysis that corroborate physiologic observations to structure. Recently, Golgi analysis has become important for developmental studies of normal and pathologic nervous systems and for the detection of structures in neurologic mutants.

By comparison, the CNS of arthropods are relatively unexplored even though they offer specific advantages, particularly with respect to the study of neuronal development, cell lineage, pattern formation, the establishment of synaptic connections, and genetic analysis. It may seem a paradox that until recently no single impregnated neuron was figured from *Drosophila* (Fig. 9-6,A–C), even though the central nervous system of this species, like that of others, can be described in terms of characteristic nerve cell geometries (see Coggshall, 1978).

Developmental and genetic studies require that the Golgi method is an effective tool for screening populations of neurons. It is therefore essential that the investigator is aware that the method may be coerced into revealing large numbers of cells. Resolution of many neurons and the detection of certain classes of neurons at the expense of others are minimal requirements for analyzing connectivities by Golgi-electron microscopy, as well as for screening neuronal forms in wild-type and mutant animals.

The presence of refractory elements can only be demonstrated after a volume of information about cell shapes and dispositions is related to general histologic procedures, thus revealing structural lacunae. Correlative studies require comparisons between Golgi preparations and preparations that show all cell profiles, such as electron micrographs or semithin serial sections stained with Mallory's azure II methylene blue. Such studies are laborious and are best performed on small and delineated regions of the brain, such as the optic neuropils. However, they are essential for defining reliable criteria that are used to evaluate the various Golgi modifications. This may be illustrated by one salutatory example, i.e., the discovery that electron microscopy could not match some pro-

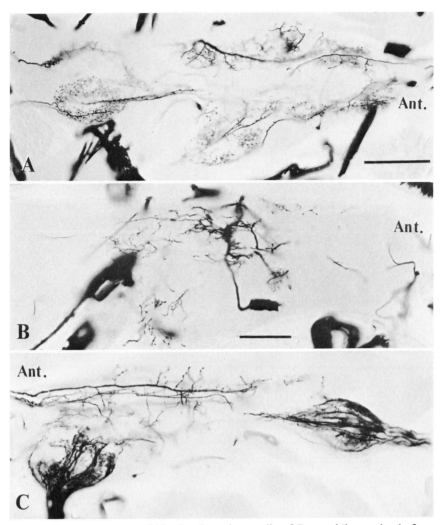

Fig. 9-6, A–C. Neurons within the thoracic ganglia of *Drosophila* resolved after PIPES-buffered glutaraldehyde-paraformaldehyde fixation and Colonnier type chromation using unbuffered sodium dichromate (18 h). They were impregnated by silver nitrate at pH 7.0, followed by a second chromation-impregnation cycle. Note the complete absence of crystalline deposits at the surface of the neuropil. Anterior (*Ant*) is indicated for each section; dorsal is upward. **B** illustrates a giant motor neuron within the mesathoracic ganglion. **A** and **C** show receptor endings in leg neuromeres.   Scale **A,C** = 100 μm; **B** = 50 μm.

files in the outer plexiform layer of the fly with Golgi-Colonnier impregnated neurons. Degeneration studies later showed that these profiles were intrinsic to the lamina because they remained intact alongside receptor terminals after the axons of interneurons and long visual fibers had been severed in the first optic chiasma and had degenerated (Campos-

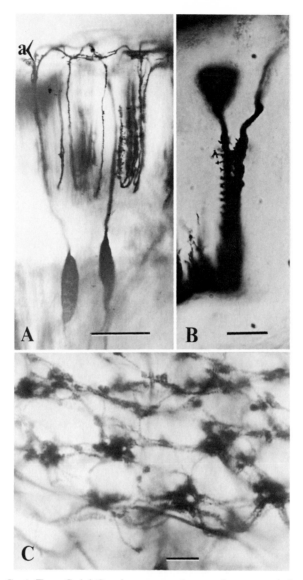

Fig. 9-7, A–C. A Two Golgi-Cox-impregnated amacrine neurons in the lamina of *Lucilia caesar*. Initial fixation was in Millonig's phosphate-buffered glutaraldehyde, followed by Weiss' modification of the Cox method. **B** An exception to the rule: Impregnation showing a receptor terminal contiguous with the monopolar cell onto which it is presynaptic. Silver influx was local, through the cut retina (*Musca domestica*). Scale = 5 μm. **C** Part of a regular hexagonal network of amacrine cell processes over the lamina surface (level *a* in **A**) (buffered glutaraldehyde fixation: Golgi-Cox procedure). Amacrine fibers are pre- and postsynaptic to each other at this level. This preparation illustrates another rare exception to the rule of nonneighbor impregnation. Scale = 5 μm.

Ortega and Strausfeld, 1973). Although the Golgi-Colonnier technique occasionally resolved cell fragments that partly fitted non-degenerated profiles it was only after application of the Golgi-Cox method in combination with aldehyde prefixation that the entire form of an amacrine neuron became apparent (Fig. 9-7, A–C). Subsequently it was found that prefixation in buffered aldehydes, followed by treatment with dichromate and osmic acid prior to Colonnier's chromation, were essential steps in revealing this type of nerve cell.

The Golgi-Colonnier and Golgi-Rapid Procedures

Criteria for Fixation, Chromation, and Metal Impregnation

Historically, improvements of the method have been founded upon intuition rather than on an understanding of the chemical reactions involved between Golgi reagents and neural tissue. However, the method is characterized by specific chemical criteria with respect to the reagents that are employed and their mode of application. For example, chromating agents are invariably used prior to metal impregnation, and aldehydes or osmium tetroxide are either employed as initial fixatives or are used in conjunction with chromation. Other oxidizing agents, such as chlorates, have been used as substitutes of osmium (Ramon-Moliner, 1957; Frontera, 1964) but do not impart any special advantage in insect material.

We may consider three interdependent stages of the procedure that are subject to manipulation: (1) fixation, (2) chromation, and (3) metal impregnation. All are carried out within a characteristic pH range. The proper execution of each stage is crucial for the success or failure of the method.

1. Fixation must be such that it preserves fine structure. Such fixatives include the monoaldehydes, acrolein and formaldehyde (but not acetaldehyde), and in particular the dialdehyde, glutaraldehyde. Also excluded as fixatives are glyceraldehyde, crotonaldehyde, and propionaldehyde. Used alone, however, the monoaldehydes fail to promote metal impregnation even when they are subsequently incorporated in a single chromation. Double chromation with formaldehyde will reveal only few neurons: Triple chromation reveals substantially more, but fewer than after dialdehyde chromation (Table 9-3). Thus, if buffered formaldehyde is used as the initial fixative, it is essential to chromate in the presence of glutaraldehyde or osmium tetroxide. The usual practice is to fix for electron microscopy, prior to chromation, using buffered glutaraldehyde or a combination of glutaraldehyde and paraformaldehyde or acrolein (see Appendix II: Methods).

In any event, fixation for the Golgi methods excludes any fixative that does not become incorporated into the structure of cross-linked dena-

Table 9-3. Dichromate and Metal Nitrate Combinations for Multiple Impregnation Cycles

Chromation Cycle I	Impregnation Metal nitrate	Exceptions	Intermediate treatment	Chromation Cycle II	Impregnation Metal nitrate
Osmium tetroxide with dichromate	Silver Nitrate		—	Osmium tetroxide with dichromate	Silver nitrate
As above	Mercury I nitrate Mercury II nitrate	Sodium, lithium, or ammonium dichromate	—	As above	Mercury I nitrate Mercury II nitrate
Glutaraldehyde with dichromate	Silver nitrate		Dichromate only	Osmium tetroxide with dichromate	Silver nitrate
As above	Mercury I nitrate Mercury II nitrate	Sodium, lithium, or ammonium dichromate	As above	As above	Mercury I nitrate Mercury II nitrate
Osmium tetroxide with dichromate	Silver Nitrate		Dichromate only	Glutaraldehyde with dichromate	Silver nitrate
As above	Mercury I nitrate Mercury II nitrate	Sodium, lithium, or ammonium dichromate	As above	As above	Mercury I nitrate Mercury II nitrate
Glutaraldehyde with dichromate	Silver nitrate		—	Glutaraldehyde with dichromate	Silver nitrate
As above	Mercury I nitrate Mercury II nitrate	Sodium, lithium, or ammonium dichromate	—	As above	Mercury I nitrate Mercury II nitrate

tured proteins and into the structure of lipids (or lipoproteins).   Therefore, the class of coagulent fixatives is not applicable to selective impregnation even though they are commonly employed for reduced silver
techniques on sectioned material or for some modifications of the block
Cajal reduced silver method.   In the latter, sites for the reduction of silver
are related to the cytoskeletal core of neurofibrillae after coagulent fixation, whereas after formaldehyde fixation the sites for silver reduction
are predominantly at the cell membrane and within the axoplasm (Fig. 9-
8,A–C).

Fixatives that are said to extract or coagulate proteins and lipids include most classic materials such as alcohols, acetone, bichloroacetic
acid, picric acid, and chloroform.   In addition, most dichromates, when
used alone, are said to act as partial coagulents, though this does not seem
to be the case for ammonium dichromate.   Secondary fixation and block
staining, as employed for electron microscopy, must be used with caution
in conjunction with the Golgi method.   Potassium permanganate wholly
inhibits impregnation of single neurons, and this may be due to its reputation as a poor preservative of lipids and that it gives rise to conforma-
tional changes in proteins (Lenard and Singer, 1968).   Buffers that are
useful for electron microscopy, such as s-collidine, but which are reputed
to extract some proteins (Luft and Wood, 1963), are mostly inapplicable

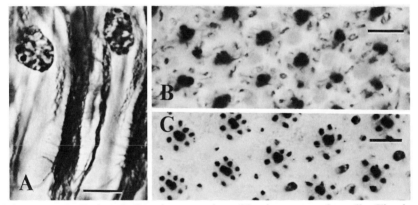

Fig. 9-8, A–C. Reduced silver impregnation of lamina monopolar cells: Fixation
in aldehydes or coagulant fixatives gives rise to different reduced silver distributions.   In **A**, tissue was fixed in a mixture of coagulant and formaldehyde fixative and reveals both the cytoskeletal matrix of a monopolar cell axon and its
dendritic spines.   In **B**, tissue was fixed in formaldehyde and reveals monopolar
cell *outlines:* The cross sections resemble Golgi-impregnated cells.   **C** is at the
same level: Material was fixed in coagulant (Carnoy) and shows the cytoskeletal
core of monopolar cells and receptor terminals (*Musca domestica*).   Scale
A–C = 5 μm.

to material that is to be treated by the Golgi methods, nor are cacodylate buffers recommended.

2. Even without chromation, partial resolution of neuron-like Gestalten can be visualized in vertebrate material after fixation in buffered formaldehyde followed by impregnation with mercuric nitrate (Bertram and Ihrig, 1959). A similar effect can be achieved when tissue is incubated in silver nitrate (after aldehyde fixation) and then placed in a fresh aldehyde solution. The second aldehyde may serve to reduce nascent silver in neurons where nuclei of metallic silver had been previously formed by reactions between Ag^+ and fixed tissue. This is similar to the block Cajal method (see p. 170).

For proper Golgi impregnation it is necessary to treat tissue with a "chromating solution." This is made by dissolving a dichromate salt, using this solution either alone or with osmium or glutaraldehyde. Before chromation, tissue is washed in buffer to remove excess fixation-aldehydes.

Various species of dichromate have a characteristic pH when dissolved in distilled water (Casselman, 1955). Chromate salts also go into solution at characteristic alkaline pH. However, zinc and cadmium chromate are only soluble in acids to give an orange "dichromate" solution. One of the earliest realizations that the pH of the chromating solution was important for obtaining optimal impregnation was by Betram and Ihrig (1957). These authors found that chromation in a solution of zinc dichromate gave rise to optional impregnation when the solution was adjusted to pH 3.1–3.3.

Casselman (1955) showed that fixation in different species of chromates gives good or poor preservation of cell structures, depending on their cations. The pH of any chromating solution (characteristic of each species of dissolved dichromate salt) is an important factor for the Golgi method. Casselman (1955) showed that, at different pHs, dichromates dissociate into different concentrations of chromate ions, base ions, and chromic acid. In the case of a 0.17 M solution of $H_2Cr_4O_7$ at pH 4 the proportion of chromate ions, etc., was approximately as follows: $CrO_4^{2-} = 10^{-4}\ M$, $H_2CrO_4 = 10^{-4}\ M$, $H^+ = 10^{-4}\ M$, $Cr_2O_7^{2-} = 10^{-1}\ M$, and $HCrO_4^- = 10^{-1.5}\ M$. When the pH is shifted to 1 the concentrations of H_2CrO_4 and H^+ are 10^{-2} and $10^{-1}\ M$, respectively, and the concentration of CrO_4^{2-} is only $10^{-7}\ M$. Shifting the pH to 6 increases the concentration of CrO_4^{2-} to $10^{-2}\ M$, whereas the concentration of chromic acid and hydrogen ions is correspondingly decreased. The concentrations of di- and hydrochromate ions are maximal at pH 4 and are symmetrically reduced at pH 1 and 6 to about $10^{-1.5}$ and $10^{-2}\ M$, respectively.

Observations of Golgi impregnated material show that at low pH a solution of potassium or sodium dichromate with osmium tetroxide (the Golgi rapid mixture) will give rise to fewer neurons than when the solution is buffered to about pH 5.5 (Nässel and Seyan, 1979). The number of neurons impregnated may be correlated with the calculated concentrations of CrO_4^{2-}. Similar results are obtained when dichromate solutions are used alone at various pHs. In the case of zinc dichromate solutions the range is 2.5–4.0. In the case of sodium or potassium the range is 3.0–5.8.

In the presence of neural tissue there is a gradual shift of the unbuffered pH of the chromating solution toward the alkaline. This is of advantage when chromating in solutions made up from a dichromate salt alone, or with added osmium tetroxide, and the tissue can remain in the solution for days or even weeks. The pH of the solution also becomes less acidic using unbuffered solutions made up from dichromates with added glutaraldehyde; however, it is a distinct disadvange to leave tissue in these solutions for too long. In unbuffered dichromate-glutaraldehyde solutions the rate of pH shift is correlated to the base component of the dichromate (ammonium, lithium, sodium, potassium, zinc, cadmium). In the case of ammonium dichromate-glutaraldehyde the time taken for the pH to shift from 3.5 to 5.5 is about 30 min. This can be restrained a little by adding glutaric or acetic acid to the solution. In the case of potassium dichromate-glutaraldehyde solutions an equivalent pH shift takes place after 36 h (at 20°C). For impregnation with silver nitrate the subjectively best Golgi preparations (many neurons, clear background) are achieved if tissue is removed from the chromating solution when it reaches pH 5.0 (Fig. 9-9).

Little is known about reactions between glutaraldehyde and constituents of dichromate solutions. However, a solution of potassium dichromate and glutaraldehyde becomes dark brown after reaching pH 5.5 and a muddy precipitate is formed. In fact, even if the solution is held at a constant pH by borax-borate buffer, it undergoes the same change, taking about twice as long. Tissue that is taken from the solution in this state either fails to impregnate with silver nitrate or only a few neurons are seen. These are accompanied by massive crystalline deposits on the tissue surface or within the tissue itself (Fig. 9-10).

What, then, is the role of glutaraldehyde as an additive? Is it merely reacting with the dichromate solution in such a way as to accelerate the shift in pH to within a range at which the CrO_4^{2-} ion concentration predominates? Or, within the working range, does glutaraldehyde promote reactions between the tissue, chromate (CrO_4^{2-}) ions, and base ions (Na^+, K^+, etc.) so that organochromate complexes are formed? There is currently no evidence to support the notion of organochromate complex formation, however.

When chromated tissue is placed in silver nitrate, precipitates are immediately formed outside the tissue and on its surface. According to Fregerslev *et al.* (1971a), X-ray powder diffraction and selected area diffraction studies of Golgi precipitates showed that black deposits inside

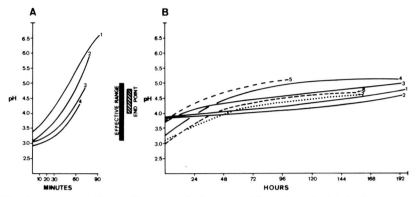

Fig. 9-9, A and B. Depending upon their composition, the pH of unbuffered chromating fluids shifts toward the alkaline at different rates, during chromation. Here pH is plotted against time. The end-point of chromation and the working range are indicated. **A**, 2.5 ammonium dichromate, 25% glutaraldehyde (4:1); *1*, initial pH 3.44, unbuffered; *2*, initial pH 3.18, formic acid buffer; *3*, initial pH 3.00, acetic acid buffer; *4*, initial pH 2.85, glutaric acid buffer. **B**, *1*, 2.5% sodium dichromate; *2*, 2% osmium tetroxide, 2.5% sodium dichromate (1:20); *3*, 2% osmium tetroxide, 2.5% potassium dichromate (1:20); *4*, 2.5% sodium dichromate, 25% glutaraldehyde (4:1); *5*, 2.5% potassium dichromate, 25% glutaraldehyde (4:1); *6*, as 5, initial pH adjusted by acetic acid; *7*, 6% zinc chromate in 17.5% formic acid, 25% glutaraldehyde (4:1).

Fig. 9-10. Crystalline deposits at the surface and within exposed tissue after silver impregnation of material that has remained too long in the glutaraldehyde-dichromate solution. Crystals appear to form at the expense of neuron impregnation, possibly depleting the chromate pool for Golgi reactions within neurons. In this case only a single element has been impregnated within a 100-μm-thick slice of the medulla of *M. domestica*. Scale = 50 μm.

impregnated cells consist of silver chromate, not dichromate. The precipitate evolved outside the tissue is not known. In a solution, adding cations that form insoluble chromates (Ba^{2+}, Pb^{2+}, Ag^+) precipitates chromate, not dichromate, formation. It would seem that the counter ion diffusion out of the tissue during the influx of silver cations is CrO_4^{2-} (and/or $Cr_2O_7^{2-}$). Yet a substantial amount of CrO_4^{2-} remains in the tissue for reaction with Ag^+. Direct observations of the black reaction in fresh sections (p. 167) describe a delay between silver entering the tissue and its subsequent visible precipitation as silver chromate in cells.

Using different species of dichromate salts suggests that the number of impregnated neurons is correlated with the base component of the dichromate, all other factors being equal. The success of the Golgi method, in terms of numbers of neurons resolved, is also correlated to the pH of the chromating solution and its duration. In our laboratory we now use buffered chromating solutions (Holmes borax-boric acid with sodium dichromate) as the standard method for sampling a wide range of neurons; silver nitrate impregnation is also carried out at a controlled pH (Nässel and Seyan, 1979). For a full description of the method, see Appendix II (p. 202). Rates of pH shift in different unbuffered chromating solutions are shown in Fig. 9-9, and optimal chromation times are shown in Table 9-4. The ratios between dichromate solutions and additives—2% osmium tetroxide or 25% glutaraldehyde—are 20:1 and 4:1, respectively. Approximately 100 vol of solution are used for 1 vol of tissue.

3. The last stage of the Golgi *cycle* is treatment of chromated tissue by silver nitrate. This is called "metal impregnation." Mercuric and mercurous nitrate can be used instead of silver nitrate (Bertram and Ihrig, 1958; Blackstadt *et al.*, 1973). X-ray powder and selected area electron diffraction studies have identified the intracellular deposits as consisting

Table 9-4. Chromation Periods

Species of dichromate		Timing with osmium (days)	Timing with glutaraldehyde (h)
Ammonium		2–3	1–1.2
Lithium		2–3	4–8
Sodium	unbuffered	2–5	24–36
Calcium		3–5	4–8
Potassium		6–10	36–60
Zinc	chromates in	5–10	48–72
Cadmium	formic acid	5–10	36–60
Sodium	buffered with borax-borate	1–6	12–36
Potassium	Holmes buffer	1–6	12–36

of Ag_2CrO_4, $Hg_3O_2CrO_4$, and Hg_2CrO_4, respectively (Fregerler *et al.*, 1971a; Blackstadt *et al.*, 1973).

Metal impregnation is carried out in the following way. The chromating fluid is decanted and the brains are washed for between 5 and 10 min in a chromate solution made from distilled water and the appropriate dichromate salt. If buffers were previously used then this solution should be appropriately buffered. Tissue is then transferred directly from this into 0.75% silver nitrate (or the appropriate concentration of mercuric or mercurous nitrate, see p. 182). A reddish or brown precipitate is evolved from the tissue. The tissue is washed until there is no more precipitate and is then placed in a fresh silver nitrate solution. In the past, it has been the practice to transfer tissue directly from the chromating bath into silver nitrate. The intermediate wash is recommended, especially after chromation with osmium. Unbuffered silver nitrate is commonly used (about pH 6.8 for a 0.75% solution in distilled water) or it is buffered in Holmes borax-borate buffer (see Appendix II, p. 202). Varying the concentration of silver nitrate between 0.5 and 1.0% seems to play no significant role in the quality of impregnation. Below 0.4% and above 2%, neurons seem to partially impregnate. Varying the pH of the silver nitrate solution is, however, critical with respect to the species of chromating solution used previously. The following observations may serve as a useful guide to habitual users of the method.

A series of experiments was set up using *M. domestica* CNS that had been fixed in PIPES-buffered glutaraldehyde-paraformaldehyde. A 24-h chromation was done in an unbuffered solution of 2.5% $Na_2Cr_2O_7$ (1 volume) added to 5 vol of 25% glutaraldehyde or 1 vol of osmium tetroxide to 20 vol of dichromate. Various results were observed using 0.75% $AgNo_3$ within a pH range of 3.3–5.4 (acetic acid-acetate), unbuffered, and pH 4.7–7.4 (adjusted with borax-borate buffer). Below pH 3.8 no neurons were resolved. At pH 3.8–4.2 some neurons were resolved as pale profiles against a completely clear background. At pH 5.5–6.4 many neurons were resolved against an almost translucent background, and with unbuffered silver nitrate (pH 5.0) or silver nitrate buffered to pH 7.0 a large number of very dark neurons appeared. Surface crystallizations occurred mainly at pH 6.8–7.2. On the other hand, after treatment with zinc dichromate [made up with formic acid according to Betram and Ihrig (1957) at pH 3.1–3.3, with the addition of osmium tetroxide] 0.75% $AgNO_3$ gave Golgi impregnation in the lower pH range but was less effective at pH 6.0–7.2. Compared with sodium chromate-treated CNS, there was a shift of 2 pH units toward the lower end of the scale for effective impregnation. Interpretations of this are difficult since the nature of chromates in tissue available for reaction with silver is unknown. If organic metallochromate complexes are formed then possibly their dissociation is correlated with the pH of the silver nitrate solution. Zinc chromate, for example, does not go into solution at pH 7.2, whereas sodium and potassium chromates do.

Mercuric nitrate is usually adjusted to pH 5.5 and mercurous nitrate is used in the pH range 1.6–2.0. Both give best results on tissue that has been treated with zinc dichromate, although Blackstadt *et al.* (1973) used potassium dichromate before mercurous or mercuric nitrate impregnation. When sodium dichromate is used mercurous nitrate gives rise to exceedingly pale neurons that are resolved as ghosts. These are best seen by dark-field illumination (Fig. 9-23,H,I). This is in clear contrast to tissue treated with zinc dichromate-osmium tetroxide (Fig. 9-28,D).

Good Golgi impregnations have been achieved with these metal nitrates (the use of mercuric chloride for the Golgi-Cox method is described on p. 181); attempts to achieve impregnation with other metal nitrates have not met with success.

Silver impregnation is traditionally carried out for several days or, when using large pieces of tissue, for weeks. However, in thin tissue slices impregnated neurons have been resolved after 15 to 30 min (see p. 167). A solution of 0.5% $AgNO_3$ was used to impregnate neurons in 100-mm^3 pieces of mouse cerebellum after only 2 h in the dark at 4°C (Ribi and Berg, 1970). Ribi and Berg achieved a pale impregnation of neurons. Their examination by electron microscopy showed that the deposits appeared as a thin layer at the inner face of the plasmalemma, as predicted by observations of the Golgi reaction (p. 189) and by the impregnation model (Appendix II). Possibly the combination of low temperature (also during chromation) and a short impregnation period slows down the rate of reactions in the tissue.

Manipulations of the pH, time, and concentration of silver nitrate impregnation suggest three further parameters for successful Golgi preparations. The first (pH control) is also correlated with the type of chromation (species of dichromate salt used, the type of additive, the temperature at which chromation took place). The second parameter, time, is difficult to judge since at 20°C impregnation in a small piece of CNS, such as of *Drosophila,* occurs within 1–2 h. The relationships between optimal timing and tissue volume are not yet known. The third parameter, temperature, is suggested by the experiments of Ribi and Berg. The concentration of the $AgNO_3$ solution possibly has little importance between 0.5% and 1.0%. In any event, criteria for a successful Golgi preparation depend on its purpose. For electron microscopy this is, ideally, few neurons with enough silver chromate deposits in them to be identifiable by light microscopy but allowing internal structures, such as vesicles or synaptic ribbons, to be resolved by electron microscopy. For general studies of cell forms and pathways it is desirable to achieve impregnation of many neurons that can be easily resolved by light microscopy.

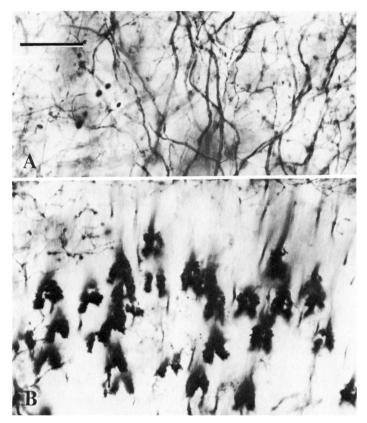

Fig. 9-11, A and B. **A** Silver nitrate can be substituted by mercuric nitrate to give mass impregnations. This figure shows numerous tangential elements of the medulla of *C. erythrocephala* after fixation in Sorensen's phosphate-buffered glutaraldehyde, postfixation in phosphate-buffered osmium tetroxide, chromation in unbuffered potassium dichromate-glutaraldehyde, and impregnation in mercuric nitrate (double cycle). **B** "Pseudoperiodic" arrays of identical profiles (L2 monopolar terminals) after Colonnier-type double impregnation with mercuric nitrate preceded by dichromate-osmium tetroxide postfixation.

Establishment of "Latent Images" for Subsequent Impregnation: Influencing Selectivity by Fixation

This section turns again to fixation and the role it seems to play in achieving various selectivities for subsequent impregnation. It is suggested later that, although only a fraction of all cells are filled with silver chromate, any cell could, in principle, be stained. Fixation and chromation alter the susceptibilities of various cells and seem to prepare "latent images" that could support reactions during metal impregnation.

It is an established fact that the type of fixation is crucial: Fixation optimal for one species may not be optimal for another (indeed, it may be hopeless). For example, PIPES-buffered glutaraldehyde-paraformaldehyde generally gives better all-around impregnation in Diptera than does Karnovsky's phosphate-buffered glutaraldehyde. In the same species, various types of fixatives will alter the susceptibility of various classes of neurons to later stages of the technique. Adding special compounds to fixatives further influences the outcome of the method.

As mentioned before, fixatives should be physiologic in that their molarity, pH, and ionic concentration are appropriate to the tissue in question. Fixation should preserve ultrastructure (also, it should facilitate the formation of organochromate complexes, presuming that these occur).

Buffered glutaraldehyde fixation gives rise to stable protein cross-links by reactions with amino groups (Richards and Knowles, 1968), and dialdehydes become incorporated into their denatured structure (Bowes and Cater, 1966). Dialdehydes donate reactive aldehydic groups to tissue and thus render it a potent reducing substrate for the formation of metallic silver or chromate. By comparison, formaldehyde-treated tissue more weakly reduces Ag^+ if the monoaldehyde is used alone in the initial fixative. The different reducing properties of fixed tissue may be evaluated by physical development of cobalt sulfide-containing neuropil that has been treated by di- or monoaldehydes. In the former, background intensification is pronounced, and there is a tendency for suppression of small-diameter fibers. After formaldehyde fixation, intensified profiles provide greater contrast against a paler background.

Although there is little evidence that osmium tetroxide interacts with proteins (see reviews by Hayat, 1970), it is reputed to be reduced by unsaturated fatty acids, thus cross-linking phospholipids and giving rise to stable diesters (Criegee et al., 1942). Valverde (1970) suggested that in the presence of dichromate, free osmium dioxide would be reoxidized to the tetroxide, thus constituting an oxidative catalyst for the conversion of diols to diketones as the sites for chromate conjugation. Also, mild oxidation of glycoproteins may release aldehyde groups for the subsequent reduction of silver ions to metallic silver. Whatever the precise nature of the reactions, it is usual that the nucleation sites for the formation of silver chromate or mercuric chromate, or mercurous oxide chromate reside within nerve cells and glia.

Modifications of the fixative (Table 9-5) are initially attempted in an arbitrary fashion. However, the design of a Golgi modification should consider possible events during the procedure, even though the sites of chromate binding, and the nature of their organic complexes, are entirely unknown. One event that may be common to both the Golgi method and the reduced silver techniques is the capacity of fixed, and in the former case, chromated tissue to reduce Ag^+ to metallic silver. It is known that

Table 9-5. Fixative Modifications[a]

	Modifications of the first fixative (F-A)	Modifications of the second fixative (F-B)	Chromation (C-1)	First metal nitrate impregnation (M-1)	Second chromation (C-2)	Second metal nitrate impregnation (M-2)
IA	Glutaraldehyde-paraformaldehyde, or glutaraldehyde alone in presence of phosphate buffer; osmolarity and ionic composition adjusted	Osmium tetroxide in phosphate or veronal buffer	pH-adjusted zinc or potassium dichromate with glutaraldehyde or pH-adjusted potassium or zinc dichromate with osmium tetroxide	Silver nitrate or mercury (I) nitrate or mercury (II) nitrate	As IA/C-1	As IA/M-1
IB		Osmium tetroxide in appropriate dichromate		pH-adjusted silver nitrate or mercury (I) nitrate	As IB/C-1	As IB/M-1
II	Glutaraldehyde-paraformaldehyde, or glutaraldehyde alone, in PIPES buffer	Osmium tetroxide in PIPES buffer	Sodium, lithium, zinc, potassium, or cadmium dichromate with glutaraldehyde	pH-adjusted silver nitrate in Borax-Borate or TES buffer	As II/C-1	As II/M-1

III	Fixation as for F-A IA,IB,II	Optional postfixation as for F-B/IA,IB,II	Potassium or sodium dichromate with chloral hydrate and/or potassium chlorate, with glutaraldehyde	Unbuffered silver nitrate preceded by incubation in dichromate alone after chlorate	As III/C-1 but chlorate usually omitted
IV	Unbuffered glutaraldehyde with silver nitrate and pyridine (see Appendix II)	Potassium or sodium dichromate alone	Potassium or sodium dichromate with glutaraldehyde	Unbuffered silver nitrate	As IV/C-1
V	Phosphate or PIPES-buffered acrolein, paraformaldehyde, glutaraldehyde	Osmium tetroxide in phosphate, PIPES, or veronal buffer	Potassium or sodium dichromate with osmium tetroxide or glutaraldehyde	Unbuffered silver nitrate	As V/C-1

	As III/M-1
	As IV/M-1
	As V/M-1

[a] Comments: Tissue is rinsed in an appropriate dichromate solution if it is transferred from an osmium-containing medium into one that contains glutaraldehyde or *vice versa* (see also Table 9-4). Postfixation (F-B) is optional for all methods. Combined fixation and chromation in dichromate-glutaraldehyde may be employed for methods I and II, omitting postfixation.

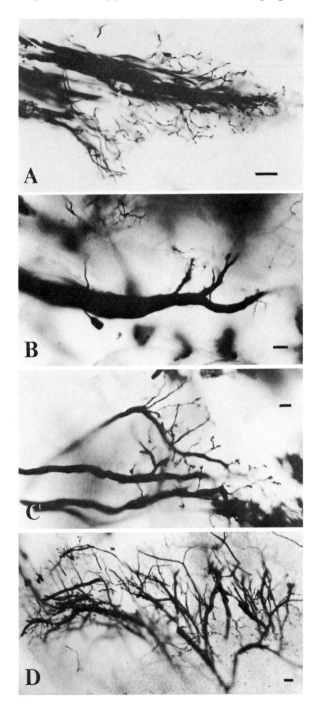

silver acts as a catalyst for reactions between Ag^+ and CrO_4^{2-} (Webster and Halpern, 1956, 1957). Thus, manipulations of the fixative that would serve to promote initial silver reduction, prior to the reaction between silver and chromate ions, or that would lay down metallic silver prior to chromation, might be of particular advantage. The incorporation of silver nitrate with the initial fixative (using nonreactive buffers) may possibly promote the black reaction in specific categories of neurons (see also Porter and Davenport, 1949; Fig. 9-12,A–D).

The use of hyperosmotic or hypoosmotic solutions may also serve to decrease or increase, respectively, permeability barriers for movement of silver ions through membranes of neurons or glia. In insects, glia is rarely revealed after chromation in a hyperosmotic solution (5 g sucrose/100 ml dichromate solution). However, small amounts of chloral hydrate are used as an additive to the chromation stage in order to shift the selectivity of silver impregnation to reactive sites in glia (Del Rio Hortega, 1928). In insect material, the addition of chloral hydrate even to hyperosmotic solutions of the chromating bath promotes impregnation of glia cells. The same effect was reported by Cajal and de Castro (1933) when chloral hydrate was added to the silver nitrate solution. In the reduced silver method, chloral hydrate shifts the selectivity to glia at the expense of the neurocytoskeleton (Figs. 9-13, 9-14A–C).

Other characteristic selectivities imparted by the Golgi rapid procedure, and combinations of osmium and aldehyde chromations, are listed in Table 9-6.

The Nature of Random Impregnation

What is the mechanism of the Golgi method? This question is asked by anyone familiar with the technique. However, it seems unanswerable because many factors are involved, most of which are unin-

---

Fig. 9-12, A–D. Selective impregnation of large-diameter, long-axoned interneurons. **A** Visual cell endings at an optic focus (*C. erythrocephala*). PIPES-buffered glutaraldehyde fixation, osmium tetroxide-dichromate postfixation, Collonier-type double impregnation with silver nitrate. **B** Unbuffered silver nitrate-pyridine-glutaraldehyde fixation, phosphate-buffered glutaraldehyde postfixation, Golgi rapid double impregnation. A terminal of one giant axon (horizontal cell) from lobula plate of *C. erythrocephala* is shown. **C** Terminals of vertical giant neurons of the lobula plate of *Syrphus elegans*. Phosphate-buffered glutaraldehyde fixation, veronal-buffered osmium tetroxide postfixation, then double impregnation by the Golgi-Colonnier procedure. **D** Impregnation of large-diameter lobula neurons (*C. erythrocephala*) after fixation with Sorensen's phosphate-buffered glutaraldehyde-paraformaldehyde and double impregnation using zinc dichromate (pH 3.2)-osmium tetroxide followed by unbuffered silver nitrate. Scales **A-D** = 10 μm.

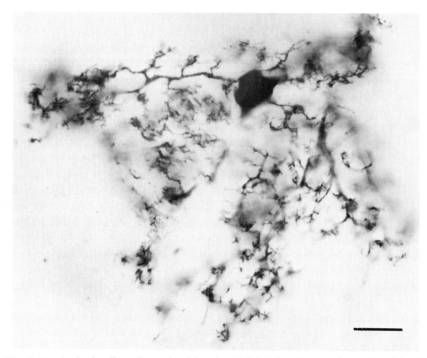

Fig. 9-13. A single glia cell resolved by the Golgi-Colonnier routine using chloral hydrate as an additive to chromation.   Scale = 5μm.

vestigated, particularly those concerning tissue preservation and molecular structure. A simple explanation of impregnation is that silver dichromate precipitates in any closed space. Observations only partly support this, and it does not explain the phenomenon of "randomness." Simple tests show that an alkaline chromating solution made from the chromate salt (not the dichromate) is insufficient for silver chromate impregnation. It seems that a proper balance between CrO_4^{2-}, $Cr_2O_7^{2-}$ ions, etc., and chromic acid is necessary. These are only three variables whose actions on tissue are not known. Certain fixations seem to promote better impregnation than others. Is this because ultrastructure is better preserved for silver chromate precipitation? Or is it because reactant sites in tissue are optimal for complexing to chromates or because tissue better reduces Ag^+ after aldehyde fixation? Again, there are no clear answers. Only the latter seems demonstrable: Aldehyde-fixed chromated tissue is capable of triggering silver reduction at specific sites in physical development. Also using physical development, profiles can be intensified shortly after Ag^+ influx, at a time before silver chromate-filled profiles are normally resolved.

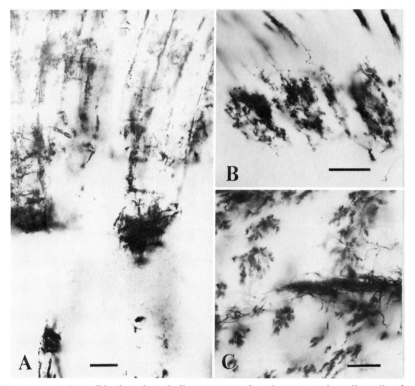

Fig. 9-14, A–C. A Block reduced silver preparation demonstrating glia cells after fixation in glutaraldehyde-chloral hydrate. **B** Golgi-Colonnier impregnated glia cells after fixation with glutaraldehyde-chloral hydrate. **C** "Gliartifacts": Intrinsic crystallite deposits, reminiscent of glia cells, incurred by prolonged exposure to chromation fluid followed by impregnation with mercurous nitrate.

This section describes these and other observations of the method and suggests that impregnation is a complex, multi-step phenomenon.

Observations of the Black Reaction in Neuronal Tissue

Distribution of Ag^+

Chromated profiles are occasionally visible prior to metal impregnation as yellow or orange arborizations against a paler background (Fig. 9-15, A,B) organized in a nonrandom fashion throughout the brain. Despite their occurrence only a fraction of the total population of neurons is revealed after metal impregnation. This phenomenon has been the overwhelming advantage of the method: However, it is offset by unpredictability. Therefore it is of some interest to observe the phenomenon of impregnation during its progress and to offer some explanations for its occurrence.

Table 9-6. Some Suggested Method Combinations for Selection of General Morphologic Classes of Neurons (Observations from *Calliphora*, *Drosophila*, and *Ersitalis*). See Table 9-5.

Neurons with large-diameter axons and dendrites including lamina monopolar cells	Giant neurons of visual neuropil and descending elements to thoracic ganglia	En masse impregnation of small diameter diffuse anaxonal neurons	Glia elements	Nonselective screen for the majority of neuron morphologies
Method IA, using zinc or potassium dichromate, and silver nitrate; F-B as IB, or short IA	Method 1B, using potassium or sodium dichromate, silver nitrate	Method 1B/F-A or F-B; IA/C-1 and M-1 [mercury (II) nitrate]; 1A/C-2; 1A/M-2	Method III omitting potassium chlorate	Method II/F-A; F-B omitted; II/C-1 (sodium, lithium, or potassium); II/M-1; IV/F-B; 1B/C-1; II/M-2
Method combination: 1B/F-A; F-B omitted; C-1; II/M-1; II/C-2 (sodium or potassium) II/M-2	Method IV/F-A; IV/FB; IB/F-B, then as IV, etc.	Method II/F-A (F-B omitted); II/C-1 (omitting zinc or cadmium); M-1 and then recycling		
Method IB/F-A, IV/F-B IB/C-1; II/M-1; IB/ C2; II/M-1	Method IV			

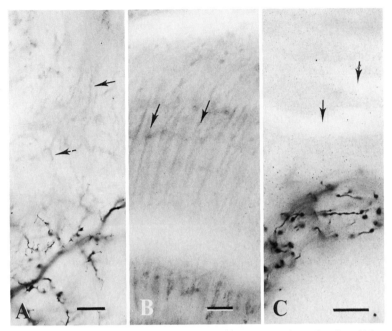

Fig. 9-15, A–C. A Pale neuronal profiles, not representative of "ghost" impregnations. These are bright orange, possibly representing organochromate deposits, some of which persist even after silver impregnation. Profiles accompany material that has undergone prolonged buffered chromation at concentrations of 5%–10% potassium dichromate. **B** Similar profiles can be detected within tissue even before metal impregnation (medulla, *Musca domestica*) and reveal the presence of nonrandom "chromophore" impregnations. **C** Prolonged fixation in osmium tetroxide (veronal buffer) can result in the formation of grainy nucleation sites for silver deposition, sparing the lumen of large-diameter elements (*arrowed*) (Calyx: *Apis mellifera*). Scale **A,C** = 5 μm; **B** = 10 μm.

It is customary to leave the tissue in silver nitrate solution for 1 to 5 days. However, brains may be removed from it at various times after initial immersion, and intact (*Calliphora*) neurons are first resolved after 2 to 3 h, whereas mechanically injured elements are sometimes impregnated after 30 min: Usually these are contiguous with crystalline deposits that extend from the tissue surface into damaged neuropil.

One method for assaying the distribution of nascent or reduced Ag^+ during impregnation is to treat the tissue with ammonium sulfide and afterward physically develop the intact brain with colloidal solutions containing hydroquinone and silver nitrate (see Chap. 20). Silver sulfide catalyzes the reduction of Ag^+ in the presence of a developer, thereby enhancing the general distribution of deposits. Brains of *Calliphora* that have been completely excised from the head capsule prior to impregnation contain silver throughout their volume after 60 min immersion in $AgNO_3$, whereas brains that reside within intact head capsules and have access to silver only through a small opening in the cuticle reveal silver-intensified AgS in neuropil beneath the site of initial AgNO$_3$ influx. If the same experiment is performed on tissue that has been immersed in silver nitrate for 3 h, silver-intensified AgS is observed throughout the tissue of both excised and encapsulated brains.

How do assays of nascent silver bring us any closer to formulating a hypothesis of impregnation? First, the apparent distribution of silvered specks after the first 30–60 min matches the distribution of impregnated cells. Normally, silver chromate-containing cells are revealed near the site of localized Ag$^+$ influx even though silver ions eventually flood the tissue. In brains that have had general access to Ag$^+$, neurons show up scattered throughout the brain. It seems, therefore, that the early distribution of Ag$^+$ is correlated with the eventual location of silver chromate precipitates.

A further observation supports this. If brains are removed from silver nitrate after 30 min, treated with sulfide, and then physically developed in hydroquinone and AgNO$_3$, occasional profiles are determined (in *Calliphora*-size brains it is rare that normal impregnation occurs so soon). Profiles detected in this manner are surrounded by a halo of tissue that contains relatively few intensified AgS specks, indicating that there has been a depletion of Ag$^+$ in the vicinity (Fig. 9-16). These profiles are

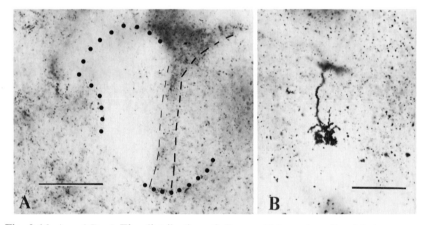

Fig. 9-16, A and B. A The distribution of silver sulfide grains after 2 h impregnation of whole excised brains, resolved by physical development (see text). Some parts of the neuropil, known to be notoriously refractive to the Golgi method, include zones of the corpora pedunculata. This figure illustrates zones of the β lobe that are deficient in silver, others that contain many silver grains. Differential diffusion of Ag$^+$ into areas of neuropil contributes a variable in the so-called "random" Golgi method. **B** A single neuron surrounded by silver-plated specks of AgS, resolved 30 min after influx of Ag$^+$. Tissue treated with sulfide, then silver intensified (see Chap. 18).

resolved because they selectively promote silver reduction in physical development. They act as catalytic supports and it could be proposed that they therefore represent an early stage in the impregnation phenomenon. Is the support silver chromate or something else? This is not known. However, one possibility is that they contain trace quantities of metallic silver, which is not identified by the diffraction methods of Fregerslev *et al.* (1971). Aldehyde-fixed tissue is known to reduce silver nitrate. Aldehyde-fixed and chromated cells retain their capacity for nucleating silver reduction by physical development, as is described later (see p. 170).

Depletion of Chromates

The silver-intensified profiles described above appear to be surrounded by an area of silver depletion. In Golgi preparations, the yellow color of the background is correlated to the number of cells impregnated. When many are impregnated, the background appears pale, indicating a depletion of the chromophore. When few are impregnated, the background is strongly colored. This is similar to the depletion of chromate reactant in Liesegang rings (see p. 172). Unlike silver ions, chromate available for reaction is contained only within the tissue. A second cycle of chromation replenishes the pool. If chromate diffuses into the tissue from a glass micropipette, this, too, will give rise to many impregnated cells (Obermayer and Strausfeld, 1980). The amount of chromate ions appears to be another limiting factor in the number of cells impregnated.

Microscopical Observation of Impregnation

The deposition of some of the products of reactions between chromated tissue and silver nitrate can be followed by direct microscopic observation. Thin slices (50–100 μm) of chromated tissue are cut by hand and put under a cover glass that also covers the tips of two oppositely placed glass micropipettes filled with 0.75% silver nitrate. Injection of $AgNO_3$ is immediately followed by a reaction at the tissue surface giving rise to crystalline deposits that increase in dimension for between 10 and 15 min. Filamentous crystallites from whole brains may achieve a length of between 100 and 1000 μm during the first 5 h of impregnation (Fig. 9-17). A visible black reaction is detectable in cut axons after an induction period of 60 to 120 s. Their outlines are initially resolved, by progressive precipitation along their length, as coarsely branched structures, some of which later assume geometries that are identical to parts of neurons previously described from intact Golgi preparations. Initial precipitation occurs during between 2 and 10 min and is followed by a period during which no further precipitation is observed (between 15 and 30 min). Thereafter, the time taken to complete their impregnation is between 120 s and 5 min (Fig. 9-18,A,D).

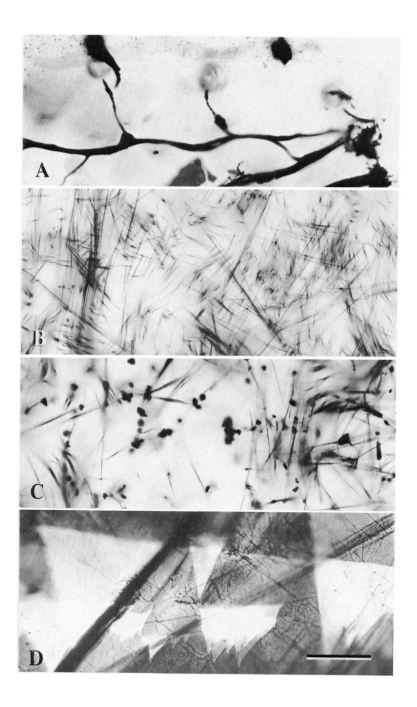

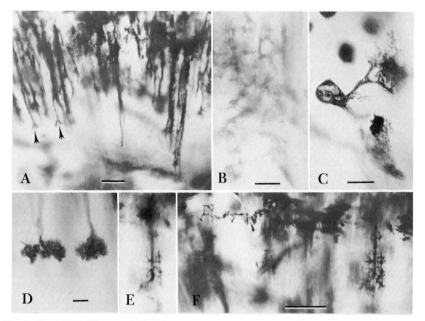

Fig. 9-18, A–F. Observed Golgi reactions (see text).   **A** Silver chromate precipitation through severed axons gives rise to rough outlines of neuronal profiles (*arrowed*) that subsequently develop to assume the shapes of recognizable cell types (**D**).   **B** The ghost-phase of impregnation of an intact neuron within the tissue (compare the form of elements after completion of the black reaction in **E** and **F**). Scale = 2.5 μm.   **C** An early stage of the observed rapid phase of silver chromate precipitation in a glia cell (cf. Fig. 9-13).   Scale **C,D** = 5 μm; **E,F** = 10 μm.

A second mode of precipitation is visible between 15 and 30 min after AgNO₃ application within undamaged cells in the chromated tissue slice. Cells are first resolved in their entirety as "shadows" that slowly darken and become recognizable shapes over a period of about 5 min (Figs. 9-18, B,C,E).   After a refractory period of 5–10 min cells appear to rapidly

---

Fig. 9-17, A–D. So-called perfect preparations should, ideally, have few deposits at the tissue surface.   **A** shows the appearance of cuticle after 16 h chromation in sodium dichromate-glutaraldehyde, followed by impregnation with silver nitrate at pH 5.5.   There are almost no silver dichromate deposits within this area of cuticle, whereas sensory cells and their axons are well impregnated.   Neural tissue that is directly exposed to unbuffered chromating solutions and afterward to silver nitrate at pH 6.8 and above is usually covered by surface deposits.   This is why it is useful to retain tissue within the cuticle, when possible, allowing Ag⁺ access through a limited area.   **B** shows characteristic hair-like crystallites after Colonnier impregnation, **C** after chloral hydrate-dichromate-glutaraldehyde impregnation, and **D** after osmium tetroxide and ·potassium dichromate impregnation. Silver nitrate was used throughout.

darken over a period of 120 sec to 5 min. After this they are identical to impregnated cells in intact brains (Fig. 9-18,E,F). Unlike damaged elements there is no evidence of a progressive extension of silver chromate deposits through the cell, as described elsewhere (see the discussion in Nauta and Ebbesson, 1970). Any cascade reaction through the arborizations of a neuron or glia is likely to be an initial and rapid event of the black reaction, which is beyond the resolution of the light microscope.

Reduction of Ag^+ to Metallic Silver in Chromated Tissue

Evidence that neurons are capable of initiating the reduction of silver nitrate to metallic silver can be simply demonstrated by physical development of chromated material that has been rapidly dehydrated and embedded in "soft" Araldite. Although this process probably removes most free chromates, substantial amounts of the chromophore remain within the tissue structure. Sections that are intensified in colloid-carried physical developer at 20°C for 8 h reveal palisades of neurons that are manifested by granular deposits of reduced silver (Fig. 9-19). Characteristically the largest-diameter elements show up more intensely than small-diameter processes due to the relative areas of reducing sites. The forms of neurons revealed by this experiment are identical to Golgi-impregnated Gestalten and not to profiles typical of reduced silver preparations.

Random Impregnation by Default in Reduced Silver Preparations

Random impregnation is not unique to Golgi methods, but can be mimicked by manipulation of the block Cajal technique, which otherwise reveals nonrandom distributions of neurofibrillar skeletons or axonal membranes, depending upon the initial fixative (Figs. 9-1C and 9-8B).

Cajal methods rely upon incubation in silver nitrate followed by its reduction with hydroquinone and pyrogallol. When the whole surface of the brain is exposed to both these steps, entire populations of parts of neurons are resolved. If, however, only a small area of the brain surface is exposed to the developer, elements in the vicinity of developer influx are best resolved. In the Cajal procedure, the reactant pool of silver nitrate is carried over in the tissue, as is the reactant pool of chromate in the Golgi method. The fixed tissue initially reduces silver nitrate to metallic silver at specific sites in cells during silver incubation (Peters, 1955a–e). These later serve to nucleate subsequent reduction of nascent Ag^+, by the developer, depleting it from the surroundings. Localized reduction gives rise to densely blackened elements. Reduction throughout excised brains reveals weakly intensified profiles within the tissue and darker elements near its surface (Fig. 9-20A).

It is common practice to transfer tissue from silver nitrate directly into the reducing solution. Usually, if silver nitrate is washed out prior to reduction, no neurons are resolved. However, a proportion of nucleation

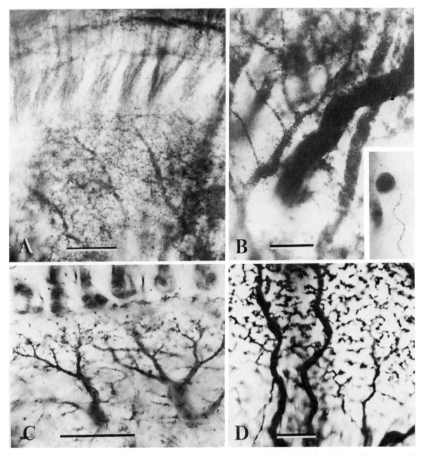

Fig. 9-19, A–D. Nucleation of silver reduction during physical development, by chromated material embedded in soft porous Araldite. Granular profiles of neurons are revealed: These can be favorably matched to the shapes of Golgi-impregnated cells (**D**) or cobalt-filled neurons (**C**). **A** shows fiber bundle projections from the medulla and a species of wide-field neuron in the lobula plate of *Calliphora* whose form matches a type of wide-field centrifugal cell (**C**). **B** illustrates arborizations of an intensified chromated neuron within the lobula plate that can be favorably compared to a normal Golgi-impregnated neuron from intact tissue (**D**). Physical development of chromated material in Araldite reveals cell bodies and their neurites (inset, **B**), unlike any of the reduced silver methods. Scale **B,D** = 10 μm.

sites can still be demonstrated if the tissue is treated by ammonium sulfide, setting up the stronger catalyst, silver sulfide. Tissue is cleared of excess sulfide and then incubated in colloidal physical developer that contains small amounts of both the reactants, hydroquinone and silver nitrate. Fragments of neurons and occasional single nerve cells are re-

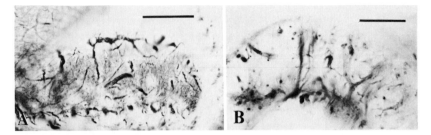

Fig. 9-20, A and B. **A** Block reduced silver preparation (*Musca*) showing dense impregnation near the surface of the tissue, at the site of initial Ag⁺ and developer influx, and pale impregnation in depth (upper left). **B** Selective intensification by Timm's method of a random cluster of neurons within the central body, after silver nitrate impregnation of glutaraldehyde-fixed but unchromated tissue (see text).

vealed randomly distributed, either throughout the brain or within the vicinity of silver-reducer influx. Possibly, randomness here is due to the competitive and autocatalytic nature of the intensification reaction and the concomitant depletion of the reactants—hydroquinone and silver nitrate: Nucleation sites that first trigger the reduction of silver develop at the expense of other elements (Fig. 9-20B).

Golgi Impregnation of Homogeneous Substrates and Artificial Fibers

Direct observation of reactions between chromates and silver nitrate within model substrates support the notion that the random element of the method is incurred after chromation, by the dynamics of Ag⁺ diffusion and the interplay between Ag nucleation and counterdiffusion of chromates. Experiments performed by Liesegang in the last century (1896; see also Stern, 1954) first led to suggestions that such phenomena might play a role in the Golgi method. Liesegang showed that if gelatine sheets were soaked in dichromate and then placed in silver nitrate, reactions between $Cr_2O_7^{2-}$ and Ag^+ were nucleated as sets of concentric bands that were subsequently revealed as granular crystalloids spaced more or less equidistantly. Their separation could be correlated to their size. Rings of silver dichromate aggregates were separated by concentric areas that appeared to be depleted of reaction products.

Liesegang rings can also be obtained in agarose blocks prepared with chromate or dichromate solutions. The rings are presumed to be obtained by counterdiffusion between silver ions and CrO_4^{2-} to precipitate Ag_2CrO_4: Nucleation occurs at supersaturated zones, and precipitation continues as long as the reactant pool of CrO_4^{2-} is undepleted. Diffusion of Ag^+ to deeper levels will take place without the formation of precipitates until regions are encountered in which CrO_4^{2-} has not been depleted.

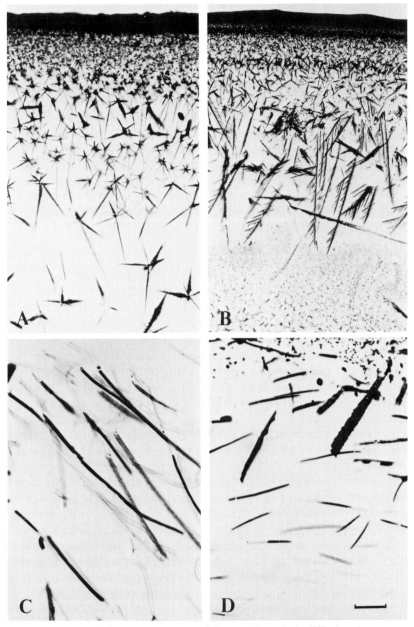

Fig. 9-21, A–D. Liesegang-type precipitates after Ag$^+$ diffusion into agarose-dichromate (acidic) (**A**), agarose chromate (alkaline) (**B**) gels. Rings are not observed in substrates that contain chromated cellulose fibers (**C,D**). Instead random distributions of precipitations are within some fibers. Others appear "unimpregnated" or are resolved as ghosts (see text).

This phenomenon, however, appears somewhat different in chromated agarose sheets that contain chromated fibrous structures embedded within them, such as cellulose fibers. These substrates are prepared so that some fibers meet the surface of the agarose, i.e., imitating the condition of a slice of chromated tissue. Silver nitrate gives rise to surface deposits and black reactions can be seen to progress along the length of some, but not all, fibers that are contiguous with the outer precipitates. Those in close proximity appear to compete for the reactant pool in that near neighbors are partially impregnated or unimpregnated. Diffusion of silver ions into the tissue results in nucleation at single fibers and then precipitation throughout their length. The precipitate is granular, unlike crystalloids of silver chromate formed in solution or in Liesegang rings (Fig. 9-21,A–C). Precipitation in one element can be observed to progress at the expense of a neighboring element when these are in close proximity and compete for the same reactant pool.

Possible Events during Silver Impregnation

The sum of neurons impregnated after two Golgi cycles is, on average, greater than twice the number of neurons impregnated after a single cycle (Fig. 9-22). This indicates that each silver impregnation increases the susceptibility of neurons to the next. Multiple chromation alone does not enhance impregnation, and after a single chromation the reactant pool is probably saturated, possibly in the form of organic-chromate complexes.

The nonlinear characteristics of neuronal impregnation thus suggests that the mechanism of the black reaction is composed of two or more discrete steps.

Some indications of this have already been described from observations of the black reaction in tissue slices: Crystalline growth is observed to extend within cut axons but not into their finer bifurcations and terminals. Impregnation products appear to develop slowly in these, after a refractory period, as they do in intact neurons. It is unlikely that the black reaction comprises crystalloid growth of silver chromate that progresses, worm-like, through the lumen of axons and dendrites from a nucleation site, unless the axon or process has been severed (see Fig. 9-18).

Further evidence that impregnation is a multi-step phenomenon is derived from electron and light microscopy of Golgi preparations. In the former, some unimpregnated profiles in the vicinity of an impregnated element appear darker (i.e., less translucent) than those more distant. In contrast, conventionally fixed material usually does not reveal dark profiles except in the case of experimentally degenerating axons. In tissue

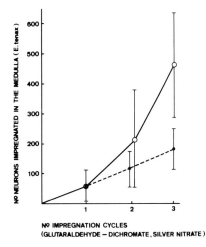

Fig. 9-22. Double impregnation does not give rise to twice as many elements as might be expected (*broken line*). Instead there is a nonlinear increase in the number of impregnated neurons under conditions where Ag$^+$ influx occurs throughout the whole tissue surface.

that has been treated with zinc dichromate, two kinds of impregnated profiles are resolved by light microscopy: Black Gestalten, characteristic of the Golgi method, and pale "ghost" profiles (Fig. 9-23,E,F).   Like the electron-opaque satellites of Golgi-impregnated cells, the ghosts reside laterally with clusters of fully impregnated elements.   After a single cycle, visible ghosts may comprise about 2% of the total number of impregnated neurons, whereas after two cycles they may comprise between 5% and 6%, and the absolute number of fully impregnated elements is increased about threefold.

We may envisage that, contrary to the orthodox view, the "black reaction" is not an all-or-nothing event that either deposits silver chromate (or the mercury equivalents) in cells or leaves them empty.   It is significant that under certain experimental conditions cells are revealed as dark profiles or as semi-opaque profiles or as barely visible ghosts.   Using mercuric nitrate, a single cell may show up as dark- and light-banded precipitates (Fig. 9-23,A,D).   Slowing reactions between Ag$^+$ and CrO$_4^{2-}$ appears to result in a thin layer of deposits on the inner plasmalemma (Ribi and Berg, 1980; Ribi, personal communication).

How could these various aspects of impregnation occur?   One possibility is that silver chromate will simply precipitate within any membrane-bound area, and that precipitation favors sheet-like structures rather than, say, nuclei.   Possibly the observed gradual development of impregnation is nothing more than a late stage of precipitation visible to the naked eye; possibly a thin layer of precipitation, invisible to the eye, does indeed grow throughout the entire cell—worm-like, as sometimes described by other workers.   If this is the case it does not explain why other enclosed spaces are usually spared by the reaction—such as mitochondria and nuclei.   Also, it does not explain why the impregnation

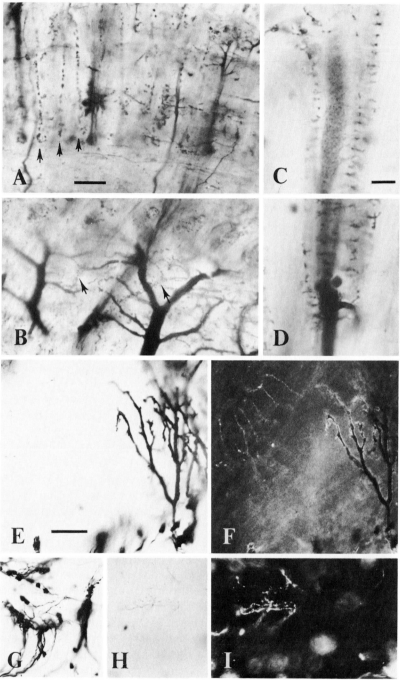

reaction seems to discriminate between closed spaces and intercellular spaces: The latter are hardly ever impregnated. Observations indicate that the predilection of the precipitate to "select" certain classes of cells is correlated with the composition of fixation or chromation. The method cannot, therefore, be called random.

It is proposed here that the formation of the silver chromate precipitate is preceded by an early event that depends on a property of the fixed cell. Direct evidence that this occurs is lacking. However, it has been demonstrated that aldehyde-fixed neurons can serve as active sites for physical reduction of silver, and that combinations of Ag^+ and aldehydes will show up neural elements. It is proposed that impregnation is related to a reduced silver reaction during the first of three phases: (A) *adsorption* of Ag^+ onto the inner plasmalemma and *reduction* of Ag^+ to metallic silver at the inner membrane (as occurs in glutaraldehyde-fixed material in block reduced silver staining) followed by (B) a *rapid catalytic propagation* of reduction over the entire inner membrane and (C) gradual *precipitation of silver chromate* onto metallic silver, the metal serving to catalyze reactions between Ag^+ and CrO_4^{2-} (Fig. 9-24). It is suggested that the random element of the Golgi method is incurred during phase A and that events B and C preclude impregnation of two adjacent neurons. Event C limits the number of visible impregnations due to depletion of the reactant pool of chromate ions. A model for this is given in Appendix I. The following sections discuss these three hypothetical phases.

Adsorption and Reduction

Diffusion of silver ions into the tissue is through a heterogeneous environment that consists of permeability barriers: These are the fixed mem-

---

Fig. 9-23, A–I. Partial impregnation is incurred by certain combinations of chromation and metal impregnation, e.g., fixation in Sorensen's phosphate-buffered glutaraldehyde, postfixation in phosphate-buffered osmium tetroxide (or dichromate-osmium tetroxide), chromation in zinc or potassium dichromate-glutaraldehyde followed by impregnation with mercuric nitrate. In **A** and **B** some elements are revealed fully impregnated, interspersed among "pseudoperiodic" arrays of ghosts (*arrowed*) characterized here by discontinuous granular precipitants. **C** shows a ghost impregnation of a lamina monopolar cell, and **D** illustrates dark to pale impregnation (*C. erythrocephala*). **E** and **F** illustrate an impregnated neuron. In **F** this is seen with a "ghost" profile revealed by dark-field illumination. The preparation was made by zinc-dichromate glutaraldehyde (postfixation in zinc dichromate-osmium tetroxide) and impregnated by silver nitrate. **G** Fully impregnated neurons within the metathoracic ganglion of *Drosophila* after mercurous nitrate impregnation. **H** "Ghost" profiles, and **I**, their resolution by dark-field illumination. Scale **C,D** = 2.5 μm; **A,B & E–I** = 10 μm.

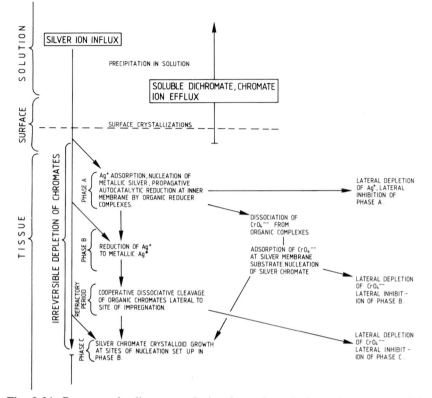

Fig. 9-24. Processes leading to exclusive formation of silver chromate precipitates (hypothetical model).

branes of neurons and glia. From the crude assays of nascent silver distribution, it seems that diffusion of silver ions is quite rapid, whereas neurons appear to become impregnated later. It is also improbable that diffusion occurs synchronously through the length of a membrane that lies parallel to the diffusion front because some parts of its fixed and heterogeneous structure are likely to be less permeable than others. Even if this is not so, it is as difficult to conceive that adsorption of silver ions onto the inner hydrophilic layer takes place simultaneously over a large area within a branched structure, rather than at a point location. Likewise, the chance of adsorption occurring at identical locations in two preparations may be considered negligible unless both preparations have undergone precisely the same treatment and are structural duplicates that present identical access to silver ion influx or unless conditions approximate those in which reactants are generated with the tissue.

Propagation of Reduction

Observations of the black reaction describe a gradual development of the impregnated Gestalt. This implies that silver chromate deposits are initiated at a site that comprises the entire inner surface of the cell. However, if we propose that the initial nucleation of the reaction product is, per se, point located, we must also propose that point locations spread laterally at the inner membrane as a very rapid event indeed. The simplest mechanism for this may be envisaged as a cooperative phenomenon in which initial point reduction catalyzes further reduction at the adjacent membrane: Diffusion, adsorption, and the reaction between Ag^+ and the reducing substrate would thus be propagated over the entire inner surface of the membrane. At locations where double membrane-bound structures, such as mitochondria, entirely plug a process, the propagation of silver reduction could be short-circuited, thereby giving rise to observable partial impregnation.

If silver chromate precipitation within the cell is preceded by the formation of metallic silver by reduction of Ag^+, then the species of fixation will play an important role. For example, aldehydes involved in protein cross linkage may have the capacity to reduce silver. Fixation in dialdehydes would hold an advantage over monoaldehydes. Postfixation in osmium may also influence the reduction properties in a class of cells and, as has been mentioned, early treatment with silver nitrate, during fixation, may lay down silver nuclei in certain neurons and thereby explain the observed differences in Golgi impregnation after this procedure. All these are variables, as are optimal pH of impregnation, diffusion rates, and so forth: Each contributes to the vagaries of the technique.

Silver Chromate Precipitation

If Ag_2CrO_4 precipitation is proportional to the mass of the initial catalytic substrate, depletion of CrO_4^{2-} will also be proportional. It might be expected that large neurons are impregnated at the expense of small elements if precipitation is initiated in both synchronously, or in the former before the latter, given that both are near neighbors (see Appendix I). Similarly, it might be expected that periodic impregnations within arrays of large-diameter elements will be spaced further apart than within an array of small-diameter elements. Such predictions are derived from two-dimensional models of impregnation, but do, in fact, fit quite well to observations of impregnated neurons within regular lattices. In the model it is assumed that the second and third phases of the black reaction are competitive ones, analogous to the autocatalytic intensification of silver grains, or metal sulfide, in the photographic process or in physical development. The simple rule is that two elements in proximity compete for reactants. It follows that initiation of the third phase of the reaction at time t_0 will tend to suppress a neighboring element in which the reaction begins at t_1, assuming that the mass of the second is equal to or less than that of the first. Depending upon their spatial separation at the initiation of the second phase, one or both elements will be fully impregnated. It may thus be envisaged that a palisade of neighboring elements that initiate phase C synchronously at t_0 could be mutually inhibitory and give rise to a palisade of pale profiles (ghosts).

A second cycle of chromation replenishes the pool of reactant, and ghost arrays would be expected to strongly catalyze reactions with silver nitrate. In periodic neuropils such as in the lamina, it is possible to achieve impregnation of almost an entire population of neurons after several cycles of the Golgi routine. Likewise, the introduction of an effectively infinite source of chromate ions to the tissue, from a glass micropipette during metal impregnation, will mediate local impregnation of as many neurons as is achieved after double impregnation (Obermayer and Strausfeld, 1980).

The role of the site of Ag^+ influx, and the competitive nature of the Golgi reaction, may be turned to advantage for screening wild-type and mutant tissue when the conditions of fixation and chromation are exactly duplicated (see Methods, Appendix II) on intact animals. Immediately prior to impregnation, openings are made in the cuticle. For the exami-

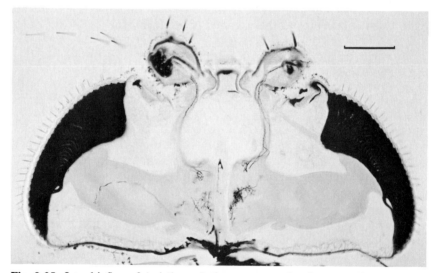

Fig. 9-25. Local influx of Ag^+ through the antennae of a chromated intact *Drosophila* head: selective impregnation of antennal sensory terminals and interneurons in the deuterocerebrum. Scale = 100 μm.

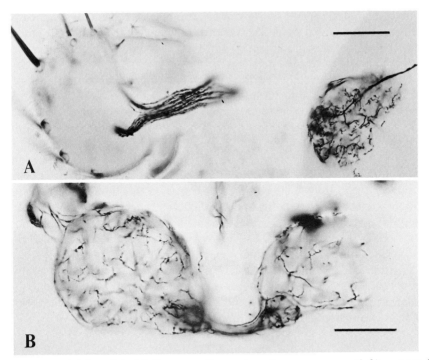

Fig. 9-26. (A) Local Ag⁺ influx, revealing antennal sensory axons (left), antennal
lobe interneurons (right), and (**B**) sensory terminals.

nation of olfactory neuropils, the site of silver influx can be duplicated
merely by removing one or both antennae (Figs. 9-25 and 9-26). A second Golgi cycle follows the first and almost duplicates constellations of
impregnated neurons in the several preparations.

Golgi Methods Using Mercury Salts

Mercuric Chloride (The Golgi-Cox Procedure)

Unlike methods that employ silver nitrate, impregnation by mercuric
chloride relies upon its reduction to mercurous chloride (Fregerslev *et al.*,
1971a) in the presence of the chromating solution. The Golgi-Cox
method is infamous with respect to its reputation as the slowest histologic
procedure. The time needed for its completion in some types of vertebrate neuropils may be as long as six months. However, insect neuropil
requires a somewhat shorter duration, and neurons are revealed after
between 2 and 8 weeks' impregnation.

Without further treatment neurons are resolved as ochre profiles against a similarly toned background. It is therefore general practice to treat the tissue with a weak alkali in order to convert the initial precipitate to a complex of metallic mercury and mercuric oxide (Fregerslev *et al.*, 1971a) in order to resolve cells black.

Nerve cells are usually distributed as bilateral near-symmetric arrangements. This is possibly because reactants behave as if they were generated within the tissue, after a prolonged induction period, and nucleation of the precipitate occurs synchronously throughout the neuropil. Commonly many hundreds of cells are resolved in a single brain.

As in other Golgi methods, the induction period is not succeeded by an all-or-nothing reaction. Rather, parts of fibers are resolved after impregnation for, say, two weeks, whereas after twice this time entire neurons may be visualized, depending upon the initial strength of the solution (see Methods, Appendix II). The first indication that the Golgi-Cox reaction has been initiated are specks of deposits that are aligned parallel to the fibroarchitecture (as in columnar visual neuropil) and in cell bodies.

The possible mechanism of Golgi-Cox impregnation, and the role of the chromate solution, are wholly obscure, and attempts to carry out the procedure on tissue slices have proved fruitless (although alkalinization after impregnation can be performed on celloidin sections).

Mercuric Nitrate Impregnation

Mercuric and mercurous nitrate can be used as substitutes for silver nitrate (Bertram and Ihrig, 1959; Blackstad *et al.*, 1973). Using the former, buffered to within a pH range of 5.0 to 6.5, no efflux of precipitate is evolved from the chromated tissue when it is first immersed in the impregnating solution. This is in contrast to the initial reaction with mercurous nitrate or silver nitrate solutions. Instead, a red granular precipitate is evolved at the tissue surface. Neurons are first resolved after impregnation for at least 10 h and impregnation periods of between 36 and 72 h are recommended.

Mercuric nitrate impregnates an abundance of neurons throughout the neuropil after chromation in potassium or zinc dichromate in the presence of glutaraldehyde (pH 3.6). Chromating solutions that omit glutaraldehyde or osmium favor incomplete impregnation of large-diameter elements. Normally, the method is used to demonstrate selectively small-diameter diffuse neurons, in particular species of anaxonal cells that are wholly refractive to other methods (Fig. 9-27). In regular lattice-like neuropil, clusters of impregnated cells usually comprise pseudoperiodic arrays which may, or may not, reflect the actual arrangements between elements within the area of impregnation. This feature of the method qualifies it as most useful for quantitative studies between neurons whose

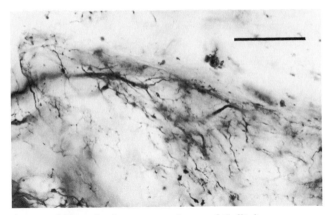

Fig. 9-27. Anaxonal fibers in the protocerebrum of *Calliphora*, accompanied by a single axon, revealed by mercuric nitrate impregnation of material fixed in Millonig's phosphate-buffered glutaraldehyde.   Zinc-dichromate osmium tetroxide chromation.   Recycled.

forms indicate a common identity.   However, arrays of this kind must be treated with circumspection with respect to interpreting real geometric relationships (see Discussion, Chap. 18).

The method shares many features that are typical of the Golgi-Cox procedure.   At least three times as many neurons may be impregnated by it as compared with double impregnation by silver nitrate, and the distribution of impregnated elements is almost bilaterally symmetric.   Neuropil is inundated with mercuric nitrate before visible reactions are initiated between mercury (II) ions and the organic chromate substrate.   It is to be expected that synchronous impregnation is initiated throughout the neuropil.   The number of impregnated elements is primarily limited by the depletion of the chromate pool, and non-neighbor impregnations are postulated on the basis of initial local depletion of the mercury (II) reactant.   It is predicted that under these conditions regular palisades of elements will be resolved, some as ghosts, some barely discernible, and others containing intense deposits of mercuric oxide chromate.   In fact, the method clearly reveals two modes of impregnation: light or dark elements within dense clusters.   Moreover, the largest-diameter axons usually show up as periodic banded patterns of intense and less intense precipitates, or alternatively, as regularly distributed clumps of precipitate (see Figs. 9-23,A–D, and Fig. 9-28,A–C).

Mercurous Nitrate [Mercury (I) Nitrate]

When chromated tissue is immersed in an acidic solution of saturated mercurous nitrate a precipitate is immediately evolved.   This phenome-

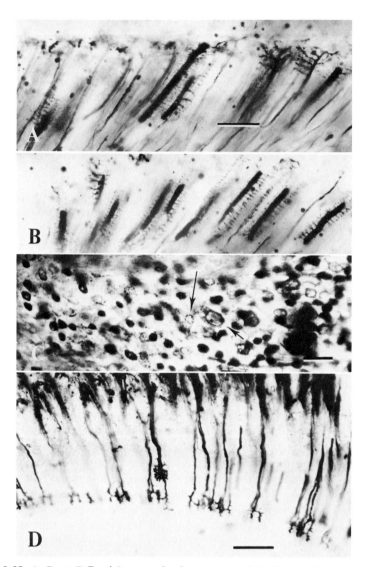

Fig. 9-28, A–D. A,B Partial mercuric chromate precipitations of lamina neurons, revealed en masse (*Calliphora;* see also Fig. 9-22, C, D).   Scale = 20 μm.   **C** Impregnation by mercuric nitrate, showing fully impregnated cell bodies and cut profiles in which precipitates appear to collect at the inner surface of the cell-body membrane (*arrowed*).   **D** Impregnations by mercurous nitrate showing a cluster of monopolar cell terminals in the medulla of *Calliphora* after local influx of mercury I ions through the cut retina.

non is reminiscent of the initial reaction between a piece of chromated tissue and silver nitrate. The two methods are in fact alike with respect to the distribution of impregnated cells as isolated clusters, or local clusters in the vicinity of mercury (I) influx (Fig. 9-28D). The timing of impregnation is identical to that of silver nitrate methods. Neurons are resolved as pale red or brown profiles containing mercurous chromate (Blackstadt *et al.,* 1973).

Conclusions

1. The Golgi method[3] resolves single neurons or glia cells. The number of impregnated cells and the intensity of impregnation have been correlated with a number of variables during the three stages of the method (fixation, chromation, and metal impregnation). All other factors being equal, modifications of the chromation stage alone, such as using various species of dichromate solutions at various pHs, with various additives, will achieve a variety of selectivities. If only one method is employed, then it may provide only a narrow range of cells for putative reactions with Ag^+.

2. Modifications of fixation and choice of silver or mercury (I,II) nitrate also bias the selectivity of Golgi methods toward some specific cell morphologies: e.g., neurons with large-diameter axons and dendrites; neurons that have large dendritic, or terminal fields, or, alternatively, diffuse anaxonal cells, and short-axoned interneurons.

3. Treatment with $AgNO_3$ can be manipulated (pH or temperature) to show different intensities of impregnation as well as to reveal few or many neurons.

4. There is no evidence that any particular species of neuron is wholly refractive to the method. An exception may be degenerating or moribund cells: Impregnation of experimentally induced degenerating nerve cells has not been achieved in insects. However, associated hypertrophy of glia cells is revealed.

5. Simulated impregnations, derived from a theoretical model of the black reaction, predict that the reaction is accompanied by an initial depletion of reactants within the surroundings of a neuron: Lateral depletion persists throughout the course of impregnation and suppresses impregnation in near neighbors. Early events in impregnation (silver reduction) would account for the phenomenon of nonimpregnation in adjacent elements, assuming that the initial phase of the Golgi reaction rapidly spreads throughout the whole cell.

6. Affinities between the Golgi method and some reduced silver tech-

---

[3] The Golgi-Cox method is not included in this summary.

niques are suggested.   But, while the latter resolve general features of cytoarchitecture, the Golgi method is reputed to select neurons randomly. The "random" element should, however, be distinguished from selectivity incurred by fixation or chromation, and the technique is better described as stochastic.

7. The Golgi method can be used on small brains so that it will consistently screen neuronal shapes of a "random" choice of elements within a nonrandomly chosen location.   This is achieved by local influx of Ag^+ into chromated tissue.

8. Neuronal arrangements that are manifested by impregnated elements are the *product* of a stochastic sampling device.   This applies even to regular palisades of impregnations.   Therefore, arrangements in Golgi-impregnated material cannot demonstrate real arrangements of nerve cells in neuropil unless the intermediate neurons are identified.   This can be achieved by combined Golgi and electron microscopy or by another

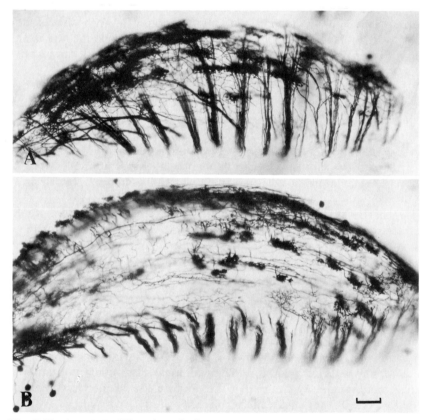

Fig. 9-29, A and B. "Pseudoperiodic" clusters of columnar interneurons (two successive sections), in the medulla of a fourth instar *Gomphocerripus rufus* (see text).   The spatial arrangements here do not truly reflect actual arrangements between nerve cells because many are unimpregnated.

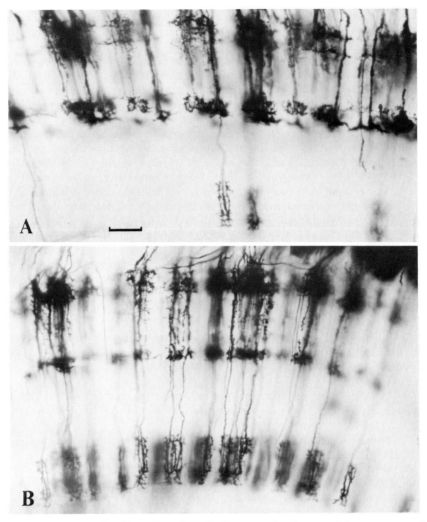

Fig. 9-30, A and B. "Pseudoperiodic" clusters of columnar interneurons and monopolar cell terminals in the medulla of *Eristalis tenax*. The distance separating each cluster or single neurons in a cluster in this orderly neuropil seems to be proportional to their mass (see text).

procedure, such as uptake of cobalt by nerve cell assemblies or correlative studies of reduced silver material such as the Holmes-Blest technique or the Bodian methods (see Chaps. 6, 7, and 8 of this volume).

Appendix I: A Model of Random Impregnation

In Golgi preparations that are made with silver nitrate, impregnated neurons are revealed as clusters of elements. The unit structure of some neuropils is such that spatial relationship between impregnated clusters of neurons can be described with reference to a hexagonal lattice. The lamina and medulla of Diptera provide such examples (Strausfeld, 1970).

Golgi-impregnated neurons in latticeworks can be classified not only according to their disposition and branching patterns but also on the basis of their dimensions and lateral extents. Thus, unit separation between impregnated neurons of various sizes can be described within a cluster. Distances separating clusters can be correlated to the dimensions of their component nerve cells and their density. When a regular array of large-diameter neurons is impregnated within a cluster, the distance separating each element is greater than that which separates two small-diameter elements of an analogous array. Some examples are shown in Figures 9-29 and 9-30. Observations of the Golgi reaction have led to a model of events that possibly occur during impregnation. This is illustrated in Figure 9-31. The model requires that impregnation be accompanied by lateral depletion of reactants that, initially, preclude impregnation of two adjacent neurons. Subsequent depletion of chromates is proposed to further inhibit neighborhood impregnation. The area of inhibition around a neuron depends, in part, on its mass, in which the early phase of the Golgi reaction has been initiated. A simple two-dimensional model has been devised to demonstrate the principle of the phenomenon of "random" impregnation by means of computer simulation that uses a program based upon the following assumptions.

1. Depletion of the reactant (CrO_4^{2-}) by a single substrate takes place according to a defined wave form that integrates over the rate of deposition of the reaction product (a sigmoid function). Only one reactant (CrO_4^{2-}) is accounted for in this model (Fig. 9-32).

2. The lateral extent of depletion of the reactant (the reactant pool) is limited within a given region. This is proportional to the mass of the catalytic substrate (i.e., the neuronal inner membrane area).

3. The deposition rate at a single substrate is a function of time and the available reactant within a symmetric boundary.

4. Two substrates in which a reaction is initiated, are independent of each other only if their associated reactant pool boundaries do not overlap.

5. Two or more elements in proximity, whose associated reactant pool boundaries overlap, compete for the reactant pool.

6. If reactant pools overlap their boundaries are not thereby extended. This is to say, two adjacent substrates, each of an identical mass, do not draw from a reactant pool whose boundaries are the same as that of a single element twice that mass.

7. Two or more elements in proximity whose impregnation is initiated at time $t = 0$ interact linearly with respect to their overlapping reactant pools. Two or more elements whose impregnation is initiated at unit time $t = 0, t = 1, \cdots$, exhibit a phase shift throughout the course of the depletion of the reactant. They thus interact nonlinearly. Thus, phase-shift and non-phase-shift conditions give wholly different patterns of impregnation with respect to the same geometric array of elements. It is therefore predicted that in a system which exhibits no phase shift, regular palisades of elements will be impregnated. This is the case for Golgi reactions with mercuric nitrate, where the refractory period between influx of mer-

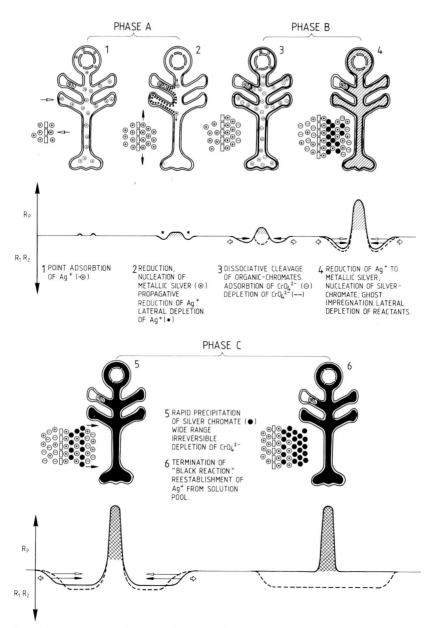

Fig. 9-31. Summary diagram of one possible sequence of events during Golgi impregnation of chromated tissue by silver nitrate illustrating the lateral depletion of the reactant pool (R_1, R_2) and successive deposition of the reaction product (R_p). Initial silver nucleation could also be modeled according to Silver's (1942) negatively charged micelle theory of colloidal silver staining. Hatched and crosshatched areas of R_p indicate silver chromate precipitation. The black inner margin (*3*) represents a layer of metallic silver. Crosshatching and black in cells *4–6* represent silver chromate precipitation.

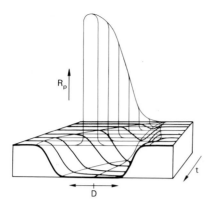

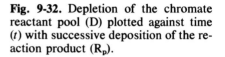

Fig. 9-32. Depletion of the chromate reactant pool (D) plotted against time (*t*) with successive deposition of the reaction product (R$_p$).

curic II ions and the initiation of their reaction with chromate ions is greatly prolonged.

Some examples from this model are shown in Figs. 9-33 and 9-34.

Appendix II: Methods

Introduction

Neural tissue is fixed in situ within the animal. Insects are best immobilized by cooling to about 4°C. Large insects are injected with fixative, smaller insects are submerged after openings are made in the ventral part of the head (at the mouth parts) and dorsally in the thorax. It is important that the tissue has been partly fixed before further dissection so as to minimize distortions of the internal organs and consequent displacement of ganglia.

Fixation of insects equivalent in size to *Drosophila* is for 2 h at room temperature. Larger Diptera and Hymenoptera should be fixed for at least 4 h and fixation overnight is not excessive for large Orthoptera. Several washes in the appropriate buffer are used to remove fixative. Thereafter tissue may be stored for up to a week in buffer at 20°C. This is particularly advantageous for field work. Prior to chromation the specimen is blotted and excess buffer is replaced by the appropriate dichromate before immersion in the dichromate-dialdehyde or dichromate-osmium tetroxide solution. Postfixation (prior to chromation proper) is either in buffered osmium tetroxide solutions or in osmium-dichromate. The tissue is washed thoroughly in buffer after postfixation and then rinsed in dichromate, to remove excess osmium.

Specimens should be submerged during all stages of the technique. At the end of chromation they may be laid out on tissue paper and their cuticle partly removed, where appropriate, before immersion in the metal nitrate solution. Specimens are kept moist by the old chromating solution or in fresh chromate. As mentioned previously, openings at specific regions of the head capsule of small insects, such as *Drosophila* or *Calliphora,* give rise to local impregnation in their vicinity. Using the Colonnier technique, removal of the cornea of the eye results in impregnation of lamina elements whereas removal of the outer part of ommatidia, to the level of R8 receptors (see Kirschfeld, 1971), resolves lamina and medulla neurons. Removal of the entire retina resolves laminal, medulla, and lobula elements as well as projections from the lobula to the midbrain. In contrast, the Golgi rapid technique resolves receptors and a few interneurons unless a second impregnation in dichromate-dialdehyde is employed.

To resolve complete interneurons that extend from the head to the body, openings must also be made ventrally in the thorax to allow inward diffusion of silver nitrate there. Any operation made on the specimen may be performed by stainless steel instruments during fixation and chromation, contrary to folklore. Chromation is performed at 20–25°C or at 4°C (Ribi and Berg, 1980) in the dark, as is metal impregnation.

Fixation

The simplest way to fix tissue is by direct immersion in the dialdehyde (or osmium)-dichromate solution. Although neurons are afterwards well resolved, the quality of fixation is not useful for electron microscopy. It is recommended that initial fixation is in buffered glutaraldehyde or in buffered paraformaldehyde-glutaraldehyde. Osmium tetroxide may be carried in the same buffer as used for initial fixation, or in veronal acetate buffer. Cacodylate buffers are not recommended. The commonest buffers for Golgi methods are Karnovsky's (1965) phosphate, Sorenson's phosphate, Millonig's (1961) phosphate, or Bauer and Stacey's (1977) piperazine-N,N'-bis(2-ethanol sulfonic acid) (PIPES). The latter is excellent for all Golgi methods that employ silver nitrate. The first three are useful for silver and mercury impregnation; their compositions are listed below.

Buffered Aldehyde and Osmium Tetroxide for Primary
and Secondary Fixation
I. Aldehyde Fixatives
A. *Low-molarity buffered glutaraldehyde-paraformaldehyde [modified from Sörensen after Glauert (1974)]*

1. Prepare stock buffer:

Solution A	Dibasic sodium phosphate $Na_2HPO_4·2H_2O$	35.61 g/liter	(0.2 M)
Solution B	Monobasic sodium phosphate $NaH_2PO_4·H_2O$	27.60 g/liter	(0.2 M)

Add 24.5 ml soln. A to 25.5 ml soln. B to obtain pH 6.8 at 25°C

30.5 ml soln. A	19.5 ml soln. B	pH 7.0
36.0 ml soln. A	14.0 ml soln. B	pH 7.2
40.5 ml soln. A	9.5 ml soln. B	pH 7.4

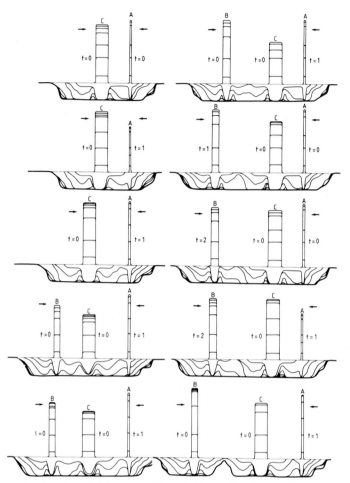

Fig 9-33. Simulated model impregnations indicate that for a substrate to achieve the final phase of the black reaction (90% impregnation is indicated by an *arrow*) several factors play a critical role. These are the time of initiation of the reaction in two or more elements, their mass, and the distance between them. Examples should be read from top to bottom of the left- and right-hand columns. Elements A and C require a minimal separation of three neurons (three of their widths) in order to reach full impregnation when reactions in both are initiated at $t = 0$. If the reaction in A occurs after C, precipitation is inhibited by depletion of its re-actant pool by C. If A and C are farther apart, then both achieve impregnation. If impregnation is initiated synchronously in C and a third element B, at $t = 0$, both B and C inhibit each other to the advantage of A even though the reaction in A is initiated later, at $t = 1$. B and C mutually inhibit unless their geometric separation or the timing of the initiation of the reaction is modified (*right-hand column*).

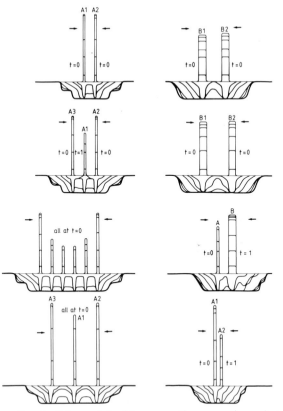

Fig. 9-34. According to this model of impregnation, reactions that are initiated synchronously in a palisade of elements should give rise to partial impregnation or ghost profiles. The *left-hand column* illustrates impregnations within clusters of two, three, six, and again three neurons of identical masses to illustrate the relationship between space and timing of impregnation within an array. In the *right-hand column* two elements of mass B (B1,B2), reaction initiated at $t = 0$, are mutually inhibited when they are close but would achieve full impregnation if more separate. The last two figures of the right-hand column similarly illustrate inhibition of close neighbors during phase-shift reactions.

2. Prepare stock sucrose solution, magnesium chloride.
   $MgCl_2$, 0.14 g/50 ml distilled water
   sucrose, 17.0 g/50 ml
   Add 1:1 (solution C)
3. Prepare stock solution of paraformaldehyde by adding 10 g to 100 ml distilled water and heating to 60°C. Then 1 N NaOH is added drop by drop until the solution clears (about 6 ml), then add 0.5 ml excess NaOH. Cool and filter.
4. Prepare the fixative by adding 20 ml solution C to 50 ml of the buffer (A + B). Then add 20 ml 10% paraformaldehyde and 10 ml 25% glutaraldehyde (contains 3 mM $MgCl_2$, 0.1 M sucrose).

5. Fix at room temperature for 2 to 6 h.
6. Wash out excess aldehydes in several changes of buffer (with solution C).
7. Wash out buffer in several changes of 2.5% dichromate solution (containing 3.4 g sucrose/100 ml).
8. Transfer to the chromating solution (the addition of sucrose is optional here).

Comment: Stages 5–8 apply to this and all following schedules for buffered aldehydes if they are succeeded for chromation with either glutaraldehyde or osmium tetroxide. Stages 7 and 8 are *omitted* if aldehyde fixation is succeeded by postfixation in osmium tetroxide buffered with phosphate, veronal-acetate, or PIPES. Stages 5–7 *precede* postfixation in dichromate-osmium tetroxide.

B. *High-osmolarity buffered paraformaldehyde-glutaraldehyde (Karnovsky, 1956)*
1. Prepare paraformaldehyde solution:
   8.0 g paraformaldehyde in 200 ml distilled water is heated to 60°C and 0.9 g granular NaOH is then added. After cooling add 3.75 g monobasic sodium phosphate and 4 ml 25% glutaraldehyde. The pH of this solution is between 7.2 and 7.3.
2. Fix for 2–6 h, then wash out excess aldehydes in phosphate buffer:
   Solution A. 0.91% monobasic potassium phosphate (KH_2PO_4).
   Solution B. 1.19% dibasic sodium phosphate ($Na_2HPO_4 \cdot 2H_2O$). 72.6 ml B is made up to 100 ml with solution A to give a pH of about 7.2.

C. *Millonig's (1961) phosphate-buffered glutaraldehyde*
1. Prepare buffer
   A. $NaH_2PO_4 \cdot H_2O$ 22.6 g/liter distilled water
   B. NaOH   25.2 g/liter
   Add 8.5 ml B to 41.5 ml A to give a pH of 7.3 at 20°C. The pH can be adjusted by varying the amount of B to more or less than this value.
2. Add 10 ml 25% glutaraldehyde to 50 ml buffer, and make up to 100 ml with distilled water. 3.4 g sucrose may be added to adjust the molarity. If added at all, magnesium chloride should not be mixed to the buffered aldehyde: To avoid precipitation the required amount is dissolved in 1 liter distilled water, which is then used to bring up the final volume of the fixative to 100 ml (0.6 g $MgCl_2$/liter: final concentration of 3 mM).

Comment: This fixative is recommended for prefixation prior to the Golgi-Cox method and Ribi and Berg's low temperature "pale" impregnation method (1980).

D. *Sorensen's phosphate buffer for glutaraldehyde-paraformaldehyde or glutaraldehyde*

1. Prepare stock buffer.
   Solution A. Dibasic sodium phosphate
   $Na_2HPO_4 \cdot 2H_2O$ 11.9 g/liter
   Solution B. Monobasic potassium phosphate
   KH_2PO_4 9.1 g/liter
   Add 24.6 ml soln. A to 25.4 ml soln. B to obtain pH 6.8 at 20°C

30.6 ml soln. A	19.4 ml soln. B	pH 7.0
36.3 ml soln. A	13.7 ml soln. B	pH 7.2
40.9 ml soln. A	9.1 ml soln. B	pH 7.4

2. Prepare 10% paraformaldehyde solution as described in I,A,3.
3. Add 10 ml 25% glutaraldehyde to the buffer or add 10 ml glutaraldehyde and 20 ml 10% paraformaldehyde solution. Add distilled water to make up fixative to 99 ml and 1 ml 3% MgCl₂ and 3.4 g sucrose to adjust ionic balance and osmolarity.

E. *PIPES [piperazine N,N'-bis(2-ethane sulfonic acid)]-buffered glutaraldehyde or glutaraldehyde-paraformaldehyde*

1. Dissolve 30.24 g of PIPES in 1 liter distilled water by the addition of very concentrated NaOH. This should be added drop by drop until the solution clears. Its pH is adjusted to between 7.0 and 7.4 by the addition of excess 1 *N* NaOH.
2. 10 ml of 25% glutaraldehyde is diluted to a final volume of 100 ml using PIPES buffer. If electronmicroscopy will be used.
   To prepare paraformaldehyde-glutaraldehyde in PIPES:
   Add 2 g paraformaldehyde and a pellet of NaOH (about 0.5 g) to 80 ml distilled water and heat to 60°C. When cool add 25% glutaraldehyde to give a final concentration of 1.5% dialdehyde (6 ml). Add 3.024 g PIPES and dissolve using concentrated NaOH: Check pH of the solution afterwards. Adjust the pH to between 7.0 and 7.4. Bring the final volume to 100 ml with distilled water.
   The buffer is 0.1 *M*.

Comment: this is the best fixative for routine Golgi impregnation either using the Colonnier glutaraldehyde-dichromate or the Golgi rapid osmium tetroxide-dichromate procedure. The fixative is, however, inapplicable to the Golgi-Cox method.

F. *Phosphate-buffered acrolein-paraformaldehyde-glutaraldehyde*

Acrolein is most toxic and is not at all recommended for routine study. In any event it should be handled under strict safety precautions. Its exceptional use is for embryonic, larval, or early pupal neuropil of holometabolous insects. It is used either with Sorensen's phosphate buffer or with PIPES:
A. To 50 ml Sorensen's buffer add 10 ml glutaraldehyde, 10 ml paraformaldehyde and 5 ml acrolein. Add distilled water to make up the solution to 99 ml. Adjust ionic balance with 1 ml 3% MgCl₂ (optional).
B. Prepare paraformaldehyde-glutaraldehyde in PIPES, as above. Before adjusting the final volume to 100 ml add 4 ml 10% acrolein.

General Comment
The pH of the initial aldehyde fixatives should be adjusted to suit the type of tissue undergoing fixation. Brains of hymenopterous insects are best impregnated after an initial fixation at pH 7.4–7.6. Diptera, Odonata, Hemiptera, and Coleoptera are fixed at between pH 6.8 and 7.2. Orthoptera and Lepidoptera are fixed at between 7.0 and 7.4.

II. Osmium Tetroxide Postfixation Solutions
A. *Phosphate buffers*

Two percent osmium tetroxide may be carried in any of the above buffers. For electronmicroscopical purposes osmium tetroxide is usually diluted 1:1 in buffer. However, this renders the tissue too dark for light microscopy of Golgi impregnations.   Instead 1 vol 2% osmium tetroxide is added to 9 vols buffer.   Postfixation of *Drosophila*-sized brains is for between 0.5 and 1.0 h at 4°C.   Brains of larger insects are postfixed for between 1 and 4 h.   Excess osmium tetroxide is removed by several changes of buffer, and excess buffer is washed out in several changes of 2.5% dichromate, prior to immersion in glutaraldehyde-dichromate or osmium tetroxide-dichromate.

B. *Postfixation in Palade's (1951) veronal acetate buffer*

> Solution A.   2.89 g barbital ($C_8H_{12}N_2O_3$)
> 1.15 g anhydrous sodium acetate
> 100.0 ml distilled water
> Solution B.   5 ml solution A
> 12.5 ml 2% osmium tetroxide
> Add 3–5 ml 0.1 N HCl to adjust pH to between 7.0 and 7.2
> Make up to 25 ml with distilled water.

This concentration is used for postfixation prior to electron microscopy. However, it is diluted 1:3 with distilled water if it is used prior to Golgi impregnation.   Fixation should be for 1–2 h at 4°C.

C. *Postfixation in unbuffered osmium tetroxide in potassium or sodium dichromate*

One volume of 2% osmium tetroxide is added to 20 volumes of 2.5% dichromate.   The addition of 3.4 g sucrose/100 ml solution is optional.   Tissue is postfixed in this solution for 6 to 24 h, after which it is washed several times in 2.5% dichromate prior to immersion in glutaraldehyde-dichromate.

Comment: Osmium tetroxide may also be used in conjunction with zinc dichromate.   This is prepared by dissolving zinc chromate in a dilute solution of acetic or formic acid (6 g zinc chromate, 100 ml distilled water, and 3–4 ml glacial acetic or formic acid); 1 ml 2% osmium tetroxide is added to 20 ml acetic acid-acetate (pH 3.1–3.4) buffered zinc dichromate.

Postfixation (Osmification)

Osmium tetroxide is used as a 2% solution diluted 1:9 in buffer or 1:20 in dichromate.   Osmium-dichromate solutions may be used as a postfixative, prior to dialdehyde-dichromate treatment, or as the first and second chromations for the recycled Golgi rapid method or in a combined method that employs a single Golgi rapid cycle followed by glutaraldehyde-dichromate chromation prior to the second silver impregnation.

Postfixation in buffered osmium is between 1 and 4 h at 4°C depending upon the size of the animal.   It is advisable to slightly enlarge initial openings made in the cuticle prior to postfixation, and it may be necessary to partly expose neuropil of larger insects directly to the solution.   Veronal acetate buffers may be used in conjunction with osmium postfixation after initial fixation in Karnovsky's phosphate-buffered paraformaldehyde-glutaraldehyde.   Osmium postfixation is not applicable to Golgi-Cox methods, but promotes mercuric or mercurous nitrate impregnation, after chromation in zinc or potassium dichromate solutions.

Chromation

Except in the case of zinc, cadmium, and lithium dichromate (Table 9-7), 2.5% solutions of dichromates are used in conjunction with osmium or glutaraldehyde. Zinc or cadmium dichromate are made up by dissolving 6 g of the chromate salt in usually 100 ml of 3.5% formic acid, after which the solution may be buffered with sodium formiate. Ammonium, lithium, calcium, sodium, and potassium dichromates are available commercially, some at exorbitant expense, and the last two are most convenient for routine use. Potassium, or better zinc dichromate, is used in conjunction with mercurous or mercuric nitrate impregnation.

Metal Impregnation

Silver nitrate is used at a concentration of 0.75% or 0.5% within the pH range 4.5–7.4. Surface crystallization is minimal in the pH range 5.0–6.0, which is usually adjusted by an appropriate buffer. Mercurous nitrate [mercury (I) nitrate $Hg_2(NO_3)_2$] is dissolved in 0.001% nitric acid to make a 1%–2% (saturated solution) whose pH is between 1 and 1.6; this can be further adjusted to 1.6 to 2 by the addition of sodium acetate.

Mercuric nitrate [mercury (II) nitrate, $Hg(NO_3)_2$] is commercially available as a saturated solution [Merck analar solutions contain 0.0162 g $Hg(NO_3)_2$/ml], and the pH of the solution is usually adjusted to the optimal range 5.5–6.5 by the addition of an appropriate buffer. Impregnation times depend somewhat on the volume of tissue; this should be borne in mind with respect to vertebrate or cephalopod material. However, for most insects the periods shown in Table 9-8, are sufficient, irrespective of size.

Additives to Chromating Fluids, Fixatives, and Silver Nitrate

Properly, osmium tetroxide and glutaraldehyde should also be classified as additives to chromating solutions since neither played a role in the original method. However, impregnation may be achieved without them if the tissue has been initially fixed in glutaraldehyde. If unbuffered (except after zinc dichromate-acetic acid-acetate) only a few neurons are partially resolved by dichromate alone: Impregnation reveals dendritic arborizations at the expense of axons and their terminals, amacrine neurons being an exception.

Additives to the chromating solutions were previously reviewed by Ramon-Moliner (1967). Those that can be applied to insect neuropil are listed in Table 9-9. In general, their routine employment holds no real advantage except in the case of 1, which selects for glia cells and 5, which shifts impregnation toward large-diameter neurons: The concentration of pyridine in 5 is critical, and this reagent can be omitted. Formic acid, osmium tetroxide, or formaldehyde, added in small amounts to silver nitrate, may also influence impregnation in terms of the types, quantities, or intensities of cells. Modified $AgNO_3$ is usually used as an intermittent step, followed soon after by a second chromation. This is succeeded by a second silver nitrate impregnation without an additive. Treatment with a weak

Table 9-7. Dichromate Solutions

Chromate	Dichromate	Distilled water	Formic acid	Sodium formiate	Glutaric acid
Cadmium 6 g	—	100 ml	3.5 ml	Xg, to pH 3.1–3.7[a]	—
Zinc 6 g[c]	—	100 ml	3.5 ml	Xg, to pH 3.1–3.7[a]	—
—	Ammonium 2.5 g	100 ml	—	—	—
—	Ammonium 2.5 g	100 ml	—	—	1 N, to pH 3.4
—	Sodium[c] 2.5 g	100 ml	—	—	—
—	Sodium[c] 2.5 g	100 ml	—	—	—
—	Potassium[c] 2.5 g	100 ml	—	—	—
—	Potassium[c] 2.5 mg	100 ml	—	—	—
—	Calcium 2.5 mg	100 ml	—	—	—
—	Lithium 3.0 g	100 ml	—	—	—

[a] Optimal pH is 3.1–3.2.
[b] Used with acetic acid.
[c] Recommended for routine use.

reducer in the presence of silver ions might be used to enhance the nucleation of metallic silver, thus giving rise to more powerful catalytic supports for subsequent silver-chromate precipitation. Addition of Ag^+ to the initial fixative may act identically.

Table 9-8. Impregnation Times

Metal nitrate	Initial treatment after chromation	Impregnation period
0.75% solution $AgNO_3$ (pH 4.5–7.4); saturated $Hg_2(NO_3)_2$ (pH 1.6)	Rinse in fresh solutions until no further precipitate is is evolved	24–28 h
Saturated soln. $Hg(NO_3)_2$ (pH 5.5–5.6)	Place directly in impregnation solution	48–96 h

Sodium acetate	Acetic acid	Sucrose	Chloral hydrate	25% Glutaraldehyde		2% Osmium tetroxide
Xg, to pH 3.1–3.7[b]	(3.5 ml in lieu of formic acid	0–4 g	—	25 ml	or	5 ml
Xg, to pH 3.1–3.7[b]	(3.5 ml in lieu of formic acid)	0–4 g or 0–5 g		25 ml (or none if buffered to pH 4.0–5.5)	or	5 ml
—	—	0–4 g or 0–5 g		—		5 ml
—	—	0–4 g	—	25 ml		—
—	—	0–4 g or 0–5 g		25 ml	or	5 ml
Borax-borate buffer: see Appendix II, p. 202		0–4 g	—	25 ml	or	5 ml
		0–4 g or 0–5 g		25 ml	or	5 ml
—	—	0–4 g or 0–5 g		25 ml	or	5 ml
—	—	0–4 g	—	25 ml		5 ml
—	—	0–4 g or 0–5 g		—		5 ml

Recycling (Double and Triple Impregnation)

Tissue may be treated several times by the chromating-impregnation routine (except where stated in Table 9-9). The metal nitrate solution is decanted, and fresh chromating solution poured over the specimens. An immediate precipitate will be formed between metallic and chromate ions (in the case of silver nitrate and mercurous nitrate). The fluid is again decanted, and the specimens are rinsed once again in the chromating solution prior to the second cycle. The second chromation is as long as the first and employs the same species of cation. Initial osmium tetroxide-dichromate may be substituted by glutaraldehyde-dichromate in the second cycle (or, rarely, vice versa). The second metal impregnation usually employs the same nitrate as the first, although mercuric nitrate may be followed by a second impregnation with silver nitrate to reveal black, partially impregnated, profiles superimposed upon pale mercuric oxide chromate profiles. Information from superimposed impregnation is only useful if these fragments can be recognized as belonging to species of nerve cells that have already been described from conventional techniques.

Table 9-9. Chromating Solution Additives

Additive	g or ml/100 ml recipient solution	Dichromate				Aldehyde fixative	0.75% AgNO$_3$ solution	Comments
		With osmium tetroxide	With glutaral-dehyde	With chloral hydrate	Alone			
1. Potassium chlorate	6 g	–	– (or +)	+	+	–	–	Wash in appropriate dichromate, prior to metal impregnation to remove soluble reaction products other than chromates.
2. Barbital	0.5 g	+	+	–	–	–	–	Use only for single cycle unless this is followed by a chromation that omits additives 1–3. 3 gives rise to an opaque background.
3. Bromal hydrate	2 g	–	–	+	+	–	–	
4. AgNO$_3$ and (optional) pyridine	0.25 g 0.04–0.08 ml	–	–	–	–	+ (AgNO$_3$ in borax-borate)	–	Wash several times in buffer before chromation.
5,6. 5% Formic acid or 5% formaline	1 ml	–	–	–	–	–	+	Impregnate at most for 1 h, followed by 2nd chromation and metal impregnation without additive.
7. 2% Osmium tetroxide	0.5 ml	–	–	–	–	–	+	

Dehydration and Embedding

The earlier practice of embedding in celloidin has been abandoned in favor of resin (araldite, Epon, Spurr's). Neural tissue can also be enclosed by soft paraffin wax and sectioned in its "wet" state on a sliding microtome. This is useful for observing the phenomenon of the black reaction, but it is not recommended for routine studies. If they must be made at all, celloidin sections of Golgi-impregnated material (including Golgi-Cox) are mounted in permount under coverslips. Within 4–5 years the precipitate leaches out of profiles and material becomes useless. The overwhelming advantage of resin sections is their permanence. They can also be removed from the slide, re-embedded, and sectioned at 1–3 μm for counterstaining, or at 500–1000 Å for electron microscopy.

A routine for resin embedding is as follows:

1. Remove from the impregnation solution and wash briefly in distilled water.
2. Dehydrate in ascending ethanol, each for 10 min, twice in dried absolute ethanol.
3. Soak twice for 20 min in dried propylene oxide.
4. Infiltrate overnight in propylene oxide-resin (1:1) in an open vessel (propylene oxide will gradually evaporate).
5. Blot and infiltrate in fresh resin for 1 h. Orient the specimens in an appropriate cast, in fresh resin, polymerize for 1 h at 60°C, 0.5 h at 80°C, 0.5 h at 90°C, and 1 h at 100°C. Cool slowly. Cut 20–100-μm sections on a sliding microtome; rinse in xylol and embed in Permount or in resin.

Three Routine Golgi Methods

The following methods are given in full. The Golgi-Cox precedure (Weiss, 1970) has been found reliable for a variety of orders (Orthoptera, Diptera, Hymenoptera, Odonata, Lepidoptera, and Hemiptera). The method should not be used for combined Golgi–electron microscopy because mercury evaporates in the electron beam with obvious consequences.

The buffered modifications of the Golgi rapid and Colonnier techniques were devised by Nässel and Seyan (1979) and are reliable. Silver nitrate dissolves easily in borax-borate buffer at 20°C and there is no precipitate.

Method 1: The Golgi-Cox Procedure

For many years this method was employed on fresh tissue: The chromating-sublimate solution served both as a fixative and as an impregnating solution. Used in this way the method does not work on insect neuropil, and prefixation in phosphate-buffered glutaraldehyde or glutaraldehyde-paraformaldehyde is essential. The fixative is washed out with buffer, and the buffer removed by washing in dichromate-chromate solutions prior to impregnation.

The schedule is modified from accounts by Ramon-Moliner (1970) and Weiss (1972). An alternative method for initial fixation is directly in the Colonnier dichromate-glutaraldehyde solution, which is adjusted by the addition of acetic acid to pH 3.1, for 6 h. Afterward it is rinsed in the appropriate solution of dichromate and chromate salt.

1. Fix in phosphate-buffered glutaraldehyde (Millonig or Karnovsky buffers) for 2–4 h. Wash in buffer and store for up to 6 days.
2. Wash out buffer with potassium or sodium dichromate and potassium or sodium chromate solution (5 g and 1 g, respectively, in 100 ml distilled water).
3. Place in the impregnating solution for between 8 days and 4 months. Full strength solutions should be used for no longer than 8 weeks. Dilute solutions are used for no less than 3 weeks. Useful dilutions are 1:4 and 1:10. The impregnation solution is made up as follows:
   Solution A. potassium or sodium dichromate 5 g
   potassium or sodium tungstate 2.5 g
   distilled water 100 ml
   Solution B. potassium or sodium chromate 1 g
   mercuric chloride 1 g
   distilled water 85 ml
   Solution B is boiled, and after cooling the volume is adjusted to the original 85 ml. Solutions A and B are added 1:1.
4. Impregnate for between 8 days and 6 weeks in the dark at 20–24°C.
5. Rinse tissue in several changes of distilled water for 30 min.
6. Immerse in: ammonium hydroxide 2 ml of lithium hydroxide 0.5 g.
   potassium nitrate 15 g
   distilled water 100 ml
   Brains are alkalinized for between 30 min and 2 h, depending upon their size.
7. Wash well in a 0.2% solution of acetic acid using several changes during 4 h.
8. Dehydrate and embed in Araldite.

Methods 2 and 3: Buffered Golgi Rapid or Golgi-Colonnier Methods (Nässel and Seyan, 1979)
Stock solutions: Holmes borax-boric acid buffer.

pH	0.2 M H_3BO_3 (12.4 g boric acid/liter H_2O)	0.05 M $Na_2B_4O_7 \cdot 10\ H_2O$ (19.0 g borax/liter H_2O)
7.0	97.50 ml	2.50 ml
7.2	95.0	5.0
7.4	90.0	10.0
7.6	85.0	15.0

Add 0.005 M $Na_2B_4O_7 \cdot 10\ H_2O$ drop by drop to 99 ml 0.2 M H_3BO_3 in order to adjust the pH in the lowest ranges from approximately pH 4.7–6.8, at 20°C.

Buffered sodium or potassium dichromate: Dissolve 2.5 g dichromate in 100 ml buffer (at pH 7.4). This will give a final pH of 5.4–5.6 at 20°C.
Buffered dichromate-osmium: Add 1 vol of 2% osmium tetroxide to 20 vol of buffered sodium or potassium dichromate. This gives a pH of 5.6–5.7 falling to pH 5.5 after 2 days.
Buffered dichromate-glutaraldehyde: Add 1 vol of 25% glutaraldehyde to 4 vol of buffered dichromate. This gives a pH of 5.3–5.4, which rises to pH 5.5 after 2 days.

Buffered silver nitrate: Dissolve 0.75 g AgNO$_3$ in 100 ml of Holmes' buffer at pH 7.0–7.4.

The Schedule
  1. Fix tissue in PIPES-buffered 25% glutaraldehyde for 3–12 h (pH 7.0–7.4).
  2. Wash tissue in PIPES buffer (several changes during 2 h).
  3. Wash tissue in Holmes' buffered chromate (several changes during 2 h).
  4. Immerse either in Holmes' buffered dichromate-osmium (Golgi rapid) or in buffered dichromate-glutaraldehyde (Golgi-Colonnier) for 12–72 h (in the dark at room temperature).
  5. Rinse several times in Holmes' buffered silver nitrate. When precipitate is no longer evolved, immerse in buffered silver nitrate for 12–72 h.
  6. Repeat stages 4 and 5 (optional).
  7. Dehydrate and embedd in Araldite, Epon, or Spurr's.

As described earlier, the first chromation can be with osmium, the second in dichromate. If this is done, tissue should be thoroughly washed in buffered dichromate after the first silver impregnation.

A modification by Ribi (1979) employs the following chromation fluid:

  2 g potassium dichromate
  2 g D-glucose
  100 ml distilled water
  5 ml 2% osmium tetroxide

The pH is adjusted to 7.2 by adding concentrated KOH. Unbuffered 0.5% silver nitrate solution is used for 4°C for 4 h (Ribi and Berg, 1980).

Acknowledgments

I am indebted to Peter Speck (European Molecular Biology Laboratory, Heidelberg) for collaboration on the design of the random impregnation model and program for Appendix I. The bulk of the technical assistance for experimental Golgi impregnation on *Drosophila* was performed by Harjit Seyan, and chromate injection experiments into intact brain, during metal impregnation, were devised and carried out by Malu Obermayer. I also thank Dr. M. Weiss for data concerning the Golgi-Cox method. Golgi studies on insects were initiated with Dr. A. D. Blest in 1967 with the advice and encouragement of Professor B. B. Boycott, F.R.S. I am indebted to both of them.

Chapter 10

Electron-Microscopic Methods for Nervous Tissues

M. P. Osborne

University of Birmingham Medical School
Birmingham, England

The first electron micrographs that we would recognize as high quality even today were produced as long ago as 1952 by F. S. Sjöstrand (Pease, 1964). Thus examination of thin sections of biological tissue at high resolution (2.0 nm or less) with the electron microscope has been possible for a quarter of a century. Since the early fifties there have been dramatic improvements in the techniques of fixation, embedding, and sectioning, and these have been paralleled by the development of greatly improved ultramicrotomes and electron microscopes. Nevertheless these techniques were more or less well established by the early sixties. They include the almost universal adoption of aldehydes as fixatives, epoxy resins as embedding media, and glass knives for production of ultrathin sections.

A detailed knowledge of cell structure, coupled with the immense strides that have been made in cytochemistry, have revolutionized our understanding of cell biology. One branch of cell biology that is the object of intense study is neurobiology, and it is not surprising that work with insect material has featured strongly in this field.

I have been engaged in the study of the fine structure of insect nervous systems since the late fifties and over the years have standardized preparative methods that I have found successful. Many of these methods are not greatly different from those which can be found in most books on electron-microscopic techniques, but what I have tried to do here is to place emphasis on those aspects of specimen preparation that are directly relevant to problems connected with investigation of insect nervous tissue and to include only those techniques with which I have had direct personal experience. I have also included many small tips and short cuts that I hope will be of real benefit to anyone contemplating work with the insect nervous system.

Fixation

Probably more confusion and mystique surrounds the process of fixation of biological tissues than any other aspect of electron microscopy. Yet if a few simple principles are kept in mind even a novice should have no difficulty in obtaining well-fixed material with his first attempts.

At present, there are two excellent general fixatives for electron microscopy. These are osmium tetroxide and glutaraldehyde, although of late formaldehyde, particularly in combination with glutaraldehyde, is rapidly gaining favor. While the aldehydes have gained the pride of place as the best currently available chemical fixatives, it is of historic interest to note here that initially osmium tetroxide was the fixative of choice in most laboratories (a possible exception was in laboratories where plant material was the main interest and here potassium permanganate was usually preferred) and that early attempts to use aldehydes met with little success. The breakthrough with aldehydes was made by Sabatini *et al.* in 1963. They obtained excellent results with several aldehydes by using them as primary fixatives and by using osmium tetroxide as a post or secondary fixative. Since this classic paper, the technique of "double fixation" has become *the* routine method for preservation of biological tissues throughout the world. In my laboratory this method has over the past fifteen years or so given excellent preservation of a variety of insect nervous and muscular tissues, and so most of this section will be devoted to describing how to fix insect tissues with aldehydes.

Several factors are critical in obtaining good preservation. They are (1) ensuring adequate and rapid access or penetration of the fixative to the tissues; (2) determining the correct osmotic pressure of the fixative; (3) determining the correct pH of the fixative.

Penetration of Fixatives

This is not too serious with insect tissues if they can be rapidly excised and placed in relatively large volumes of fixative in which they are agitated for several minutes, and then left to stand in this fixative for the remainder of the fixation time.   Fixation of ganglia and nerves in situ poses more of a problem.   All extraneous tissue such as fat and muscle must be dissected away so that the tissues to be preserved are directly exposed to the fixative.   It is important, too, to see that all traces of hemolymph are washed away from the tissues; otherwise it may clot and form a barrier preventing access of the fixative.   This is best accomplished by cutting open the insect and pinning it out flat on wax in a small dish.   After all tissues to be fixed are well exposed, a continuous stream of the fixative (in a fume cupboard) is poured over the tissues for about 1–2 min before flooding the dissection with a large pool of fresh fixative in which the animal remains for the rest of the fixation period.   Some workers dissect insect tissues under saline solution or wash tissues before fixing them.   I do not recommend this practice unless you are sure of the saline solution, since decidedly inferior fixation may result.

If after these precautions good fixation is still not obtained using glutaraldehyde, and penetration of the fixative is known to be the problem, then formaldehyde or formaldehyde/glutaraldehyde mixtures can be tried since formaldehyde is known to penetrate biological tissues more rapidly than the larger molecule, glutaraldehyde (see below).

Osmotic Pressure of Fixatives

That the osmotic pressure of the fixative, particularly when using aldehydes, is important is no longer a question of debate, but what is open to argument is exactly what the correct osmotic pressure should be.   Some authors consider that the osmotic pressure contributed to the fixative by the aldehyde can be ignored and that the critical factor is the osmotic pressure contributed by the rest of the constituents of the fixative.   These are the buffer and added salts or carbohydrates such as sodium chloride and sucrose.   Other authors disagree and suggest that some measure should be taken of the osmotic pressure of the fixative agent although it is not precisely clear as to how much this should be.   What must be realized at this point is that if no account is taken of the osmotic pressure of the aldehydes then most fixatives in current use are osmotically hypotonic.   On the other hand if the aldehydes do contribute effectively to the total osmotic pressure then most are hypertonic.   Moreover, making the fixative isotonic with the body fluids does not always give satisfactory preservation.   For detailed discussions about osmotic pressures of fixatives see Sjöstrand (1967) and Glauert (1975).   In attempting to calculate the osmotic pressure of fixatives there is no straightforward procedure that can guarantee success, and it has to be admitted that even with competent investigators, good preservation of tissues is achieved as much by trial and

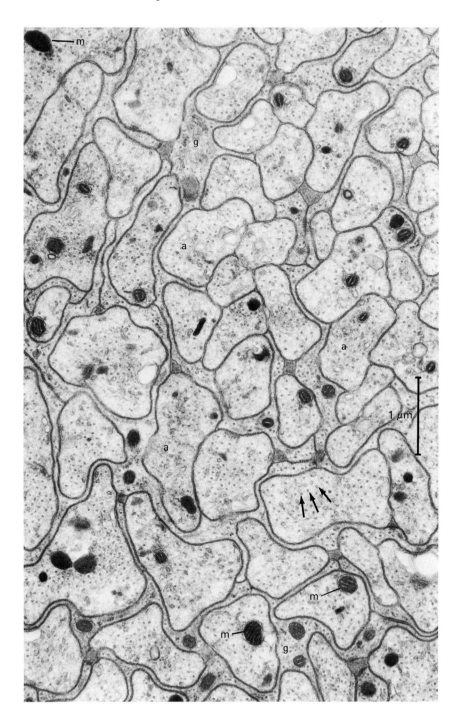

error as it is by logical means.   However, for the beginner, I can pass on a method for adjusting the osmotic pressure of fixatives that has been in use in my laboratory for some time and that has reliably given good preservation of nervous tissue in a wide variety of insects (Fig. 10-1).   The method has the virtue of being very simple and is as follows: Assuming that the osmotic pressure in osmoles of the hemolymph of the insect is known, subtract from this value, first, the concentration of glutaraldehyde to be used divided by 10 and, second, the molarity of the buffer (not its actual osmotic pressure).   The figure that remains will give the amount of sucrose in moles which has to be added to the fixative to raise its osmotic pressure to the "correct" value.   Thus for the stick insect, *Carausius morosus,* the calculation is worked out in the following manner: The osmotic pressure of the hemolymph $\simeq 0.4$ osmol.   Subtract the concentration of glutaraldehyde in the fixative divided by 10.   Thus for a fixative containing 2.5% glutaraldehyde, this is $0.4 - (2.5/10) = 0.15$.   If the molarity of the buffer to be used is 0.05 M, subtracting this value leaves 0.1.   Thus 0.1 is the molarity of sucrose required to adjust the osmotic pressure of the fixative to a suitable value.   Thus, to each 100 ml of fixative, 342 (mol. wt. of sucrose) $\times 0.1 \times (100/1000) = 3.42$ g of sucrose are added.

The osmotic pressure of the postfixative, osmium tetroxide, is equally important.   Calculation of the "correct" osmotic pressure for this fixative is also simple.   Again, referring to the example of the stick insect and using a buffer of the same molarity of 0.05 M as for the aldehyde fixative, all one need do is subtract the molarity of the buffer from the osmotic pressure of the insect's blood to leave the molarity of sucrose needed to be added to the fixative.   Thus for the osmium tetroxide fixative, $0.4 - 0.05 = 0.35$ M sucrose $= 11.97$ g sucrose are added to each 100 ml of fixative.   In practice the osmotic pressure contributed to the fixative by osmium tetroxide can be ignored.   Osmium is usually used at a concentration of 1%–2%, and in our laboratory there is a slight preference for the higher concentration.

Salts such as sodium chloride can be used instead of sucrose for adjusting the osmotic pressures of fixatives.   I do not in general recommend this

---

Fig. 10-1. Transverse section of a nerve within the corpus cardiacum of the stick insect, *Carausius morosus.*   The material was fixed in 2.5% glutaraldeyde in a phosphate buffer followed by postfixation in 1% osmium tetroxide also in a phosphate buffer.   In both fixatives sucrose was added to adjust the osmotic pressure.   The section was stained in uranyl acetate and lead citrate.   Preservation is generally good, there being little evidence of swelling or shrinkage of any of the cellular elements or intercellular spaces.   Good preservation of the axoplasm has been achieved, the mitochondria (*m*) and microtubules (*arrows*) being particularly well preserved.   The combination of uranyl and lead stains gives high contrast and excellent tonal quality to the electron image.   *a*, axons; *g*, glial cells ×20,000.

technique for two reasons. First, the *actual* osmotic pressure contributed by salt solutions to the fixative is not easily calculated because the degree of dissociation of the salts is not known, and in practice can only be determined by direct measurement. Second, salts do not appear to give such good preservation of insect tissues as does sucrose. Other sugars such as glucose can be used for adjusting osmotic pressure and as far as I am aware glucose gives results similar to those of sucrose.

pH of Fixatives

The pH of the fixative can be important in obtaining good preservation of biological tissues, but in general is not regarded as such a critical factor as osmolarity. Most fixatives are buffered to a pH value of between 7.0 and 7.4. However, some authors prefer a pH of 8.0 or more and others prefer a distinctly acid fixative buffered to about pH 6.0 (see Glauert, 1975). The point to be borne in mind here is that seldom can the pH of the fixative be blamed for poor preservation, and the beginner can rest assured that fixatives in the pH range of neutrality will suffice for insect nervous tissue. However, if the pH of the insect hemolymph is known it might be worth trying a fixative with an identical pH value. Far more important in my opinion is the choice of the buffer that is used to control the pH. Several buffers are in routine use and formulae for these are given in Appendix I. The most commonly used buffers are phosphate and cacodylate buffers, although veronal acetate, citrate, borate, maleate, collidine, and Tris buffers are also employed, some of these especially when specific staining and cytochemical techniques are used. At one time, particularly when osmium tetroxide was used almost exclusively as a fixative, veronal acetate buffers were commonly employed. However, it has since materialized that veronal acetate is not a good buffer at physiologic pH, and its use with aldehydes is to be discouraged since it reacts with these compounds (see Glauert, 1975). My choice has nearly always been for some variant of phosphate buffer, and although I have had good success with cacodylate I would strongly urge the beginner to start with the former. This same buffer can be used both for the aldehyde and osmium fixatives.

Problems with Fixation

In some cases difficulties with the quality of preservation of certain insect tissues may arise. Poor fixation can often be corrected by changing the osmotic pressure of the fixative simply by altering the amount of sucrose that is added. A guide to whether the osmotic pressure should be raised or lowered can usually be obtained by studying certain features of electron micrographs. Swollen mitochondria and Golgi apparati and distended nuclear and plasma membranes coupled with a leached appearance of the cytoplasm indicate too low an osmotic pressure. General shrinkage of cells and cellular organelles allied with highly convoluted or crenulated cell and organelle membranes and enlarged intercellular spaces suggest too high an osmotic pressure of the fixative. Effects of varying the osmotic pressure of the fixative upon the structure of the subsynaptic reticulum of neuromuscular junctions from the body wall muscles of the blowfly larva are illustrated by Fig. 10-2, A–C. At relatively high osmotic pressures (Fig. 10-2C) the extracellular spaces of the subsynaptic reticulum are much wider than they are at lower osmotic pressures (Fig. 10-2, A, B) due presumably to shrinkage of the cellular component of the subsynaptic reticulum in the fixative of the higher osmolarity.

Fixatives of 0.3–0.35 M (Fig. 10-2, A, B) apparently give "better" preservation of the subsynaptic reticulum. It is pertinent to note here that the osmotic pressure of the hemolymph is $\simeq 0.42$ osmol. Thus adequate fixation of a particular organelle may pose a problem despite the fact that other organelles may be very well preserved. It cannot be overemphasized that in the final analysis, what constitutes good fixation is subject to personal preference and that some tissues regardless of the fixative used do not give good preservation of all cell constituents.

It is sometimes apparent that, in spite of altering the osmotic pressure of the fixatives, mitochondria are not well preserved by aldehyde fixation. Preservation of mitochondria and membrane systems in general can be improved by the addition of 1–3 mM calcium chloride to the fixative. Indeed it is probably wise to add calcium routinely to fixatives to minimize membrane artifacts, although when phosphate buffers are used 1 mM calcium should be regarded as the upper limit to avoid precipitation of calcium phosphate compounds.

If membrane preservation is of prime importance it is becoming increasingly popular to treat the tissue after osmication with a third fixative, uranyl acetate. This is usually dissolved at a concentration of 0.25%–1% in a sodium maleate buffer at pH 5.0. Phosphate or cacodylate buffer is not used because precipitation occurs. It therefore follows that if these two buffers are used in the aldehyde and osmium fixatives, the tissues must be washed in the maleate buffer before being transferred to the uranyl acetate fixative. Uranyl acetate solutions should be used as quickly as possible after making up, and it is recommended that they be kept no longer than one week and even then should be stored in the dark (see Glauert, 1975). Incidentally, I am a firm believer in using freshly made up fixatives whatever their chemical nature. Glutaraldehyde and osmium fixatives can be stored in a refrigerator at about 4°C for up to 1 week, but we have had poorly fixed material that in some cases has been traced to the use of old fixatives. My advice is when in doubt always make up fresh fixatives. With aldehyde fixation it is often advocated that

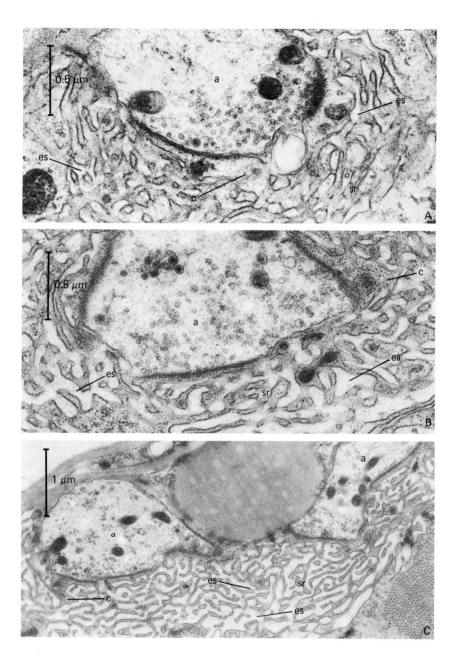

a buffer wash should be given between the aldehyde and osmium fixation. I have never, as a general rule, used this intermediary buffer wash and transferred tissues directly from the aldehyde to the osmium fixative without, as far as I am aware, deleterious results. I admit that the buffer wash can serve as a temporary phase, which is convenient for storing fixed tissues for say overnight periods, but I find it just as convenient to store tissues during the dehydration procedure in 70% alcohol at 4°C (see p. 217).

It can happen that although the outer layers of tissues are well fixed the inner zones are poorly preserved. Clearly this is due to poor or delayed penetration of fixatives. Preservation of large masses of tissue can be improved if the fixation is carried out in the cold at about 4°C. In general, though, this type of problem can be solved by improving access of the fixative to the tissues (see p. 206) or by prolonging the fixation time. Ordinarily, 0.5 to 1 h in an aldehyde is sufficient, but we have had occasions when prolonging this time to 3–4 h, or even longer, has produced an improvement. The duration of postfixation in osmium tetroxide is not usually critical, but since it penetrates much more slowly than aldehydes, prolonged soaking in osmium, even overnight, may bring about better preservation especially where tissue blocks exceed 2–3 mm in thickness. Usually, though, 0.5 to 4 h in osmium is sufficient. Persistent trouble with penetration of glutaraldehyde can often be overcome by switching to formaldehyde or formaldehyde/glutaraldehyde mixtures. The latter is made up by dissolving 2.0 g of paraformaldehyde powder in 29 ml of distilled water and heating to 60–65°C in a fume cupboard. A few drops of $1 N$ sodium hydroxide are added until the solution clears. Take 50 ml of 0.2 M phosphate or cacodylate buffer (pH 7.3–7.4) and add to it the 29 ml of paraformaldehyde solution, 10 ml of 25% glutaraldehyde in water, and 20 ml of distilled water. This fixative contains 2% formaldehyde and 2.5% glutaraldehyde in 0.1 M buffer. To this fixative sucrose

---

Fig. 10-2, A–C. A Transverse section of a nerve muscle synapse from a muscle of body wall of the blowfly larva, *Lucilia sericata*. Fixation was in 2.5% glutaraldehyde in a phosphate buffer at 0.3 M. × 36,000. **B** Field similar to that of **A** with identical fixation except that the molarity was raised to 0.35 M. Note that the extracellular spaces are generally wider and the postjunctional cytoplasm is more electron opaque than in **A**. The axon shows little change. × 36,000. **C** Field similar to those of **A** and **B**. The molarity of the fixative is higher at 0.4 M. Note a further relative increase in the width of the extracellular spaces, of the subsynaptic reticulum, and the electron opacity of the postjunctional cytoplasm. The axons with their organelles show little differences compared with **A** and **B**. × 18,000. *a*, axon; *c*, cytoplasm of postjunctional region of muscle cell; *es*, extracellular spaces; *sr*, subsynaptic reticulum. The micrographs were kindly supplied by Dr. S. N. Irving.

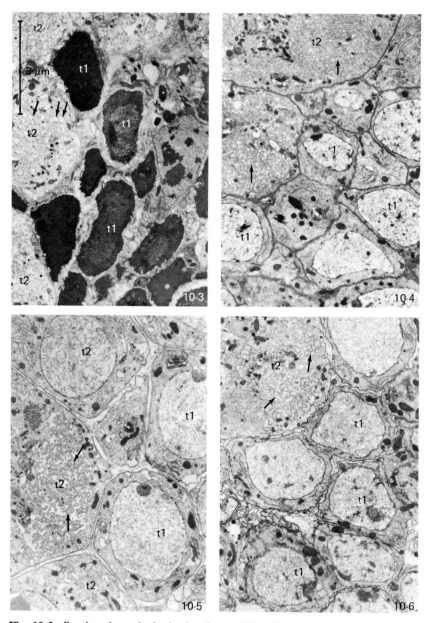

Fig. 10-3. Section through the brain of an aphid. Material is fixed in glutaraldehyde and postfixed in osmium tetroxide. Two types of neurons are shown. Type
1 (*t1*) has an electron-opaque nucleus and cytoplasm. Type 2 (*t2*) is larger and has
by comparison and electron-lucent cytoplasm and nucleus. The endoplasmic reticulum shows areas of distension (*arrows*). This figure and Figs. 10-4, 10-5, and
10-6 were kindly supplied by Dr. J. Hardie.

or glucose can be added to adjust its osmotic pressure. If required, 1.3 mM calcium chloride can also be added.

Finally, one of my former research students, Dr. J. Hardie, recently brought to my attention a problem he had in fixing central nervous tissues in aphids. With glutaraldehyde one type of neuron (Fig. 10-3) appears shrunken with an electron-opaque nucleus and cytoplasm. Another type (Fig. 10-3) exhibits a swollen endoplasmic reticulum. With a glutaraldehyde-formaldehyde (Karnovsky, 1965) fixative, the first type of neuron is not so electron opaque, but the nuclei appear leached. In the second type of neuron areas of distension of the endoplasmic reticulum are still present (Fig. 10-4). The use of osmium as the primary fixative again prevented shrinkage and the appearance of an electron-dense cytoplasm and nucleus in the first type of neuron; however, in the second type, severe distension of the endoplasmic reticulum is realized (Fig. 10-5). The best general overall fixation of both types of neurons was achieved with a mixture of glutaraldehyde (2%) and osmium tetroxide (1%). Neurons of the first type show good preservation of both nucleus and cytoplasm, and the endoplasmic reticulum of the second type shows good form with evidence of excellent preservation (Fig. 10-6).

It may be that other workers have had similar problems with insect nervous tissues, and in these cases I think that the glutaraldehyde/osmium fixative is worth trying (see also Hirsch and Fedorko, 1968). This mixture is unstable at room temperature and therefore should only be made up immediately before use. Even so, the mixture is best made up and used at a temperature not exceeding 4°C. Fixation times longer than 1 h are not recommended. Mixtures of osmium tetroxide, glutaraldehyde, and formaldehyde have also been used with success (Pollard and Ito, 1970).

Fig. 10-4. The same two types of neurons as in Fig. 10-3, but fixed in glutaraldehyde-formaldehyde and postfixed in osmium tetroxide. The cytoplasm of the first type of neuron (*t1*) is not so electron opaque but has a leached nucleus. The cytoplasm of the type 2 neuron (*t2*) still exhibits swollen areas of the endoplasmic reticulum (*arrows*).

Fig. 10-5. A similar field to the previous figure but fixed in osmium tetroxide. Type 1 neurons (*t1*) show better preservation than in Fig. 10-4 but the cell appears swollen with poorly preserved nucleoplasm. The nucleoplasm of type 2 cells (*t2*) is also poorly preserved and the endoplasmic reticulum is grossly swollen (*arrows*).

Fig. 10-6. Types 1 and 2 neurons fixed in a mixture of glutaraldehyde (2%) and osmium tetroxide (1%). The general preservation of both neuronal types compared with the three previous figures seems much better. The cytoplasm in particular seems well preserved with the endoplasmic reticulum of the type 2 cell showing good structural detail (*arrows*).

Embedding

Compared with the early days of electron microscopy when only methacrylate (Plexiglas) was available for routine embedding of biological material, there is today an almost bewildering choice of embedding media. I only mention methacrylate out of historic interest because with the exception of one or two specialized problems, it is no longer in wide use. It has been supplanted by the epoxy resins in one form or another, although in some laboratories the polyester resins are used to yield excellent results. However, if epoxy resins are used in combination with the latest embedding techniques the beginner need have little fear that the quality of preservation will be improved by changing to a different type of embedding medium.

The major problem with epoxy resins is caused by their relatively high viscosity, which can make some specimens difficult to impregnate. However, some resins such as Epon 812 and notably that of Spurr (1969) do have a relatively low viscosity and are to be recommended in cases where tissues have been found difficult to infiltrate. In dealing with insect nervous tissue, there is, apart from cuticular sense organs, little problem with impregnation, and almost any epoxy resin can be expected to give satisfactory results. Most of my experience is with the epoxy resins Araldite and Epon with a preference for an Epon 812 formula developed by Manton (1964), which differs somewhat from the Epon 812 mixtures recommended by Luft (1961), which are in such wide use today. Certainly I would strongly urge those about to start electron microscopy to use some form of Epon embedding medium, as it has a fairly low viscosity and gives excellent preservation.

Epoxy resin embedding media are made up by mixing a variety of components. The casting resin itself such as Epon and Araldite is mixed with one or more hardeners, the type and amount of which control the hardness of the embedding mixture. To this resin/hardener complex is added an accelerator to initiate hardening or polymerization and sometimes a plasticizer, which modifies the hardness and hence the cutting properties of the embedding medium. Four epoxy resin embedding mixes are given below which have been used successfully in our laboratory (Table 10-1). Although there are many other formulations available I have little personal experience with them, but the interested reader should consult Glauert (1975) for these. A list of suppliers of these embedding components is given in Appendix II.

Most embedding media are not miscible with water and hence, before impregnation can commence, the tissues must be dehydrated. This is usually carried out in ethanol or acetone; ethanol is more widely used because pure acetone is difficult to keep in a dehydrated form since it takes up water from the atmosphere. After fixation and postfixation tissues can

Table 10-1. Formulae of Epoxy Resin Embedding Media[a]

Author	Epoxy resin	Hardener	Plasticizer	Accelerator
Luft (1961)	Epon 812 (4.5 ml)	DDSA (2.7 ml) + MNA (2.8 ml)		DMP 30 (0.25 ml)
Manton (1964)	Epon 812 (3.7 ml)	DDSA (5.3 ml) + MNA (1.0 ml)		DMP 30 (0.25 ml)
Glauert and Glauert (1958)	Araldite CY212[b] (5 ml)	DDSA (5 ml)	DBP (0.5 ml)	DMP 30 (0.25 ml)
Mollenhauer (1964)	Epon 812 (2.5 ml) + Araldite 506 (3.3 ml)	DDSA (4.2 ml)	DBP (0.15 ml)	DMP 30 (0.25 ml)

[a] *DBP*, dibutyl phthalate; *DDSA*, dodecenyl succinic anhydride; *DMP 30*, 2,4,6-tridimethyl-amine methyl phenol; *MNA*, methyl nadic anhydride.
[b] Araldite CY212 = Araldite M.

be transferred directly to 70% ethanol and then run through 90% and two changes of absolute ethanol. In our laboratory we also pass our material through dried alcohol (over anhydrous copper sulfate). This step can be omitted if propylene oxide (see below) is used as a transition solvent since this compound is miscible with water and can be used to complete the final stage of dehydration. The times the tissues spend in alcohol are not critical and indeed the 70% stage can be used to store the tissues if the embedding schedule has to be interrupted for overnight periods, etc. A minimum of 10 to 15 min in each alcohol is quite adequate for small pieces of tissue, e.g., < 1 mm^3. Above this size 30 min in each concentration of alcohol is probably sufficient. If the tissues are to be embedded in an epoxy resin they should be transferred through two changes of propylene oxide (Luft, 1961), which is an epoxy resin of very low viscosity, of about 15–30 min each. Infiltration of the embedding medium can now commence. This can be done by soaking the tissues in ascending concentrations of the embedding medium in propylene oxide. Three mixtures are usually sufficient with ratios of embedding medium:propylene oxide of 1:3, 1:1, and 3:1, and tissues are left in each of these mixtures for about 1 h with constant agitation before placing in the pure embedding medium. The tissues are steeped in the embedding medium for about 1–2 h, again with agitation, before placing in fresh embedding medium in small containers such as gelatin or plastic capsules before polymerization in an embedding oven.

Some specimens, such as ganglia with associated nerves, are better embedded in flat containers. Suitable molds can be bought for this purpose, but any small, flat container will do, such as plastic snap-tops from specimen vials.

A far better method for infiltration, which cuts down drastically on embedding times and which eliminates impregnation problems even with very large chunks of tissue, is that of vacuum impregnation. This was shown to me 12 years ago by Dr. D. E. Philpott at the Ames Research Center in California. Tissues in propylene oxide are placed in a mixture comprised of equal volumes of propylene oxide and resin for 30–45 min. The resin in this mixture and that used for subsequent embedding are "outgassed" before use by placing under vacuum in a vacuum-embedding oven at room temperature for about 1 h. The vacuum is produced by an ordinary, inexpensive rotary pump. Do not use a water aspirator as this will introduce water into the epoxy resin and prevent its proper polymerization. After soaking in the Epon/propylene oxide mixture the tissues are placed in fresh (outgassed) resin in embedding containers. The containers are put into the vacuum oven, which at this stage should be at room temperature. The oven is evacuated and at the same time the heating elements are switched on. As the air pressure is reduced and the temperature in the oven raised, bubbles will probably form in the embedding medium where it surrounds the tissues. If bubbling is too vigorous the tissues may be thrown out of the embedding container. This can be prevented by slightly reducing the hardness of the vacuum. However, if tissues are ejected, air is let into the oven, the tissues returned to their embedding containers, and the whole vacuum process restarted. The important step in this technique is to increase progressively both the hardness of the vacuum and the temperature of the embedding medium over 30 min. The oven thermostat should be set at 60°C or whatever temperature is required to polymerize the resin. After this 30-min period, air is let into the oven and the resin is left to polymerize overnight or longer.

The rationale behind this method is that by reducing the atmospheric pressure and raising the temperature of the embedding medium, any propylene oxide left in the tissues after soaking in the resin/propylene oxide mix will be boiled off (the boiling point for propylene oxide at normal atmospheric pressure is 33–36°C) within the 30-min period. When the embedding mix is returned to atmospheric pressure, resin will be forced into spaces vacated by propylene oxide, thus ensuring uniform and complete impregnation of the specimen.

Problems with Epoxy Resins

As previously hinted, infiltration of tissues by the resin can be a problem, but this is completely eliminated in my experience if vacuum impregnation is employed. Where we have had problems with vacuum impregnation, they have usually been caused by the heated resin being left under vacuum for periods in excess of 30 min. In this instance partial polymerization of the resin can take place before it is returned to atmospheric pressure leaving "holes" in the tissues.

Another source of trouble is that resins are left too long at room temperature (several hours) before using them. The resin mix becomes more viscous as a result of partial polymerization and does not impregnate the tissues properly. In the techniques mentioned here, all the soaking mixtures of resin include the accelerator, and the "shelf" life of these mixes is therefore limited. For this reason I advocate the use of only freshly prepared resin mixes. Some workers suggest making up large quantities of the embedding medium and storing it in airtight containers at $-20°C$ where it can be kept for several months. I do not recommend this practice. Each time resin is needed it has to be warmed up to room temperature before use. If the seal is broken before the resin has equilibrated to room temperature water may condense on the resin and impair polymerization. Moreover, unless a careful record is kept there is no way of knowing how often the resin mix has been warmed up, or for how long it has been kept in total at room temperature. One way around these problems is to store the resin in glass ampules of about 10 ml capacity. Thus each ampule is only returned to room temperature once, and any remaining resin is discarded. In spite of this, I cannot emphasize too strongly that it is safer to use a freshly prepared mixture for each embedment.

Many components of epoxy resin mixtures take up water from the atmosphere and while some resins such as Araldite are not greatly affected by this, others such as Epon will not polymerize properly. Even gelatin capsules, which are used as containers for embedding, can imbibe enough moisture from the air to prevent hardening of the resin. Therefore, we dry our capsules over phosphorous pentoxide or other desiccants before use. This problem can of course be completely eliminated by using plastic capsules of the BEEM type. It is important to keep bottles or tins that contain embedding components tightly stoppered when not in use, but in any case, we only keep our resins for three months and exchange all the components after this time even though they may still be producing good blocks.

It is important for good embedding to mix the constituents thoroughly; this is particularly so for small-volume components such as plasticizers and accelerators. Araldite is fairly viscous and some authors recommend warming the components to improve mixing. We always mix our resins at room temperature in glass specimen vials that have snap-on plastic tops. Mixing is carried out with a magnetic stirrer, and the small bar magnet that is used as the agitator is placed directly in the vial with the resin mix. We measure out and introduce the resin components into the mixing vial with plastic disposable syringes. The resin and hardener(s) are first added and thoroughly mixed before adding the plasticizer (if needed) and the accelerator. Both plasticizer and accelerator are added from a 1-ml syringe, and in adding these, we squirt them from the syringe below the surface of the mix while it is being stirred as vigorously as possible. This method ensures rapid and even dispersion of these components throughout the whole of the resin mixture. After mixing is completed the bar magnet can be removed and cleaned in alcohol for future use. The use of disposable syringes and specimen vials means that no cleaning of glassware is needed after each embedding. Some resin components can cause skin irritation so we wear disposable plastic gloves when measuring out and mixing the resins.

Special Epoxy Resin Mixtures

Sometimes large-area sections > 1 mm^2 may have to cut. Epon, particularly if glass knives are used, may prove difficult to section at these larger areas. In this case I can recommend the Araldite/Epon mixture of Mollenhauer (1964) as shown in Table 10-1.

Insect cuticle, being hard, is not easy to embed and section. I have found that Araldite CY212 (Glauert and Glauert, 1958) holds or binds to cuticle far better than Epon because it appears to cure to a harder block. I have successfully cut sections of insect epidermis and cuticle using this particular epoxy resin. Thus if problems in sectioning insect cuticle are encountered with Epon, I certainly think it is worth trying Araldite as an alternative embedding medium.

Embedding Small Specimens

On occasions tiny pieces of nerve or small sense organs will have to be embedded. This is best done by fixing, dehydrating, and impregnating the tissues in situ before cutting them free for final embedding. Removal of small pieces of nerve prior to impregnating them with resin will almost certainly produce curling or other mechanical distortions, which makes orientation of these small specimens for sectioning almost impossible. Tiny pieces of tissue are best left tagged to small pieces of cuticle that can then serve as a convenient ''jig'' for holding the tissues straight and as a ''handle'' for transferring them from one solution to another without getting lost. I have successfully embedded insect stretch receptors using this technique that are essentially connective tissue strands some 1 mm long by about 15 μm diameter. Even after these precautions, sectioning desired area from tiny specimens can prove difficult. In this case I embed them in small flat containers of about the same size and depth as a microscope slide, where they can be orientated precisely under a binocular stereomicroscope before polymerization of the embedding medium. If tissues are vacuum-impregnated, final orientation or their separation from bits of cuticle is best completed after the vacuum treatment. Once polymerization is complete, the flat ''slide'' of embedding medium can be placed directly under an ordinary light microscope and tissues can be scrutinized to determine the exact areas to be sectioned. For example, some time ago, I had to section individual nerve cells that lay alongside peripheral nerves. This method enabled me to locate and section these cells with little difficulty.

Sectioning

This is probably the single most difficult stage in preparing material for examination in the electron microscope. A wide choice of ultramicrotomes is available for producing ultrathin sections, and many laboratories no doubt have several types available. Broadly speaking, these fall into two major groups, those with a mechanical advance mechanism (e.g., Porter-Blum, Huxley, and the recently introduced Reichert Ultracut) and those with a thermal advance system (e.g., L. K. B., and older Reichert models). Every worker has his favorite type of microtome, and although all types currently available are capable of producing excellent sections my own personal preference is for one with mechanical advance because I find them less complicated yet more flexible in use. If a new microtome is to be purchased I would recommend one with a mechanical advance. In any event they are usually much less expensive than the types with thermal advance.

Whatever microtome is purchased, it should be situated on a sturdy bench or table that stands on the floor in a part of the building that is free from mechanical vibration. It should also be housed in an area that is free of draughts or thermal air currents. In other words do not place the microtome near a door, window, or heating system. If no suitable room is available a part of the electron microscope laboratory can be enclosed to form a small cubicle with a floor area of about 4 × 4 ft. The top of the cubicle can be left open for ventilation, but the bottom of the structure should extend to the floor to minimize convection currents. If space permits several such cubicles can be built side by side, to house individual microtomes that can preferably be of different types. Another desirable feature of the cubicles is to fit them with fluorescent lighting whose brightness can be adjusted via a dimmer switch to suit the requirements of each operator.

Glass knives are used routinely throughout the world for cutting sections and are perfectly satisfactory for almost all work. Diamond knives are probably better for producing very thin sections (i.e., below 50 nm) for high-resolution work, for producing long unbroken ribbons of serial sections, and for cutting large-area sections. However, diamond knives are very expensive and for this reason alone are not suitable for beginners since the cutting edge is very easily damaged with careless use. Furthermore these knives need to be cleaned after use, and this can be quite a delicate operation for which reference should be made to the manufacturer's instructions. I make the assumption that the beginner will be using glass knives, and for producing these I cannot do better than recommend the L. K. B. Knifemaker. I know of no better piece of equipment for making glass knives and consider it indispensible in any electron microscope laboratory.

Before the glass knife is mounted in the microtome, a trough must be mounted on it for holding the liquid on which the sections will float after cutting. Various types of troughs are used, but we prefer one made from adhesive tape (Scotch Tape, 4890-33 from L. K. B.). The tape is stuck to the knife and sealed with dental wax as shown in Fig. 10-7. Care must be taken when doing this not to touch the knife edge. The tape is best cut with a razor blade at an angle so that it only contacts the front of the knife where the cutting edge intersects it. This will leave an area between the tape and the front of the knife for sealing the edges of the tape with dental wax (Fig. 10-7). Dental wax is best applied in the form of small chips about 1 to 3 mm³ in volume that have been cut from the wax sheets. Two or more of these chips are placed on the heel of the trough, and these are melted with the aid of a hot metal probe so that the wax runs and seals all the joints (Fig. 10-7). During this operation the wax must not be allowed to run onto the knife edge. Sometimes, despite adequate sealing, the tape springs away from the sides of the knife. This of course always happens during sectioning and the trough fluid leaks out, carrying the sections with it. This can be prevented by warming the adhesive tape by passing it fairly rapidly over a spirit flame before sticking the tape to the knife.

Once the trough has been constructed the knife is clamped into the knife holder on the microtome, care being taken not to disturb the wax seals. The most important point to observe in clamping the knife is to obtain the correct clearance angle (Fig. 10-8). For most sectioning, this should be as small as possible and certainly no larger than 5° (see below).

The preparation of the specimen block for sectioning deserves some comment. It cannot be overemphasized that the smaller the size of the section the easier it is to cut. In order to achieve this the blocks must be trimmed to an area of 1 mm² or even less if possible. Furthermore it is

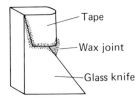

Tape

Wax joint

Glass knife

Fig. 10-7. Glass knife with tape trough showing details of wax seal between the edge of the tape and glass knife.

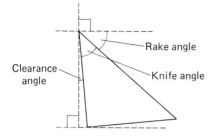
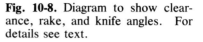

Fig. 10-8. Diagram to show clearance, rake, and knife angles. For details see text.

better to trim all embedding material away from around the specimen. This avoids difficulties caused by differences in hardness between the embedding medium and the specimen. Leaving a zone of embedding medium around the specimen not only unnecessarily increases the area of the sections, but can cause frustration later on when the sections are examined and prove mainly to be devoid of one's specimen.

Most people trim blocks to the form of a truncated pyramid (Fig. 10-9), and this is the shape that I would recommend since it gives best support to the specimen during sectioning. The only disadvantage is that as sections are cut they become progressively larger. However, this is not really a nuisance unless very large numbers of sections are needed. Some workers overcome this by trimming the block in the form of a mesa (Fig. 10-9). This is fine for large block faces, but is clearly not satisfactory for small specimens because the raised mesa will be prone to vibrations causing chatter (see below) in the sections.

If orientation of the specimen is not critical, the blocks can be removed from their gelatin or plastic moulds, placed directly in the microtome chuck, and trimmed to the desired size. I do not recommend dissolving away gelatin capsules with warm water, particularly if Epon has been used, as the water can soften the blocks making them impossible to section. A better way is to cut them free of the capsules with a scalpel. If orientation of the specimen is critical, it is best to cut a chunk of embedding medium containing the specimen from the block with a fine saw. The remaining stump of the block can be sawed flat and used as a holder on which the specimen can be mounted at the correct angle. Two

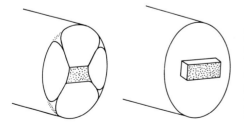

Fig. 10-9. Diagram to show two common shapes to which specimen blocks are trimmed for production of ultrathin sections. Block on the *left* has been shaped to a truncated pyramid. Block on the *right* has the form of a raised mesa.

methods have proved useful in sticking the specimen to this block.   The first involves the use of sealing wax that can be obtained from Post Offices or stationery shops in the U.K. and from stationery shops in the United States.   However, if sealing wax is unobtainable, a mixture of carnauba wax and paraffin wax can be used, the proportion of carnauba being 25%– 33% (Pease, 1964).   A bed of molten wax is formed on the sawed face of the block and the piece of embedding material containing the specimen block is placed in this wax and held at the correct angle until the wax sets firmly.   This wax joint can be further strengthened by adding more molten wax.   This is done with the aid of a warm, metal rod.   The advantage of this method is that is is quick, and if mistakes are made the wax is easily remelted and the specimen repositioned.   Even if the joint is broken it is a simple matter to remake it.

The second method is similar to the first except that a quick-setting epoxy glue is used instead of wax.   This may give a stronger bond, but if mistakes in orientation of the specimen are made it is a more time-consuming process to rectify.

Specimens that have been flat-embedded can be sawed from the bulk of the embedding material, and if the saw cuts are parallel, the specimen block can be mounted directly in a vise-type specimen holder.

Trimming of specimen blocks is best done with a razor blade.   If the two-edged variety is used it is better to break the blade in half and use each edge separately to avoid trimming one's fingers.   The blade is easily broken while it is still in its protective paper wrapper.   The condition of the blade is not critical for rough trimming of the block, but the last few facing cuts of the truncated pyramid should be made with a new blade. This ensures that the edges of the pyramid are perfectly straight and un-contaminated with bits of embedding material.   Grease from a new blade can be removed by dipping it in a solvent such as xylene.   The final shape of the block face should be made rectangular or trapezoid (Fig. 10-9).

Some manufacturers recommend the use of their microtomes to trim the specimen block to the correct shape.   I must confess, however, that most people in our laboratory trim their block, free-hand, with a razor blade.   One word of warning can be given in connection with mechanical aids to trimming.   If glass knives are used to trim the block, extra care should be taken in making the final cuts.   These should be made with a new glass knife and should be very thin cuts.   If an old knife is used, or if thick cuts are made with a new knife, small pieces of glass may break off the knife and embed themselves in the specimen block.   Thus when cutting ultrathin sections, these pieces of glass can damage the knife edge and ruin the sections.

Once a suitable specimen and knife have been mounted in the microtome, sectioning can begin.   The precise setting up of the microtome, and the method of aligning the block face with the knife edge, will vary according to the design of the microtome, and for these procedures it is best to consult the instruction book. Do ensure, however, that the block face of the specimen is exactly parallel with the knife edge, and with a rectangular-shaped face, the longer sides of the rectangle are usually set horizontally. With a block that has a trapezoid face, the two parallel sides are also set horizontally, but the longer of the two parallel sides is arranged so that it will contact the knife edge first. Thus as each section is cut, it will press evenly upon the trailing edge of the previous section and detach it cleanly from the knife edge with a minimum of wrinkling. Check that there is no grease or dust, or other contaminant, on the block face or knife edge, as these may damage the edge or encourage ''wetting'' of the block face by the trough fluid. Check, too, that all clamps holding the specimen and knife are correctly tightened. When adding the fluid to the knife trough, raise the specimen well clear of the knife edge before adding it. This reduces the risk of wetting the face of the block.

The composition of trough fluid is a matter of personal preference and to some extent experience. Some workers use distilled water, others a mixture of 10%–20% alcohol or acetone in distilled water. We routinely use a 10% alcohol solution for sectioning epoxy resins with glass knives. The fluid is easily added from a clean syringe. The level of the fluid is adjusted so that it fully wets the knife edge. This generally entails adding excess fluid so that its surface is distinctly convex and raised above the level of the knife edge. Following this the fluid level is lowered until it has a slightly concave surface and is more or less level with the edge of the knife. Final slight adjustments to the fluid level, microtome, and light and binocular microscope are made so that a silvery-white reflection is obtained from the surface of the trough fluid particularly at the knife edge. This surface reflection is not only important for judging the quality of sections as they are cut, but also for assessing their thickness. The thickness is estimated from the interference colors of the sections (Peachey, 1958). Sections with colors of silver to pale gold (60–90 nm in thickness) are adequate for most work, with gray to silver sections (<60 nm) being more suitable for high-resolution work and gold to deep gold sections (>90 nm) being useful for survey or low-magnification work. These interference colors are a much more reliable guide to section thickness than the setting indicated on the microtome. Once sections are being cut, the section thickness control of the microtome should be adjusted to achieve the desired interference color, with little notice being taken of the section thickness indicated by the microtome advance mechanism.

When all adjustments to the knife, specimen, and trough fluid are satisfactorily completed, the next job, which is probably the most tricky, is to slowly advance, in increments of 0.5 μm or less, the knife toward the specimen while cycling the microtome, until the first section is cut. This will undoubtedly be "thick" by ultramicrotome standards and will probably not show an interference color. If it does not extend across the full face of the block continue to advance the knife by small amounts (0.1 μm or less if possible) until the whole face of the block is cut. Then set the microtome to automatic advance and ultrathin sections should be produced. The first few sections may not be of even thickness, but after 5–10 have been cut the microtome system should settle down and produce sections of constant thickness. If these are not of the correct thickness the advance mechanism should be adjusted to achieve this. Once enough sections have been cut, the microtome is stopped and the sections are "expanded" by holding a piece of filter paper soaked in chloroform over the surface of the trough fluid for several seconds. This should remove any compression stains or wrinkles that are present in the sections.

Problems in Sectioning

Most problems encountered in ultrathin sectioning are generally easy to correct, but unfortunately, the correct solution is often a matter of experience. With epoxy resins, if the clearance angle of the knife is small, i.e., 1–2°, and the knife edge is good, the sectioning speed is probably at fault. In general it has been my experience that most microtome manufacturers have tended, in the past, if not now, to recommend too high a cutting speed for these resins, and changing to a slower cutting speed, e.g., at or below 1–2 mm/s may bring about an improvement in sectioning. Sometimes, small alterations of the angle that the trough fluid makes with the knife edge may also help. If sections are cut alternately thick and thin, then either the knife is of poor quality or the section advance may be too fine. Increasing the microtome feed may solve this difficulty. If all this fails, and assuming that the epoxy resin is cured properly and the knife edge is good, it may simply be that the block face is too large for the type of specimen being cut. Difficult material can often be successfully sectioned after reducing the area of the block face.

Wetting of the block face or dragging of sections over the knife edge during the cutting stroke is generally caused by contamination of the knife edge or block, but can be equally due to too slow cutting speed or too high a level of the trough fluid. A wet block face can be dried by touching it with a piece of filter paper, and risk of further wetting can be reduced by lowering the fluid level of the trough at least until the first few sections have been cut.

Variations in thickness within individual sections may be caused by too large a block face, a poor knife edge, or a poorly embedded specimen. It may also be due to loosely tightened specimen or knife clamps, although in this case the fault is often accompanied by considerable variation in thickness between individual sections. Regular variations in thickness within individual sections, if these are not related to structural features of the specimen, can be caused by vibrations reaching the microtome from outside sources. Unevenness in individual sections may not be a problem, if for example the tissues are cut at an acceptable thickness and the surrounding embedding medium at another. This may occur when sectioning small nerve fibers. Sometimes the specimen block is too soft or has areas of reduced hardness within it. The simple solution to this is to fix and embed fresh material. However, a poor block may be saved by further heating in the embedding oven. If this does not work and the material is irreplaceable or unique, soft blocks can be sectioned by placing in a refrigerator at 5°C for 0.5–1 h (not mounted in the microtome chuck!). After this treatment the blocks are speedily mounted and sectioned without delay, and it may be possible to produce sufficient sections before the block warms up and becomes too soft again. I have even heard of people sectioning very soft blocks with the aid of a cryoultramicrotome. Occasionally during very hot spells the air temperature may rise above 30°C and under these conditions we have found difficulty in sectioning epoxy resins. The answer is to have adequate air conditioning in the laboratory, but failing this, cooling the blocks as indicated above may help. Chatter, that is, variation in section thickness caused by high-frequency vibrations between the specimen and the knife, is usually traced to too high a cutting speed, too large a clearance angle of the knife, too soft a specimen, or a combination of all three. Try reducing the clearance angle and cutting speed before blaming the specimen. Scratches in the sections or rough-surfaced sections are caused by a poor knife. The remedy for this is obvious.

Collection of Sections

Once sufficient sections have been cut and expanded they can be mounted directly onto specimen grids.

Some microtomes have special attachments for holding the grids while sections are collected. However, grids can be picked up by holding their edge between the points of a pair of stainless watchmaker's forceps. The forceps should be equipped with a rubber ring that will provide the correct pressure for gripping the grid while sections are being gathered. Forceps with straight tips can be used, but I prefer those with curved tips.

The most convenient way, but not necessarily the best, to pick up sections is to raise the level of the trough fluid so that it has a distinctly convex surface. The sections are moved with an eyelash mounted on a matchstick so that they occupy the highest point on the liquid surface, and then the grid is dabbed (supporting film downward if one is used) gently onto the sections. The grid is then lifted away from the trough fluid and the excess liquid blotted off, whereupon the sections should be found adhering to the supporting film or naked grid. A disadvantage of this method is that it can cause considerable folding of the sections. A better way is to place the grid vertically into the trough fluid until only the edge of the grid help by the forceps is above the surface. A ribbon of sections can then be pushed with the aid of a mounted eyelash, so that one end of it sticks to the rim of the grid where it lies above the liquid. The grid is then slowly lifted from the trough and the ribbon of sections should adhere to the grid and lie evenly on its surface. If sections are not in the form of ribbons, but form a randomly orientated collection on the trough fluid, the vertical raising technique cannot be so easily used. In this situation the grid is placed horizontally (film-side up) just beneath the surface of the trough fluid and the sections are maneuvered to lie above it (it may be necessary to bend the rim of the grid that is held by the forceps to hold the grid level beneath the trough liquid). The grid is then lifted, still in its horizontal position, straight up through the surface of the liquid, carrying the sections with it. The surplus water on the grid is gently blotted so that the sections dry down evenly onto the supporting film.

Whatever method is used to pick up sections, it is essential that all remaining fluid on the grid and tips of the forceps should be removed before releasing the grid from the forceps. If this is not done, surface tension forces may pull the grid between the forceps or cause it to stick to one of the forceps tips with the risk of contaminating the sections. To prevent this we push a piece of filter paper between the forceps blades immediately behind the point where they grip the grid. This effectively completely dries the area where the grid is held, facilitating its release once the grip of the forceps is relaxed. The grids with their sections can then be stored in grid boxes prior to staining or examining in the electron microscope.

It is of prime importance to ensure when picking up sections from the surface of the trough fluid that the forceps tips are perfectly clean. If not, any dirt or grease on them can so severely contaminate the sections as to render them useless. The remedy is to degrease the forceps tips before each grid is picked up by dipping them into chloroform.

Supporting Films

Although sections can be mounted directly onto electron microscope grids, I recommend that a supporting film be used particularly if grids with large or hexagonal holes are used. I prefer the hexagonal type because it gives the largest transmission area combined with maximum support for the sections. Several types of plastic films are available, of which we prefer Formvar in our laboratory. A recently developed film Pioloform F (Whacker-Chemie) seems promising and gives tougher films than Formvar, but we have gained the impression that this film deteriorates more rapidly with storage or after being irradiated in the electron microscope than does Formvar.

We make Formvar films from a solution of 0.5%–1% of the plastic in chloroform. The solution is stable for many months if kept out of direct sunlight. In casting the films, the solution is poured to a depth of about 3–5 cm in a measuring cylinder whose internal diameter is such that a 76×26 mm standard microscope slide can be dipped directly vertically into the solution. The slide, which should be a new one, is wiped free of fluff and dust with a clean, dry cloth or tissue. A crocodile clip, to which a length of cotton thread or twine is attached, is fixed to one end of the slide, which is then lowered by means of the thread into the Formvar solution, to a depth of 2–3 cm. The slide is held vertically in this solution for several seconds, and then raised slowly at an even speed out of the solution and held above it in the measuring cylinder for 15–30 s, until the excess chloroform has drained from the slide. The slide is next slowly removed from the measuring cylinder and allowed to dry in the air. This should take about 30 s. Once dry the film on both sides of the slide is scored with a coarse needle or forceps tips. The score lines should be made about 1 mm away from and parallel to each of the three slide edges that the film covers. Next the film is ''loosened'' by breathing on it so that it becomes frosted. The slide is then held vertically, and the narrow edge of the slide that is coated with the film is lowered onto the surface of distilled water contained in a lotion bowl or similar vessel. When this end of the slide has contacted the water surface, waggle it gently to detach the edges of the film from both sides of the slide. The detached film will be seen as a silvery coating on the surface of the water when viewed by reflected light. Daylight is best for this purpose, but failing this, light from a fluorescent tube is almost as good. Once the bottom edges of the two films are free the rest of the film should follow as the slide is pushed further beneath the surface of the water. When both films are floating freely, they can be moved away from the slide by blowing or teasing with a fine needle. The slide can then be removed from the water. Specimen grids with their light or matt side downward can be placed on the supporting film with the aid of fine forceps. It is better to bend the grids slightly by bowing them between forceps tips and to put the convex side down on the supporting film than to place them on perfectly flat. This will prevent the film from being pulled through the grid holes and broken when it is removed from the water surface. This is done with another clean microscope slide by touching it onto the supporting film, and pushing it down-ward below the surface of the water carrying the film with it. The slide is then moved through an arc and brought up through the surface of the water with the side to which the film is stuck emerging first. This process is done with one continuous motion and with practice is far easier to ac-complish than describe. If done correctly, the film should lie flat on the surface of the slide, covering the grids and holding them firmly in place. Excess water on the slide and film is blotted off with filter paper. When dry the coated grids can be left on the slide until they are required. Alter-natively, the film bearing the grids can be removed from the surface of the water by laying a piece of filter paper over it. The paper is then carefully pulled away from the water surface carrying the film and grids with it and laid aside to dry.

Carbon films are advised for high-resolution work and as an additional coating to plastic films to stabilize them. This minimizes drift and creep of plastic films when they are in the electron microscope. I do not gen-erally use carbon films or carbon-coated plastic films for two reasons. First, sections do not adhere as well to carbon films as they do to plastic. Second, unless the carbon films are carefully made they may be contami-nated with carbon chips. In any case drift and movement of plastic films in the electron beam are of little importance these days if relatively fast shutter speeds of 0.5–1 s are used. Nevertheless carbon films have to be used for special problems, and if they have to be made, consult Kay (1963) for appropriate methods of manufacture.

Semithin Sections for Light Microscopy

It is general practice to cut sections from specimen blocks of the order of 0.5–2 μm for examination by light microscopy. This enables one to select the correct area in the block for ultrathin sectioning and to ascertain if the material is oriented correctly. In addition these sections can give a reasonable guide to the quality of preservation of the tissues.

Semithin sections are cut on a dry, glass knife on an ultramicrotome and transferred from the knife with a fine paint brush to a droplet of water on a clean microscope slide. Several sections can be placed on this droplet, which is next warmed by passing the slide over a spirit flame. As the water warms up it softens the plastic sections, allowing them to spread out and remove all wrinkles and folds. The warming process is continued until all the water has evaporated, leaving the sections firmly stuck to the slides. All one need do now is cover the sections with a droplet of immer-sion oil and examine them directly under a phase contrast microscope.

Usually, sufficient detail can be seen without further treatment to assess adequately the orientation and quality of preservation of the material. In spite of this some workers prefer to stain their sections for this type of examination. This is easily achieved by drying the sections as before, covering them with 1% toluidine blue in a 0.1 M phosphate buffer at pH 7.0–7.4, and heating the slide for 5–8 min at 85°C. The stain is washed away with tap water and the sections are dried before covering them in immersion oil for examination by light microscopy.

Details of other stains currently in use for semithin sections are described by Hayat (1975).

Staining Ultrathin Sections

Material embedded in epoxy resins has very low contrast in the electron microscope. Contrast can be enhanced after fixation by soaking the tissues in a salt solution of a heavy metal such as uranyl acetate (see p. 211), a method known as "en-block" staining. Even where this method is employed it is usual to impart even greater contrast by staining the material once it has been sectioned. If en-block staining is not used then staining of ultrathin sections is a necessary and routine procedure for obtaining maximum contrast and resolution in electron micrographs. A fairly wide range of "electron" stains for sections is currently available and all are salts of heavy metals. Two in particular are widely used either singly or sequentially as a "double" stain. These are uranyl acetate and lead citrate.

Staining can be a major source of contamination of sections and indeed can be so bad as to make the sections useless. By adopting a technique developed by Locke and Collins (1965) we have found that contamination is reduced to a very low order. This involves wedging grids bearing sections into slits cut into rings made made from 2- to 4-mm slices of polyvinyl chloride tubing (bore and wall thickness 5 × 1.5 mm from Gallenkamp Ltd.). Three vertical slits are made at 120° intervals in the inner surface of each ring to a depth of half the wall thickness (Fig. 10-10). Each slit is loaded with a grid by compressing the ring with forceps at positions adjacent to and opposite to the slit. This forces it to gape open allowing a grid to be inserted (Fig. 10-10). The pressure from the forceps can now be relaxed, allowing the plastic ring to return to its natural shape and close up the slit which then grips the rim of the grid tightly (Fig. 10-10). Once all three grids are loaded into each ring the latter serves as a "handle" for transferring the grids through the staining solutions.

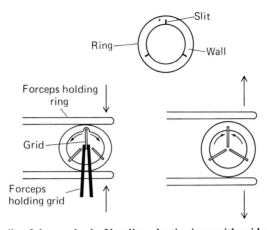

Fig. 10-10. Details of the method of loading plastic rings with grids for subsequent staining of sections with uranyl acetate and lead citrate. *Upper portion* of the figure shows a plastic ring with three slits cut into the inner wall. *Bottom left* figure shows how the ring is compressed with a pair of forceps (*vertical arrows*) to spring apart a slit (*curved arrows*) so that a grid can be inserted with a pair of watchmaker's forceps. The *bottom right* illustrates that when forceps holding ring are released (*vertical arrows*) the plastic ring returns to its original shape closing the slit (*curved arrows*) to grip the specimen grid.

Staining Procedure

A stock solution of 30% uranyl acetate in methanol (Stempak and Ward, 1964) is made up and this can be stored for several months and used repeatedly. The rings holding the grids are placed in a 10-mm-diameter glass filter tube (H/3790/02 from R. W. Jennings and Co.). One end of the glass tube is connected by plastic tubing to a disposable syringe. By means of the syringe the uranyl acetate solution is drawn up through the filter to cover the grids. Staining is carried out for 3–7 min and then the uranyl acetate is forced back through the filter into the stock bottle. Next the grids are washed 5 times for several minutes, each time in clean methanol that has been drawn up through the filter. After the fifth wash the grids are air dried and then stained in lead citrate (Reynolds, 1963). This is made by dissolving 1.33 g of lead nitrate [$Pb(NO_3)_2$] and 1.76 g of sodium citrate ($Na_3C_6H_5O_7 \cdot 2H_2O$) in 30 ml of carbon dioxide-free distilled water. The mixture is shaken vigorously for a full minute and then allowed to stand with intermittent shaking for 30 min. Then, 8 ml of 1 N sodium hydroxide are added, and the mixture is made up to a total volume of 50 ml, again using carbon dioxide-free distilled water. The pH of this solution should be 12 and it should be perfectly clear. This stain will keep for several months if stored in an air-tight bottle in a refrigerator.

For staining in lead citrate, the rings containing the grids are placed horizontally on slabs of clean dental wax in a petri dish. A few pellets of sodium hydroxide are also placed in this dish. A clean pipette is used to disburse the lead stain from the stock bottle into the center of each ring, enough solution being added to completely cover the grids (Fig. 10-11). The top of the petri dish is put on and the grids left in the lead citrate for 3–7 min. The purpose of the sodium hydroxide is to remove carbon dioxide from the air, thus minimizing the formation of lead carbonate, which is insoluble and will precipitate over the sections. For this reason care must be taken during lead staining not to breathe over the staining solution. When staining is complete, each ring is picked up in clean forceps and washed in distilled water from a wash bottle for 1 min. It is then washed in the same way for 10 s in 0.001% NaOH in distilled water and in distilled water only for 1 further minute. Finally the grids in their rings are allowed to dry on filter paper before storing in grid boxes. Material stained in this way is shown in Fig.10-1.

For high-resolution work I do not use lead stains because they have a rather coarse grain that is evident at higher magnifications. Neither do I use methanolic uranyl acetate because I suspect that it tends to attack the embedding medium. Instead I use a saturated solution (8%) of uranyl acetate in distilled water (Sjöstrand, 1967) and float the rings containing the grids in this solution. Staining is carried out for 2–3 h at a temperature of 60°C. Following this the rings are transferred to distilled water at 60°C to prevent precipitation of the stain. Finally the grids are washed in a stream of distilled water (at room temperature) from a wash bottle for 1–2 min. This technique allows very good resolution of cell membranes (Fig. 10-12).

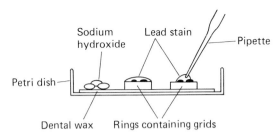

Fig. 10-11. Details of procedure for lead staining. The rings holding grids are placed on a piece of dental wax in a Petri dish. The stain is pipetted into the rings so that it completely covers the grids. Sodium hydroxide pellets are placed on the dental wax before the lid of the petri dish is put on to help prevent formation of insoluble lead carbonate in the stain.

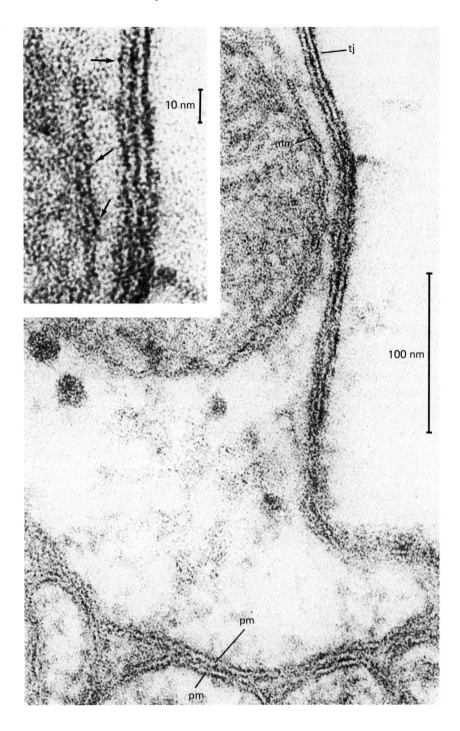

Appendix I: Buffer Solutions

Note: H_2O refers to distilled water.

1. 0.2 *M* Borate Buffer

   Solution A, 1.24 g boric acid in 100 ml H_2O
   Solution B, 1.9 g borax ($Na_2B_4O_7 \cdot 10\ H_2O$) in 100 ml H_2O

Soln. A (ml)	Soln. B (ml)	pH
90	10	7.4
85	15	7.6
80	20	7.8
70	30	8.0

2. 0.2 *M* Cacodylate Buffer

   Solution A, 4.2 g sodium cacodylate [$Na(CH_3)_2ASO_2 \cdot 3H_2O$] in 100 ml H_2O
   Solution B, 0.2 *M* HCl, i.e., 1.72 ml conc. HCl in 100 ml H_2O

Soln. A (ml)	Soln. B (ml)	pH
50	18.3	6.4
50	13.3	6.6
50	9.3	6.8
50	6.3	7.0
50	4.2	7.2
50	2.7	7.4

3. 0.2 *M* Citrate Buffer

   Solution A, 4.2 g citric acid in 100 ml H_2O
   Solution B, 5.9 g sodium citrate in 100 ml H_2O

Soln. A (ml)	Soln. B (ml)	pH
55	45	4.5
46	54	4.8
40	60	5.0
35	65	5.3
30	70	5.5
24	76	5.8
19	81	6.0

Fig. 10-12. High-resolution electron micrograph from the central nervous system of the blowfly larva. The material is fixed in glutaraldehyde followed by postfixation in osmium tetroxide. Staining is in an 8% aqueous solution of uranyl acetate. Note the detail of the plasma membranes (*pm*), the mitochondrial limiting membrane (*mm*), and the tight junction (*tj*). The *inset* shows that the resolution is limited by the size of the stain particles (*arrows*), which can easily be seen at this magnification. Material embedded in Epon. × 420,000.

4. 0.2 M Collidine Buffer

Solution A, S-Collidine (pure) 2.67 ml in 50 ml H_2O
Solution B, 1 M HCl, i.e., 8.6 ml conc. HCl in 100 ml H_2O

Soln. A (ml)	Soln. B (ml)	pH
50	11	7.2
50	10	7.3
50	9	7.4
50	8	7.5
50	7	7.6
50	6	7.7
50	5	7.8

Add H_2O to make up to a total volume of 100 ml for a 0.2 M buffer.

5. 0.2 M Maleate Buffer

Solution A, 2.32 g malearic acid + 20 ml of N (4%) NaOH in 50 ml H_2O
Solution B, 4% NaOH

Soln. A (ml)	Soln. B (ml)	pH
50	1.0	4.8
50	1.8	5.0
50	2.8	5.2
50	4.0	5.4
50	5.8	5.6
50	7.6	5.8
50	10.0	6.0

Add H_2O to make up to a total of 100 ml for 0.2 M buffer.

6. 0.2 M Sodium Phosphate

Solution A, 2.76 g monobasic phosphate ($NaH_2PO_4 \cdot H_2O$) in 100 ml H_2O
Solution B, 5.36 g dibasic sodium phosphate ($Na_2HPO_4 \cdot 7H_2O$) in 100 ml H_2O

Soln. A (ml)	Soln. B (ml)	pH
68	32	6.5
57	43	6.7
45	55	6.9
33	67	7.1
23	77	7.3
19	81	7.4
16	84	7.5
10	90	7.7

7. 0.2 *M* Phosphate Buffer (Sorensen's)

Solution A, 3.39 g dibasic sodium phosphate ($Na_2HPO \cdot 2H_2O$) in 100 ml H_2O
Solution B, 2.59 g monobasic phosphate (KH_2PO_4) in 100 ml H_2O

Soln. A (ml)	Soln. B (ml)	pH
40	60	6.6
50	50	6.8
60	40	7.0
70	30	7.2
80	20	7.4
90	10	7.7
95	5	8.0

8. 0.2 *M* Buffer (I have found this particular phosphate buffer to give excellent preservation of insect tissues)

Solution A, 2.72 g potassium dihydrogen orthophosphate (KH_2PO_4) in 100 ml H_2O
Solution B, 0.8 g NaOH in 100 ml H_2O

Soln. A (ml)	Soln. B (ml)	pH
78.1	21.9	6.5
72.5	27.5	6.7
67.6	32.4	6.9
65.8	34.2	7.0
59.5	40.5	7.3
56.8	43.2	7.5
55.0	45.0	7.7
53.2	46.8	8.0

9. 0.2 *M* Tris-Maleate Buffer

Solution A, 5.8 g maleic acid + 6.1 g tris(hydroxymethyl)aminomethane in 100 ml H_2O; approx. 1 g charcoal is added, and the solution is shaken, allowed to stand for 10 min, and then filtered
Solution B, N (4%) NaOH

Soln. A (ml)	Soln. B (ml)	pH
40	10.5	6.0
40	15.0	6.4
40	18.0	6.8
40	19.0	7.0
40	20.0	7.2
40	22.5	7.6
40	24.2	7.8
40	26.0	8.0

10. Veronal Acetate Buffer

Solution A, 1.94 g sodium acetate ($CH_3COONa\cdot3H_2O$) + 2.94 g sodium veronal in 100 ml H_2O
Solution B, 0.1 M HCl, i.e., 0.86 ml conc. HCl in 100 ml H_2O

Soln. A (ml)	Soln. B (ml)	pH
20	36	4.9
20	32	5.3
20	28	6.1
20	26	6.8
20	24	7.0
20	22	7.3
20	20	7.4
20	16	7.7
20	12	7.9
20	8	8.2

The buffer is made up to a volume of 100 ml by adding H_2O.

Appendix II: List of Suppliers

Most materials for fixation, embedding, sectioning, and staining of biological tissues can be obtained from the suppliers listed below. However, if requirements cannot be met by these firms, alternative sources of supply can be found by consulting Glauert (1975), Hayat (1975), Kay (1965), Pease (1964), and Reid (1975).

(a) United Kingdom

Agar Aids for Electron Microscopy
127a Rye Street
Bishop's Storford
Hertfordshire
(Also U.K. agents for Ladd)

EMscope Laboratories (EMscope)
99 North Street
London SW4 OHQ

Graticules Ltd.
Sovereign Way
Tonbridge
Kent
(Also U.K. agents for EFFA)

Polaron Equipment Ltd. (Polaron)
4 Shakespeare Road
Finchley
London N3 1XH
(Also U.K. agents for Polysciences)

Taab Laboratories (Taab)
52 Kidmore End Road
Emmer Green
Reading
Berkshire

(b) United States

Ernest F. Fullam Inc. (EFFA)
P.O. Box 444
Schenectady, New York 12301

Polysciences Inc. (Polysciences)
Paul Valley Industrial Park
Warrington, Pennsylvania 18976
(Also U.S. agents for Polaron)

Ladd Research Industries Inc. (Ladd)
P.O. Box 901
Burlington, Vermont 05401
(Also U.S. agents for Agar Aids)

Extech International Corp. (Extech)
177 State Street
Boston, Massachusetts 02109
(Also U.S. agents for Taab)

The following is a list of firms from which specialist supplies are bought chiefly for carrying out some of the techniques described in the text.

Gallankamp Ltd.
Frederick Street
Birmingham B1 3HT
England
(Plastic tubing for making rings to hold grids during staining)

LKB Instruments Ltd.
232 Addington Road
South Croydon
Surrey CR2 8YD
England
(Tape for making troughs on glass knives)

Jencon Scientific Ltd.
Mark Road
Hemel Hempstead
Herts
England
(Sintered glass filters for staining)

Smethhurst - High - Light Ltd.
420 Chorley New Road
Bolton
Lancs BL1 5BA
England
(Grids for mounting sections)

Whacker - Chemie GMBH
8000 - Munchen 22
P.O. Box 1
West Germany
(Pioloform F for making supporting films)

Bibliography

The following is a list of books that I have found excellent for describing the techniques of electron microscopy and should be consulted for methods that go beyond the scope of this book.

Glauert, A. M.: Fixation, dehydration and embedding of biological specimens. In: Practical Methods in Electron Microscopy (A. M. Glauert, ed.), Vol. 3, Am. Elsevier, New York, 1975

Hayat, M. A.: Positive Staining for Electron Microscopy. Van Nostrand-Reinhold, New York, 1975

Kay, D.: Techniques for Electron Microscopy. Blackwell, Oxford, 1965

Lewis, P. R., Knight, D. P.: Staining methods for sectioned material. In: Practical Methods in Electron Microscopy (A. M. Glauert, ed.), Vol. 5, Part 1. Am. Elsevier, New York, 1977

Pease, D. C.: Histological Techniques for Electron Microscopy. Academic Press, New York, 1964

Reid, N.: Ultramicrotomy. In: Practical Methods in Electron Microscopy (A. M. Glauert, ed.), Vol. 3, Part 3. Am. Elsevier, New York, 1975

Sjöstrand, F. S.: Electron Microscopy of Cells and Tissues, Vol. 1. Academic Press, New York, 1967

Chapter 11

Methods for Special Staining of Synaptic Sites

Friedrich-Wilhelm Schürmann

I. Zoologisches Institut der Universität
Göttingen, F. R. G.

Functional pathways in nervous systems can be traced by electrical recording and by means of morphologic techniques using both light and electron microscopy. Morphology demonstrates the shapes and extensions of nerve cells and the sites where they converge to form functional contacts. The functional significance of such junctions cannot be proposed from a single morphologic investigation. Rather, comprehension of the complexity of functional connections and relationships between nerve cells must be derived by correlating different special morphologic techniques with careful inference from electrophysiologic studies. It should be borne in mind that there is much that functional analysis cannot tell us, and similarly many details revealed in fine structures are still subject to speculation. The most sophisticated morphologic research can give no more than a framework for inferring the possible interactive sites between nerve cells, the synapses.

Structures that are interpreted as being synapses are specialized sites between adjacent nerve cells and between nerve and muscle cells that can only confidently be resolved by electron microscopy. Although synapses have been occasionally detected by light microscopy, none of their fine structural details was resolved.

Neuronal membrane specializations, interpreted as sites of functional

241

contiguity, were first detected about 20 years ago in vertebrates and invertebrates (De Robertis, 1955; De Robertis and Benett, 1955; Couteaux, 1961; see Palay and Chan-Palay, 1975). Since then, electron-microscopic investigations of connectivity, parallel to or directly combined with special light-microscopic techniques (and electrophysiology), have become one of the most essential tools for the analysis of neuropil. Even though recent studies report nonfunctional "synapses" (Mark *et al.,* 1970), the pursuit of "synaptology" as a specialized study is indispensable.

One basic aim of high-resolution synaptic morphology is to describe and classify synaptic contacts and to discriminate them from other specializations of neuronal and glial plasmalemmas (for a review of nonsynaptic membrane specializations, see Staehelin, 1974; insect membranes: Satir and Gilula, 1973; Lane, 1974; Skaer and Lane, 1974).

The structure of synapses in the central and peripheral nervous systems of various insect groups has been described by several authors (in the central nervous system by, among others, Trujillo-Cenoz, 1965, 1969; Osborne, 1966; Smith, 1967; Lamparter *et al.,* 1969; Schürmann, 1971, 1972; Dowling and Chappell, 1972; see also Strausfeld, 1976; at nerve-muscle junctions by Osborne, 1975; Usherwood, 1974). All the described junctions belong to the class of the so-called chemical synapses that consistently show the presence of synaptic vesicles. These are supposed to contain chemical transmitter substances. The ultrastructure of an electrical synapse in insects has not yet been described, although there is good evidence that such contacts exist in locust brain (O'Shea and Rowell, 1975).

Electron-microscopic and electrophysiologic studies indicate that synapses within insect ganglia are normally restricted to neuronal arbors that form the integrative neuropil. No chemical synapses have been demonstrated between fibers of peripheral nerve bundles in insects, but are known to be present in Crustacea (Atwood and Moran, 1970) and have recently been demonstrated in a spider leg nerve (Foelix, 1975). Also synapses appear to be absent in the perikaryal layer (the rind of somata that encases neuropil regions) typical for arthropod ganglia. Some nerve cell bodies in insect nervous systems may characteristically show several short processes emanating from them, thus appearing multipolar in Golgi preparations (Schürmann, 1973; Strausfeld, 1976). No synapses have been found between these processes, which are restricted to the cell body rind. This apparent lack of morphologic synapses at the cell body fits into previous findings that nerve cell bodies in arthropod central nervous systems are not involved in integration (Sandeman, 1969; Hoyle, 1970), although one exception was reported (Crossman *et al.,* 1972). Occasional lack of the glial wrapping around somata (glial windows) associated with the mushroom bodies of ants was tentatively suggested as the structural basis for electrical coupling between perikarya (Landolt and Ris, 1966), but experimental evidence for this is lacking.

This chapter describes the structure of insect synapses in order to present some criteria for their identification. Arguments are presented as to why these specializations should be regarded as indicative of functional sites for contiguity. Although what are termed synapses in insect central nervous systems, as revealed by electron microscopy, have not yet been experimentally verified as the exclusive loci for chemical transmission, there is good circumstantial evidence that they are. The main criteria for assuming the morphologic synapse as a functional site are listed below.

1. In appearance there are many analogies between the insect synapse and the vertebrate synapse, the latter being known longer and better investigated in its functional context. Like their vertebrate counterparts most types of insect synapses show a clear morphologic polarity (Figs. 11-1–11-14, 11-20–11-22). The profile of the "presynaptic" neural element exhibits an accumulation of vesicles associated with osmiophilic substances. These sometimes are assembled as an apparatus that projects from the presynaptic membrane into the fiber lumen (Figs. 11-1, 11-3, 11-4, 11-6–11-9, 11-11, 11-13, 11-14, 11-22). Such presynaptic figures (e.g., bars or columns) may be surmounted by a plate or disc, as in flies or bees, and often, but not always, located opposite several postsynaptic profiles, so that the same presynaptic site communicates with more than one postsynaptic element. Some of these three-dimensional arrangements were described by Trujillo-Cenoz (1969), Burckhardt and Braitenberg (1976), and Schürmann (1971). Electron-dense deposits other than the presynaptic figure may, however, line the presynaptic membrane. Structures reminiscent of presynaptic figures can also occur in neural elements at sites that are opposite certain types of glia (Boschek, 1971).

The extracellular space between two profiles at synaptic sites is normally expanded into a synaptic cleft (200–300 Å) and filled with electron-opaque material (Figs. 11-1–11-3, 11-5, 11-6–11-12). However, a marked cleft is absent in some types of synapses (Figs. 11-13 and 11-14). Postsynaptic elements can usually be recognized by band-like electron-dense material (in transverse sections through synaptic membranes) attached to the subsynaptic membrane (opposite the presynaptic membrane and its apparatus), thus denoting the so-called subsynaptic regions (Figs. 11-1–11-3, 11-5, 11-6–11-12). Subsynaptic fibrous paramembranous material shows a lateral extension corresponding to the length of the widened synaptic cleft, whereas presynaptic paramembranous components are restricted to small localized spots. Subsynaptic material projects up to 500 Å into the postsynaptic fiber and has an irregular outline at its interface with the surrounding cytoplasm (Figs. 11-1, 11-2, 11-5, 11-11, 11-12). At some loci, subsynaptic densities characteristically appear to be absent or poorly developed (Figs. 11-13 and 11-14).

Three-dimensional reconstructions of subsynaptic fibrous material have not yet been described, whereas, for the presynaptic specialization, a cross-like configuration (or T-shaped ribbon) has been demonstrated in

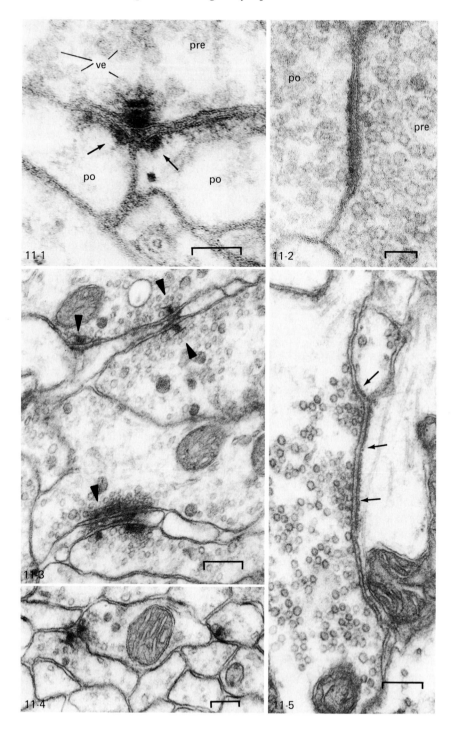

Diptera (Trujillo-Cenoz, 1969; Toh and Kuwabara, 1975; Burckhardt and Braitenberg, 1976). In Hymenoptera, a vertebrate-like regular presynaptic grid, with a hexagonal pattern of figures, was found in some synapses (Schürmann, 1971).

Morphologic "polarities" of insect synapses, useful for tracing neuronal pathways, are not so much indicated by presynaptic accumulations of vesicles, close to synaptic sites, as signified by asymmetric arrangements of intracellular paramembranous material. The occurrence of vesicles alone (translucent or dense core) may indicate functional contiguity but cannot be assumed as such: In particular such an interpretation should be avoided when vesicles are detected without associated paramembranous material within two adjoining fibers that are not separated by glial lamellae.

As in vertebrates, synaptic contact between two nerve fibers is usually established by a number of sites within the convergent arborizations. Synaptic sites are often established *en passant* between fibers that project in parallel, and this may be detected even in fascicular bundles where there are no prominent fiber arborizations (Schürmann, 1971, 1972, 1973). Thus, inferring or postulating the occurrence and position of synapses from light-microscopic criteria (e.g., swellings, blebs, and so forth) must be confirmed by electron microscopy.

2. In insect neuropil pre- and postsynaptic "specializations" can be selectively demonstrated by cytochemical methods previously applied to

---

Figs. 11-1–11-5. Different aspects of insect synapses from aldehyde-osmic acid-fixed brains.

Fig. 11-1. Dyad-synapse typical for insects. A presynaptic element (*pre*) with clear synaptic vesicles (*ve*) and a presynaptic figure, adjacent to the presynaptic membrane and surrounded by vesicles, is connected with two postsynaptic fiber profiles (*po*). Note the widened synaptic cleft, filled with amorphous substances not as electron opaque as the prominent subsynaptic membrane appositions (*arrows*). *Apis mellifera.* Brain: protocerebrum. Scale 0.1 μm.

Fig. 11-2. Synapse between two fibers, both filled with vesicles. At the postsynaptic side (*po*) subsynaptic paramembranous substances are well developed whereas presynaptically (*pre*) the membrane coating with electron-dense substances is exceedingly thin and a presynaptic figure is not present. *Acheta domesticus.* Brain: protocerebrum. Scale 0.1 μm.

Fig. 11-3. "Plate"-covered presynaptic figures (*arrowheads*). Note the widened clefts and the different views of presynaptic figures. *Calliphora erythrocephala.* Brain, optic neuropil: medulla. Scale 0.2 μm.

Fig. 11-4. Synapses with scarcely developed subsynaptic densities. *Calliphora erythrocephala.* Brain: stalk of mushroom bodies. Scale 0.2 μm.

Fig. 11-5. Extensive synapse with prominent subsynaptic densities (*arrows*) and small presynaptic membrane appositions. Structural polarity is obvious, as in Fig. 11-2. *Aeschna cyanea.* Brain: protocerebrum. Scale 0.2 μm.

vertebrate and other invertebrate phyla.   These include the following techniques: bismuth-iodide impregnation and phosphotungstic acid staining (see below).   Both indicate that similar components contribute to synaptic organelles in vertebrates and insects.

3. Vesicles and presynaptic figures have been shown in axon terminals of insect neuromuscular junctions (see Osborne, 1975).   These are similar or identical to the presynaptic apparatus within insect ganglia.   In addition, the neuromuscular synapse is known as the site of chemical transmission (see Usherwood, 1974).

4. Finally, the classification of pre- and postsynaptic components from morphology is supported by studies of degenerating neurons.   Electron microscopy of sensory fiber anterograde degeneration into ganglia reveals that vesicle aggregates and associated membrane specializations reside within axon terminals that contact nondegenerated elements: These, in turn, are equipped with subsynaptic densities (Boeckh et al., 1970; Lamparter et al., 1967; Schürmann and Wechsler, 1970; Campos-Ortega and Strausfeld, 1972).

In summary, the similarity between synaptic topology in vertebrates and insects, as well as histochemical, neuromuscular, and degeneration criteria, provides good evidence that the structures termed synapses should be considered as at least one class of functional contiguity.

Synaptic fine structure can be demonstrated using conventional fixation procedures (aldehyde fixation; osmic acid postfixation) and by a combination of aldehyde fixation with successive different heavy metal-complex impregnations.   Particularly helpful for the characterization of synapses are: Bismuth iodide impregnation (the BIUL technique of Pfenninger et al., 1969); the ethanolic phosphotungstic acid method (the EPTA method of Bloom and Aghajanian, 1966, 1968) for paramembranous synaptic substances; and the zinc iodide osmium tetroxide (ZIO) stain for synaptic vesicles (Akert and Sandri, 1968).   Other special techniques, such as Wood and Barrnett's (1964) triple fixation and the potassium permanganate stain for the demonstration of adrenergic vesicles (Hökfelt, 1968), have either been found unsatisfactory for insect nervous systems or did not yield novel information, as did, for example, fixation with osmic acid and unbuffered glutaraldehyde, followed by uranyl acetate impregnation (Kanaseki and Kadota, 1969).

Methodology

Successful application of the BIUL, EPTA, and ZIO methods is heavily dependent on the preceding fixation.   A survey of publications clearly shows the superiority of aldehyde solutions (for review, see Strausfeld,

1976) whereas fixation with osmic acid alone may allow the identification of synapses but consistently results in poor preservation of paramembranous substances that are essential for the classification of a synaptic type. Potassium permanganate fixation has been used for insect ganglia (Lamparter, 1966; Landolt, 1965) but gives poor preservation of structure in particular membrane specializations. Simultaneous fixation with glutaraldehyde and osmic acid (Franke *et al.*, 1969) has been successfully applied to the insect nervous system (F.-W. Schürmann, unpublished) and provides good resolution of plasmalemmas, which, by other methods, often appear to be tightly apposed; however, synaptic paramembranous appositions are not as well preserved. Glutaraldehyde-hydrogen peroxide fixation (Perrachia and Mittler, 1972) does not seem to be advantageous for synaptic contacts, but gives good resolution of membranes (Castel *et al.*, 1976), whereas osmic acid treatment following aldehyde fixation gives a poorer preservation of membrane proteins due to solubilization. Aldehyde fixation alone reveals proteins in the gross membrane that are obscured or removed by osmication (Hake, 1965; McMillan and Luftig, 1973).

One formula will not give satisfactory results on all groups of insects since these may differ markedly in ionic composition of body fluids and show striking differences in the thickness of the neurilemma. Thus it is advised that the worker check the osmolality of a species hemolymph (see Prosser, 1973) prior to defining the composition of the fixation fluid. The osmolality of aldehyde fixatives should be adjusted so that the solution is slightly hypertonic (compare Peters, 1970, for vertebrate nervous tissue). Phosphate or cacodylate-buffered aldehyde solutions or formaldehyde-glutaraldehyde mixtures (Karnovsky, 1965) are used (for preparation, see Chap. 10) prior to osmication and prior to special impregnations. Aldehyde-fixed tissues are briefly washed in buffer solution and then transformed either to 2% osmic acid solution or to the special impregnation solutions as described below.

Phosphotungstic Acid Technique

Phosphotungstic acid (PTA) was first applied for investigations of fine structure in the vertebrate brain by Gray (1963), who was able to show clear presynaptic projections. The method was further developed as a cytochemical stain by Bloom and Aghajanian (1966, 1968) who used an ethanolic solution of the anionic heavy metal complex for aldehyde-fixed, nonosmicated nervous tissue. They showed a strong, more or less predominant affinity of the impregnation to paramembranous synaptic material, whereas membranes remain unstained. Bloom and Aghajanian employed the new technique for visualizing synaptic development and

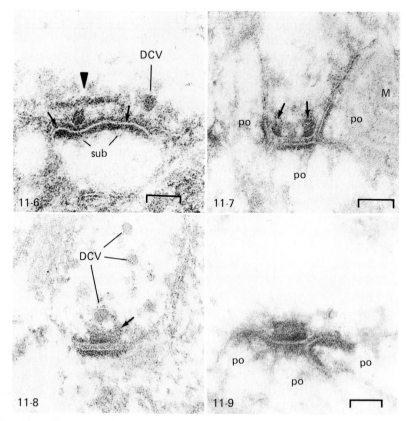

Fig. 11-6. Synapse stained with EPTA. Note the presynaptic "plate" (*arrowhead*), which is not fully attached to the presynaptic figure. Membranes remain unstained and appear as white lines. The presynaptic membrane is only partially covered with electron-dense substances. Dense subsynaptic appositions (*sub*) and cleft material (*arrows*). There is a positive reaction of the contents of dense-core vesicles (DCV). *Calliphora.* Brain: protocerebrum. Scale 0.1 μm.

Fig. 11-7. Synapse stained with EPTA showing three postsynaptic elements (*po*), two figures in the presynaptic element (*arrows*), and mitochondrion (*M*). *Calliphora.* Brain: protocerebrum. Scale 0.1 μm.

Fig. 11-8. Synapse stained with EPTA showing presynaptic figure (*arrow*), dense-core vesicles (*DCV*), and dense subsynaptic appositions. *Apis mellifera.* Brain: protocerebrum. Scale 0.1 μm.

Fig. 11-9. BIUL-impregnated synapse. A presynaptic element with presynaptic figure is connected with three postsynaptic fibers (*po*). The presynaptic figure shows inhomogeneous electron density. *Apis mellifera.* Brain: protocerebrum. Scale 0.1 μm.

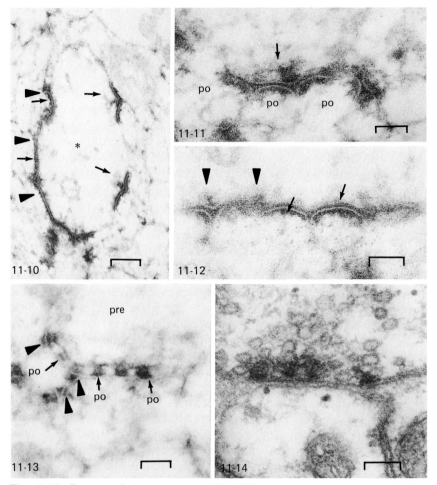

Fig. 11-10. Presynaptic bouton (*asterisk*), BIUL stain. Note the dyad-synapses (*arrows*) and the extensive synaptic sites with electron-dense cleft material and broad subsynaptic densities (*arrowheads*). *Apis mellifera.* Brain: mushroom body calyx. Scale 0.2 μm.

Fig. 11-11. BIUL-stained synapse with a presynaptic figure (*arrow*) covered by a "plate" and spots of presynaptic membrane appositions, postsynaptic elements (*po*) with broad subsynaptic appositions. *Apis mellifera.* Brain: mushroom body calyx. Scale 0.1 μm.

Fig. 11-12. BIUL-stained synapse. The presynaptic membrane is partially covered with electron-dense substances. Note a narrow white midline in the synaptic cleft (*arrows*) and the broad subsynaptic appositions and parts of presynaptic figures (*arrowheads*) opposite two postsynaptic fibers. *Apis mellifera.* Brain: protocerebrum. Scale 0.1 μm.

Fig. 11-13. BIUL-stained synapse. "Plate"-covered presynaptic figures; the synaptic cleft is not widened and shows small electron-dense patches (*arrows*); but note intercellular densities where three fibers meet (*arrowheads*): subsynaptic densities are missing (postsynaptic fibers = *po*). *Apis mellifera.* Brain: upper stalk of mushroom body. Scale 0.1 μm.

Fig. 11-14. Similar synapse as in Fig. 11-13: aldehyde-osmium preparation. The "plate" is not clearly visible, the synaptic cleft is not widened, and subsynaptic densities are not present. *Apis mellifera.* Brain: α-lobe of mushroom body. Scale 0.1 μm.

synaptic density per volume of neuropil (Aghajanian and Bloom, 1967; Bloom and Aghajanian, 1968). The ethanolic phosphotungstic acid stain (EPTA) was subsequently applied to synapses of Diptera and Hymenoptera (Trujillo-Cenoz, 1969; Lamparter *et al.*, 1969), with similar results, and on other phyla (e.g., Mollusks; Jones, 1968; Annelids; Aghajanian and Bloom, 1967; Günther and Schürmann, 1973; Onychophora; Schürmann, 1978). Rees and Usherwood (1972) reported a positive reaction of the EPTA stain for intracleft substances in insect neuromuscular junctions. The present author's experience confirms the usefulness of the method for Hymenoptera, Diptera, Orthoptera, and Odonata.

Staining Procedures

After aldehyde fixation, small trimmed pieces of specimens, or total ganglia, are washed briefly in buffer (10 min) and then dehydrated to 96% ethanol. Block impregnation in EPTA solution is performed for 1–1.5 h. The impregnation solution consists of 1% phosphotungstic acid dissolved in absolute ethanol (Bloom and Aghajanian, 1968). For 10 ml of solution 1–5 drops of 96% ethanol are added. In this manner the water content of the impregnation medium can be varied to influence the staining affinity of the synaptic material (Bloom and Aghajanian, 1968). No modification of the original method is needed for insect neuropil. Purified phosphotungstic acid and other purified chemicals should be used (Merck, $P_2O_5 \cdot 24WO_3 \cdot 44H_2O$, analytical grade). Impregnation is followed by washing in alcohol, exchange with propylene, and finally embedding in Araldite or Epon. It is necessary to double stain ultrathin sections with uranyl acetate (Watson, 1958) for 5 min, followed by lead citrate (Reynolds, 1963), to give sufficient contrast. Medium-thick sections (a yellow to white diffraction hue) are recommended for the visualization of synaptic apparatus which might otherwise appear hardly prominent in faintly stained structures.

Results

With the EPTA method, the paramembranous components of insect synapses stand out prominently against the rest of the neuropil. Presynaptic projections and membrane appositions (see Figs. 11-6–11-8) appear electron dense. Plasmalemma, vesicles, and often membranes remain unstained and are seen as translucent outlines that pass through the dense intra- and intercellular synaptic substances. The naked unit membrane (about 75 Å in diameter) is EPTA-negative. The size of presynaptic material shown with EPTA impregnation or in osmicated material is similar, but EPTA reveals profiles of presynaptic configurations more clearly.

Electron-opaque material, covering parts of presynaptic membranes, is also better demonstrated by the EPTA method.

The present findings show that presynaptic vesicles have possible access to the presynaptic membrane [for comparison, see figures in Trujillo-Cenoz (1969) and Lamparter *et al.* (1969)]. The synaptic cleft is filled with EPTA-positive material that is less opaque than pre- and subsynaptic paramembranous substances, but is considerably darker than in combined aldehyde–osmic acid-fixed material. The lateral extent of the strongly EPTA-positive subsynaptic densities (web) corresponds to the length of the widened synaptic cleft (see Figs. 11-6–11-8).

All positive reacting intra- and extracellular synaptic components exhibit the strongest electron density in the insect neuropil. Other membrane specializations and other organelles show a weak electron density after EPTA stain; e.g., desmosomes, contents of dense-core visicles, nucleus, nucleolus, and glial lacunar systems. Thus, the EPTA method cannot be considered a selective synaptic stain in the strict sense.

Chemical Nature

The cytochemical action of phosphotungstic acid (PTA) is far from clear, and descriptions are controversal (see Pfenninger, 1973; Hayat, 1975). Some authors consider PTA as a carbohydrate stain (for the "greater membrane") while others give reasons that it exclusively stains basic proteins (see Pfenninger, 1973; Luft, 1976). Pfenninger emphasizes that PTA bonding is dependent on pH and states that variations of pH would explain the effects quoted by various authors. He considers the anionic heavy metal complex PTA as a good candidate for a reaction with basic groups, and this view is supported by digestion experiments with trypsin (Bloom and Aghajanian, 1968; Pfenninger, 1973). It is still unclear which action groups bind PTA, and systematic experiments with insect synapses are lacking.

The Bismuth Iodide-Uranyl Acetate-Lead Citrate Technique

Bismuth iodide ($BiIO_4$), introduced into electron microscopy by Pfenninger *et al.* (1969), is valuable for the characterization of paramembranous synaptic material. Using this method, many aspects of the assembly of presynaptic networks have been shown in some vertebrate synapses.

The bismuth iodide method is well suited for the study of synaptic

structure in insects and has been applied to Diptera, Orthoptera, Blatto-
idea, and Hymenoptera (Lamparter *et al.,* 1969; Boeckh *et al.,* 1970;
Schürmann, 1971; Schürmann and Wechsler, 1970). A positive reaction
for paramembranous presynaptic and intercellular material has been observed in insect nerve-muscle junctions (F.-W. Schürmann, unpublished).

Staining Procedure

Impregnation with bismuth iodide follows aldehyde fixation after a brief
rinse in buffer. The original method of Pfenninger *et al.* (1969) can be
used without modification for the insect nervous system; it gives satisfactory results, provided that tissue is fixed for a long period in aldehyde
(overnight or as long as 20 h).

A modification by Pfenninger (1971a) involves transfer to the acidic impregnation solution through graded steps with decreasing pH in acetate
buffer and is in fact a gentler treatment of tissue.

The medium for the reaction of bismuth salt with potassium iodide is
2 *N* formic acid and is prepared as follows: 0.5 g $(BiO)_2CO_3$ and 2.5 g KI
are dissolved in 50 ml 2 *N* formic acid. The mixture is heated to 50°C and
filtered. Tissue blocks are impregnated for 12–18 h at 40°C. After ethanolic dehydration, specimens are embedded in Araldite or Epon.
Thicker sections (diffraction hue: yellow to white) are useful for lower
magnifications. Thin sections must be double contrasted with uranyl acetate and lead citrate to obtain outstanding synaptic structures against a
faintly stained background. For observation by light microscopy as a
preliminary to EM studies, 1- to 2-μm sections from EPTA- or BIUL-
treated tissue can be stained with toluidine blue, but contrast is weaker
than in sections from osmicated tissue. Synaptic sites cannot be detected
by light microscopy.

Results

The effect of bismuth impregnation (see Figs. 11-9–11-13,11-15, 11-16)
and of postcontrast with uranyl acetate and lead citrate is similar to the
EPTA treatment, but synaptic structures exhibit a slightly enhanced electron density with the BIUL stain. Paramembranous synaptic material is
electron dense, and the polarized structure of the synapse becomes apparent (Figs. 11-9–11-13). The synaptic cleft material appears to be more
electron dense than in aldehyde-osmic acid-treated specimens. Membranes and electron-translucent vesicles remain unstained, while dense-
core vesicles show electron-opaque contents. A medial white line in the
cleft (Fig. 11-12) can be detected in insect synapses at high resolution (cf.
vertebrate synapse: Pfenninger, 1973). Presynaptic figures are more

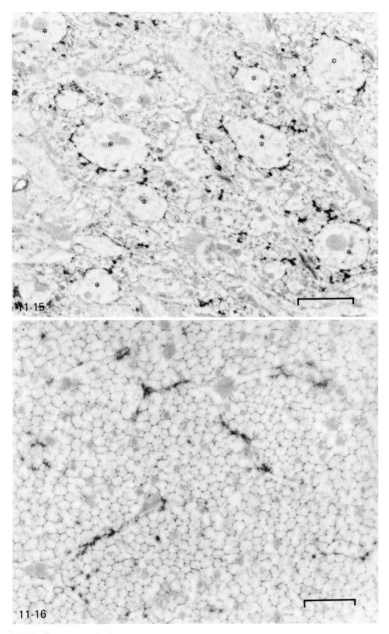

Fig. 11-15. Presynaptic boutons (*asterisks*) stained with BIUL. Synaptic sites can be detected at low magnification. *Apis mellifera.* Brain: mushroom body calyx. Scale-1 μm.

Fig. 11-16. Synapse distribution revealed by BIUL impregnation at a lower magnification. Note the staining of nonsynaptic extracellular substances between intrinsic fiber elements (Kenyon cells). *Acheta domesticus.* Brain: stalk of mushroom body, transverse section. Scale-1 μm.

clearly demonstrated. In the bee, presynaptic "pillars" are revealed as not wholly contiguous with the plate surmounting the presynaptic apparatus (Figs. 11-11 and 11-13). Presynaptic electron-dense membrane appositions are not continuous but appear interrupted, revealing the membrane as directly accessible to vesicles (Figs. 11-9, 11-11, 11-12) Paramembranous synaptic structures show the strongest electron density, but as with the EPTA stain, the BIUL technique will show up organelles other than synaptic components (see Pfenninger, 1973). For example, electron-dense extracellular substances between nonsynaptic membranes have been detected in the stalks of mushroom bodies of the cricket (Fig. 11-16). The contents of neurotubuli appear moderately opaque, and the extracellular lacunar system of glial cells (see Lane, 1974) appears electron dense in locust neuropil. Positive reactions by other cell components are described by Pfenninger (1971a,b).

Cytochemical Specifity

The heavy metal complex $BiIO_4$ was originally used in chemistry for the demonstration of amines (see Pfenninger, 1973). Pfenninger (1971a,b, 1973) studied intensively the bonding capacity of the BIUL stain to cell organelles. BIUL does not react with membranous lipid components and the staining of external glycocalyx material remains weak. Organelles and structures rich in protein and nucleic acid show high affinity to impregnation (e.g., chromatin, nucleoli, ribosomes, mitochondrial matrix). *In vitro* experiments demonstrate that the $BiIO_4$ complex has a high affinity for basic polyamino acids, but does not react with acidic or neutral protein groups or with sialic acid. Blocking acidic polar protein groups by carboxymethylation weakens contrast, and contrast disappears completely when basic and acidic groups in proteins are blocked together by carboxymethylation and acetylation. Thus, basic components of proteins can be specifically demonstrated when the reaction of acidic groups is suppressed by blockage. Without blocking, the BIUL technique is suitable for the demonstration of all polar groups. Carboxyl groups seem to represent the majority of acidic components within the external membrane coats and synaptic cleft (Pfenninger, 1971a,b, 1973). Studies on vertebrate synapses indicate that extra- and intracellular paramembranous components do not have identical compositions. Pfenninger's cytochemical results are consistent with biochemical investigations on synaptosomes (e.g., Cotman and Mathews, 1971; Cotman *et al.*, 1974; Wang and Mahler, 1976). Though systematic research on the cytochemistry and biochemistry of insect synapses is still lacking and though the external membrane coats might have different composition, there is no reason to assume a basically different staining reaction for the insect neuropil by the BIUL.

Zinc Iodide-Osmium Tetroxide Stain

Iodide salt-osmium tetroxide solutions were first used for the demonstration of autonomic nerve fibers by light microscopy (Champy, 1913; Maillet, 1962). Akert and Sandri (1968) discovered that zinc iodide-osmic acid (ZIO) treatment reveals electron-opaque contents of synaptic vesicles that normally appear translucent after aldehyde-osmic acid fixation. The method was thereafter used for various synaptic areas of the nervous system in vertebrates in order to define different types of vesicles (Akert *et al.*, 1969; Barlow and Martin, 1971; Nickel and Waser, 1969; Dennison, 1971; Kawana *et al.*, 1969; Stelzner, 1971). Martin *et al.* (1969) introduced prefixation with glutaraldehyde and obtained better preservation of tissue. The ZIO technique has been successfully applied to insect nervous tissue by several authors (Lamparter *et al.*, 1969; Boeckh *et al.*, 1970; Schürmann, 1970, 1974; Schürmann and Wechsler, 1970; Schürmann and Klemm, 1973; Scharrer and Wurzelmann, 1974; Lane and Swales, 1976; Tyrer and Altman, 1976).

Staining Procedure

After glutaraldehyde fixation and a brief washing in buffer (10 min), specimens are impregnated for 15 h in zinc iodide-osmic acid solution. Perfusion with cacodylate-buffered formaldehyde, followed by a phosphate-buffered glutaraldehyde perfusion and by a subsequent ZIO stain gave negative results for the vertebrate nervous system (Vrensen and De Groot, 1974). The solution is prepared as follows: 15 g zinc powder and 5 g of iodine crystals are vigourously shaken in 200 ml of distilled water. The solution is allowed to stand for 1 h and fresh filtered. For insects a modified mixture of ZIO solution is recommended (6 ml filtrate and 3 ml 2% osmic acid solution in 0.2 M phosphate buffer: cf. Akert and Sandri, 1968; Martin *et al.*, 1969). Subsequent treatment is performed as usual (brief rinse in buffer, ethanolic dehydration, embedding). Thin sections are double contrasted with uranyl acetate and lead citrate for optimal visualization of synaptic structures other than vesicles.

Results

The zinc iodide-osmic acid impregnation stains mainly vesicles in insect neuropil, but other cellular elements may also react positively (Lamparter *et al.*, 1969; Lane and Swales, 1976). Some Golgi complex sacculi and vesicles, parts of the smooth endoplasmic reticulum of the GERL (Golgi-endoplasmic reticulum-lysosome) complex and in nerve fibers, multivesicular bodies, cytosomes and gliosomes, consistently show a strong positive reaction (heavy electron-dense deposits).   Mitochondria are also often affected and extracellular impregnation deposits are frequently found in the membrane system of perineurial cells (Fig. 11-18).

In the central nervous system the reaction is used mainly to reveal contents of the clear vesicles (300–600 Å) whereas dense-core vesicles (diameter up to 1800 Å) are not normally impregnated (Fig. 11-17).[1] ZIO-positive vesicles are spherical, but flattened vesicles also occur, for instance in the protocerebrum of locusts (Figs. 11-19 and 11-21) (see Strausfeld, 1976; Lane and Swales, 1976).   The result of impregnation in a single nerve fiber, with the same class of vesicles (which are translucent in aldehyde-osmic acid fixation), can vary: All vesicles in one fiber profile

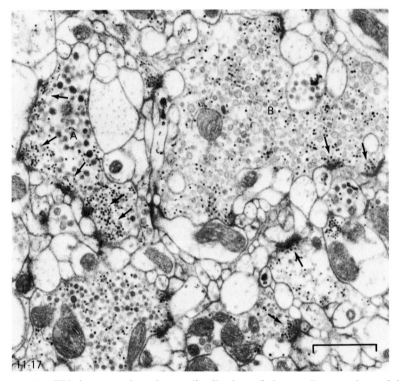

Fig. 11-17. ZIO impregnation shows distribution of the small synaptic vesicles that appear as *black spots,* and their accumulation near synaptic sites *(arrows).* Different dense-core vesicles in fiber profiles A and B remain unstained. *Acheta domesticus.*   Brain: protocerebrum.   Scale-1 μm.

[1] A comprehensive classification of the different vesicle types in the insect nervous system is still not established.   Dense-core and translucent vesicles can occur together in the same nerve fiber in different concentrations.   Translucent vesicles appear attached to presynaptic sites; dense-core vesicles are not usually distributed near the presynaptic sites.

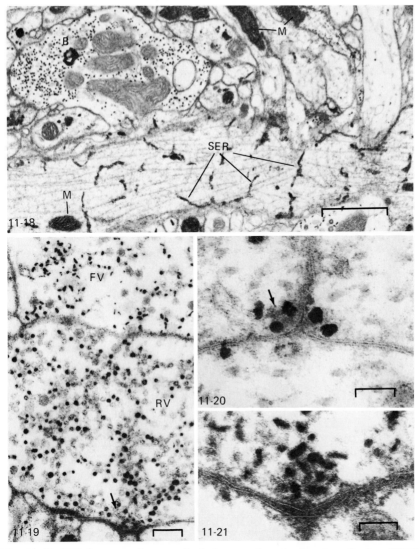

Fig. 11-18. ZIO impregnation may include smooth endoplasmic reticulum (*SER*) and some mitochondria (*M*). Presynaptic bouton (*B*) filled with synaptic vesicles and mitochondria. *Acheta domesticus*. Brain: protocerebrum. Scale 2 μm.

Fig. 11-19. ZIO impregnation of flattened (*FV*) and round vesicles (*RV*) that only partially reacted positively to the impregnation. Synaptic site (*arrow*). *Acheta domesticus*. Brain: protocerebrum. Scale-1 μm.

Fig. 11-20. ZIO impregnation showing a few vesicles at synaptic sites of two presynaptic fibers. Note presynaptic substances between the vesicles (*arrow*). *Acheta domesticus*. Mushroom body stalk. Scale 0.1 μm.

Fig. 11-21. ZIO impregnation of flattened vesicles at a dyad synapse. *Acheta domesticus*. Brain: protocerebrum. Scale-0.1 μm.

might react positively, whereas, in another, vesicles may react partially or not at all (Figs. 11-17–11-22).

Despite these variations the ZIO stain is useful for detecting clear vesicle populations within different fibers and neuropil areas. The distribution of small synapses, such as in very dense neuropil, is often better dis-

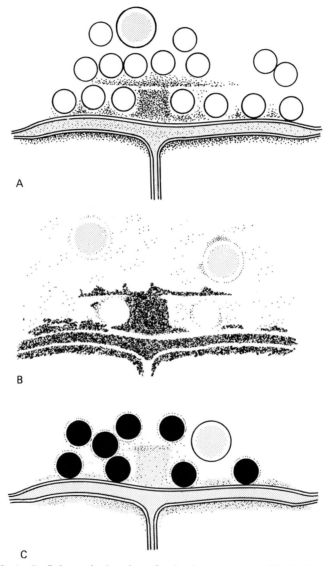

A

B

C

Fig. 11-22, A–C. Schematic drawing of a dyad synapse. **A** Aldehyde-osmic acid fixation; **B** EPTA or BIUL impregnation after aldehyde fixation; **C** ZIO impregnation after aldehyde fixation.

cerned with the ZIO method than by any other (see Schürmann, 1970). Translucent vesicles in neuromuscular junctions of the musculus levator and retractor tibiae in *Locusta migratoria* and *Acheta domesticus* are stainable with ZIO, but the method is not suitable for light-microscopic demonstrations of neuromuscular junctions in general (F. W. Schürmann, unpublished).

The ZIO technique is not limited to electron-microscopic studies but can be useful to detect synaptic areas by light microscopy. Under favorable conditions vesicle accumulations and synapses can be deduced from ZIO-stained 1- to 2-μm sections by Nomarski interference contrast (Barlow and Martin, 1971), by phase contrast (Akert and Sandri, 1968; Tyrer and Altman, 1976), or in toluidine blue-stained sections (Schürmann, 1973; Schürmann and Klemm, 1973). Tyrer and Altman (1976) have used the ZIO method to count the approximate number of synaptic sites in thicker branches of identified motor neurons in thoracic ganglia of locusts after serial sectioning and light microscopy. ZIO-positive vesicle accumulations appear as dark spots in phase contrast images. With the toluidine blue stain, ZIO-positive vesicle clusters exhibit a deep blue color against a lighter bluish surrounding (Schürmann, 1970, 1973).

Using light microscopy and ZIO impregnation to discriminate vesicle-containing areas and to detect synaptic sites, the limitations of the method and possibilities of error must be considered (see Tyrer and Altman, 1976). Electron-microscopic examination of the same neuropil area is indispensable. Difficulties in interpreting the ZIO application include:

1. Synapses with very few positive-reacting vesicles cannot be detected by the light microscope. Synapses of this type are abundant in the insect neuropil.

2. ZIO-positive vesicle clusters may be artifactually dislocated in the nerve fiber and become shifted against membranes, thus simulating synaptic contacts.

3. Glial profiles, as well as mitochondria, can react positively to ZIO stain and imitate synaptic sites in the light microscope.

4. Degenerating fibers have strong osmiophilic reactions (see Chap. 12). Degeneration, therefore, sometimes cannot be distinguished from ZIO-positive vesicle clusters in the light microscope. It should be remembered that degeneration can be age dependent in the insect imago (Herman *et al.*, 1971).

5. As noted before, ZIO stain of translucent vesicles can remain incomplete. There are some indications that clear synaptic vesicles of different neuron classes are not similarly affected with ZIO impregnation.

6. Finally, an apparently trivial but in fact an important point: fine-fiber profiles of less than 0.2 μm, which are abundant in the insect nervous system, cannot be confidently resolved by light microscopy.

The cytochemical background of the ZIO stain is not known. The impregnation mainly affects clear synaptic vesicles and includes several other cell structures and is thus not selective. The method is unsuitable for the selective demonstration of cholinergic vesicles (Pfenninger, 1973; Matus, 1970). Clear vesicles in neuromuscular junctions in insects react positively to ZIO (F. W. Schürmann, unpublished) and must be considered as noncholinergic since glutamate and γ-aminobutyric acid represent the putative transmitters of these synapses (Usherwood, 1974; Gerschenfeld, 1973). Kawana *et al.* (1969) suggested that the different affinities of the ZIO stain are due to different physiologic states of vesicles. According to Pfenninger (1973) the demonstration of clear vesicle populations associated with different transmitters might be based on ubiquitous, unidentified substances in the vesicular lumen.

Conclusions and Outlook

Some criteria for defining structural features of insect synapses can be derived from conventional electron microscopy and special histochemical techniques. These were first used on vertebrate tissues and have proved valuable for insect nervous system, with or without modification.

Comprehensive application of the different morphologic methods (standard techniques, cytochemical staining, freeze etching) is needed for the insect synapse. Precise characterization of synaptic fine structure, as well as definition of criteria for different types of insect synapses, demands the application of several procedures. However, the EPTA and BIUL techniques are particularly useful in that they show a reaction with paramembranous structures that indicates chemical similarity between components in insects and vertebrates. Intra- and intersynaptic substances show a strong affinity for EPTA and BIUL impregnations, are strongly electron opaque, and are more completely and consistently resolved than by conventional techniques (aldehyde fixation, followed by osmication and uranyl acetate-lead citrate staining). EPTA and BIUL establish structural if not functional polarity, and because of the high contrast of stained material with respect to its electron-translucent surroundings, synapses revealed by BIUL and EPTA may be susceptible to semiautomated counting devices (see Vrensen and De Groot, 1973). The enhanced demonstration of pre- and postsynaptic organelles, as well as cleft material, might be due to superior preservation of membrane proteins (see McMillan and Luftig, 1973). The ZIO stain is specially useful for detecting "small" synapses associated with few vesicles (often encountered in insect tissue) and for the distinction of synaptic areas in neuropil. If possible these techniques should be used in parallel and combined with light-microscopic procedures that reveal entire neurons.

Addendum

Wood *et al.* (1977) describe two types of synaptic contacts in the cockroach central nervous system using the bismuth iodide method and freeze fracture.

A report by Reinecke and Walther (1978) deals with the zinc iodide-osmium tetroxide stain of nerve-muscle junctions in locusts. The authors investigate the ZIO stain of presysnaptic vesicles after stimulation and suggest that the stain demonstrates intravesicular protein with SH-groups.

Chapter 12

Experimental Anterograde Degeneration of Nerve Fibers: A Tool for Combined Light- and Electron-Microscopic Studies of the Insect Nervous System

Friedrich-Wilhelm Schürmann

I. Zoologisches Institut der Universität
Göttingen, F. R. G.

In 1850 Waller observed that the distal parts of glossopharyngeal nerve fibers—separated from their nerve cell bodies—were subjected to structural changes leading to degeneration. His experiment and discovery were the basis for the subsequent development of special techniques for tracing nerve fibers and their synaptic projections in the central nervous system. The so-called experimental Wallerian degeneration (anterograde or orthograde degeneration)[1] has since become a most valuable tool for neuroanatomical studies at the level of light and electron microscopy (see Gray and Guillery, 1966; Ebbesson, 1970; Heimer, 1970).

Light-microscopic techniques that rely mainly on silver stains have been invented and improved by vertebrate neuroanatomists. However, their adaptation for arthropod nervous systems has been successfully employed for some insects and Crustacea (see Pareto, 1972). Recent use of experimental degeneration has clearly demonstrated that Wallerian degeneration can take place in sensory cells as well as motor neurons and interneurons of the insect nervous system. Structural changes during

---

This chapter is dedicated to Professor Dr. B. Rensch, University of Münster, on the occasion of his eightieth birthday.

[1] Anterograde or orthograde means degeneration of the fiber away from its soma, retrograde toward the soma of the affected neuron.

and after the establishment of degeneration have been visualized by classical staining procedures for light microscopy (e.g., Vowles, 1955; Pareto, 1972; Hess, 1958a; Farley and Milburn, 1969; Spira *et al.*, 1969) or with special silver stains, such as the Fink-Heimer technique (e.g., Lund and Collett, 1968; Rehbein, 1973).

Visualization of degenerative changes has been achieved by other light-microscopic techniques using fixations for electron microscopy, often combined with ultrastructural studies (Hess, 1960; Lamparter *et al.*, 1967, 1969; Melamed and Trujillo-Cenoz, 1963; Rowell and Dorey, 1967; Boulton, 1969; Schürmann and Wechsler, 1970; Boeckh, 1973; Boeckh *et al.*, 1970; Rees and Usherwood, 1972; Campos-Ortega and Strausfeld, 1972, 1973; Edwards and Palka, 1974; Palka and Edwards, 1974; Griffiths and Boschek, 1976; Griffiths, 1979). No general advantage can be attributed to any one of these methods, and the choice of staining is usually governed by the initial experimental procedure.

Massive effects of degeneration for tracing tracts or extensive sensory inputs have been described by means of conventional paraffin histology without special staining techniques for degeneration products (e.g., Rehbein, 1973; Pareto, 1972; Ernst *et al.*, 1977). However, detailed information about the spatial relationship of small groups of fibers or single neurons, with respect to other neural and glial elements, cannot normally be expected from these methods: Synaptic contacts can only be documented by electron microscopy.

Lamparter *et al.* (1967) were the first to bridge the gap between light and electron microscopy in their study of Wallerian degeneration in insects. They followed the sensory projections of ant prothorax leg nerves in the first ventral ganglion neuropil by phase contrast microscopy of unstained plastic sections, using it as their initial marker for subsequent observations of synapic junctions of degenerating fibers by electron microscopy. The procedure of taking serial 1- to 4-μm plastic sections with occasional ultrathin slices for high resolution study allows a qualitative description of the distribution of afferent fibers and their synaptic connectivity. This technique—subsequently applied to the nervous systems of various insects by several authors—represents one useful method for tracing experimentally selected fibers and defining their synaptic contiguities.

This chapter presents some basic instructions about degeneration by experimental lesions in the insect central nervous system, which are suitable for light-microscopic neuronal mapping and simultaneous survey of synaptic contacts by electron microscopy. Interpretation of these results is critically examined as a tool for neuroanatomy.

In addition, this chapter offers some comments about experimental procedures, the factors influencing the results of lesions, and their structural interpretations. Parts of an identified neuron and of similar neurons in the same or different species when separated from their somata show different temporal reactions to a lesion (Clark, 1976a,b; Edwards and Palka, 1974). It is left to the investigator to vary and test parameters in order to determine suitable conditions for any particular species or subsystem.

Choice of Animal

The use of animals that have been reared under controlled and constant conditions, from a well-known stock, is recommended. It is also important to know the exact age of the animal, to make sure that insects chosen for experiments do not carry a disease that could damage the nervous system and to select animals with "normal" behavior rather than to take mutants with nervous system defects. The thorough studies of Herman *et al.* (1971) demonstrated that the insect nervous system is subject to structural alterations correlated with aging. In *Drosophila* signs of normal degeneration (such as shrinkage, vascularization, deposition of dense bodies) become increasingly apparent in the brain with progressive age and can be correlated with the decline of some stereotypic behavior. It has also been suggested that the decline of mating behavior and negative geotaxis is closely related to gerontologic structural changes of the neuromuscular apparatus. The sensory apparatus (e.g., sensory hairs on the antennae and cerci) is continuously exposed to injury that can cause transient or persisting changes to afferent pathways of the nervous system.

Normal gerontologic degeneration affects the whole nervous system and must be taken into account when the results of experimental Wallerian degeneration are interpreted, especially when the induced degenerative effect is not expected to be a massive one.

Experimental Lesions to the Nerve Cord: Controlled Surgery

Axons are deprived of their perikarya either mechanically or by cautery. The choice of method depends on the degree of precision and delicacy required and the accessibility of selected parts of the nervous system. Studies on afferent pathways can often be achieved by cutting a nerve with fine scissors, split razor blades, or microknives, by induced autotomy (Horridge and Burrows, 1974), or by crushing a nerve with forceps (though this procedure leaves some uncertainty as to the success of the lesion), or after complete extirpation of sense organs (e.g., Lamparter *et al.,* 1967). The arrangement of sensillae is such as to allow their selective ablation and to trace their input into the central nervous system.

Controlled dissection of tracts or the tract-like connectives of the ventral nerve cord has been performed with microknives or scissors introduced via small windows in the cuticle under microscopic control (e.g., Farley and Milburn, 1969; Ernst *et al.,* 1977). In fact, soma of an identified neuron has been separated from its fiber within a ganglion by splitting the nerve cord with a fine knife (Clark, 1976a,b). Anterograde degeneration of nerve cell groups or of single neurons caused by thermo- or electrocautery—sometimes with the help of a micromanipulator—has been studied by several authors (e.g., Vowles, 1955; Goll, 1967; Campos-Ortega and Strausfeld, 1972), though coagulation by a heated probe is a crude method and is likely to cause undesired damage: Tissue adheres to the needle causing disruption when it is withdrawn. The extent of lesions by cautery often remains unknown.

Injuries are uncompletely closed by wound healing, and operational holes in cuticle have to be sealed (see Campos-Ortega and Strausfeld, 1973). Operated animals should be kept under the same conditions as untreated ones at constant temperature. Large surface wounds require the use of a damp chamber.

Controlled and selected anterograde degeneration by means of topically administered or incorporated chemical components has not been reported for the insect nervous system. However, the application of false transmitters such as 6-hydroxydopamine and 5,6-dihydroxytryptamine intoxicates catecholaminergic and serotonergic axon terminals in sympathetic, peripheral and central nervous systems in vertebrates and causes dose-dependent selective anterograde degeneration (Tranzer and Thoenen, 1968; Tranzer *et al.,* 1969; Baumgarten *et al.,* 1972) and thus can be used as a tool for identifying and mapping these adrenergic neuron classes.

The mode of action of these false transmitters on the insect nervous system is not known. Biogenic monoamines accumulate in the central nervous system of insects usually as distinct distributions in well-defined neuropil areas: These are putative transmitters (for review see Klemm, 1976). 6-Hydroxytryptamine is taken up by the brain of the locust *Schistocerca* (Klemm and Schneider, 1975). Experiments with false transmitters in insects could be rewarding in the search for specific neuron classes.

Timing of Postoperative Survival

Operated animals are killed after a survival period that allows the development of structural degenerative changes in a neuron up to a state suitable for its identification by both light and electron microscopy. The gradual disappearence of severed fibers in time after various periods of survival often provides valuable information on the gross pathways of large-diameter fiber elements of a neuron (Farley and Milburn, 1969; Spira *et al.*, 1969). To obtain more detailed and complete knowledge, fibers subjected to degeneration must be fixed and examined at a transient stage when their synaptic components and their spatial relationship with surrounding neuronal and glial elements are more or less intact. The optimal timing of postoperative survival is critical and not fully under the control of the experimenter. Between 1 and 6 days seems to be a reasonable period for achieving structural reactions, judging from several publications that embrace a range of insect orders (Table 12-1). The development of anterograde degeneration does not necessarily proceed linearly (in time) from the site of the lesion, nor do different segments of a fiber necessarily undergo the same sequential alterations, nor do similar fibers react synchronously. Farley and Milburn (1969) and Spira *et al.* (1969) independently found different survival times for fibers of the interneuronal ascending giant fiber system in the ventral nerve cord of *Periplaneta*. Edwards and Palka (1974) reported differences in survival for sensory fibers in severed cercal nerves. Hence, anterograde degeneration of fibers cannot be detected if postoperative periods are too short, whereas long postoperative survival leads to their complete disappearance. Series of experiments with different survival times and controls of normal untreated tissue have to be carried out.

Rees and Usherwood (1972) made a careful study of degeneration behavior of two axons of identified excitatory motor neurons in the metathoracic ganglion of *Schistocerca* that innervate the femoral parts of the retractor unguis muscle. The thicker axon (ax 1; diameter, 8–10 μm) was found to degenerate faster than the smaller axon (ax 2; diameter 5 μm) at a rate of 5 mm/day compared with 3 mm/day at 30°C. The rate of degeneration is reduced to 1 mm/day in ax 1 when the temperature is decreased to 20°C. Griffiths and Boschek (1976) found an amazingly fast rate of degeneration in retinula cell axons in optic lobes of *Musca* (5 μm/min), and a different reaction of the long visual receptor cell component in ommatidia. These authors noticed typical structural signs for anterograde degeneration in the subretinal synaptic neuropil of the lamina and the medulla as early as 2 min after ablation of the sensory somata.

The same events appear considerably later in the corresponding receptor cell system of *Calliphora* (10–15 min after surgery). Interestingly, this anterograde degeneration is first seen in the more distal parts of the retinula cell axons, starting in the lamina and spreading progressively centrally toward the medulla. Griffiths and Boschek's observations of differing local reactions of fibers, isolated from their perikarya, correspond well to those of Clark (1976a,b). In an elegant study, using the cobalt chloride back-filling technique for an identified motorneuron (CDLM) in

Table 12-1. Periods of Postoperative Survival in Various Insect Orders

Group	Genus	Part of nervous system	Postoperative survival	Authors
Hymenoptera	Apis	Brain, mushroom bodies	3 days	Vowles, 1955
	Formica	Interneurons		
	Formica	Brain, mushroom bodies, interneurons	3 days	Goll, 1967
	Formica	Prothoracic ganglion, leg-nerve afferents	1–35 days	Lamparter et al., 1967
	Apis	Brain, deutocerebrum, sensory axons	1–9 days	Pareto, 1972
Diptera	Calliphora	Brain, deutocerebrum		Boeckh et al., 1970
Diptera	Musca	Brain, optic lobes, retinula cells	16 h–6 days	Campos-Ortega and Strausfeld, 1972
	Musca	Brain, optic lobes, interneurons	1 day	Campos-Ortega and Strausfeld, 1973
	Phormia	Prothoracic ganglion, leg-nerve afferents	6 h–8 days	Geisert and Altner, 1974
	Drosophila	Brain, thoracic ganglion, sensory axons	12–96 h	Stocker et al., 1976
	Musca	Brain, optic lobes		Griffiths and Boschek, 1976
	Calliphora	Retinula cells	2 min–6 days	
Lepidoptera	Sphinx	Deutocerebrum, sensory axons	1–7 days	Lund and Collett, 1968
Blattoidea	Periplaneta	Brain, various regions, incl. mushroom bodies	2–90 days	Drescher, 1960
	Periplaneta	Ventral nerve cord, connectives, and leg nerves	2–30 days	Hess, 1958a, 1960
	Periplaneta	Ventral nerve cord, connectives, cercal nerve, afferent fibers	2–36 days 2–10 days	Farley and Milburn, 1969

	Genus	Structure	Duration	Reference
	Periplaneta	Ventral nerve cord, connectives, giant interneurons	4–60 days	Spira et al., 1969
	Periplaneta	Sensory axons	6 h–4 days	Boeckh et al., 1970
	Periplaneta	Brain, deutocerebrum		Ernst et al., 1977
Orthopteroidea	*Schistocerca*	Brain, optic lobes, deutocerebrum	1–7 days	Lund and Collett, 1968
	Laplatacris	Ventral nerve cord, interneurons	30 min–88 h	Melamed and Trujillo-Cenoz, 1963
	Schistocerca	Ventral nerve cord, connectives, interneurons	Up to 10 days	Rowell and Dorey, 1967
	Schistocerca	Ventral nerve cord, connectives, interneurons	1–23 days	Boulton, 1969
	Locusta	Brain, deutocerebrum, sensory axons	1–11 days	Schürmann and Wechsler, 1970
	Locusta *Tettigonia* *Decticus* *Gryllus*	Ventral nerve cord Tympanal receptor cells	3 days	Rehbein, 1973
	Schistocerca	3rd thoracic ganglion, leg nerve, 2 motor axons	2–8 days	Rees and Usherwood, 1972
	Acheta	Terminal ganglion, cercal nerve, sensory axons	8 h–7 days	Edwards and Palka, 1974
	Teleogryllus	3rd thoracic ganglion, motorneuron, CDLM	Up to 168 days	Clark, 1976a,b
	Locusta	Sensory axons and interneurons	6 h–17 days	Ernst et al., 1977

the metathoracic ganglion of the cricket *Teleogryllus,* and by splitting the fiber from its soma, Clark demonstrated that the proximal fiber trunk with its arborization in the neuropil can remain structurally and functionally intact for 63 days postoperatively and may exceptionally survive for 168 days. In contrast, he found that the distal part of the nerve fiber (axon) is destroyed in a few days by cutting the nerve root close to the ganglion. His study demonstrates that in some neurons there is a long-lasting independence of some fiber components from their somata. Cell bodies are generally regarded as the trophic and metabolic "center" of a neuron: Clark's experiments illustrate that the site of severence can be important for the induction and location of degenerative changes. In addition, his study demonstrates a remarkable variance of degenerative processes between individual animals with respect to timing, which was also reported by previous authors (Bodenstein, 1957; Guthrie, 1967; Rees and Usherwood, 1972).

Preparation for Light and Electron Microscopy

Aldehyde-osmic acid fixation is a prerequisite for combined light- and electron-microscopic studies. Excellent preservation of tissue is necessary for correct interpretation of aspects of degeneration. Convenient fixation procedures and subsequent preparation steps are dealt with elsewhere in this volume. Serial plastic sections (1–5 μm thick) can be cut from small and even large blocks to include nerves, parts, or whole ganglia, or brains. The gross effect of anterograde degeneration in terms of its passage in the three dimensions can be judged by light microscopy of unstained 1-μm sections observed with phase contrast illumination (Melamed and Trujillo-Cenoz, 1963; Lamparter *et al.,* 1967) or from sections stained with toluidine blue (Schürmann and Wechsler, 1970) or methylene blue (Boeckh, 1973).

Structural Aspects of Anterograde Degeneration

Structural changes in isolated axons due to degenerative reactions at various times after surgery were described extensively by Hess (1960), Lamparter *et al.* (1967, 1969), Rees and Usherwood (1972) and confirmed for different insect groups by other authors (see Table 12-1). It is not within the scope of this article to describe the entire gamut of changes occurring up to the disappearance of the nerve fiber but rather to emphasize features that allow its identification and, possibly, its synaptic contacts and to discuss factors that might obscure a clear-cut distinction between a degenerating and nondegenerating element.

Light Microscopy

The main criteria for detecting degeneration—shrinkage and darkening of the axoplasm—are usually observed between 1–10 days postoperation, and they can be resolved by light microscopy. Dark droplets and fiber profiles seen by phase contrast microscopy, and dark blue aggregations in stained sections (Figs. 12-1–12-4), correspond to osmiophilic electron-dense degenerative axonal segments, or endings, as revealed by electron microscopy (Figs. 12-6–12-8, 12-10–12-12). This darkening proceeds to the fiber terminal after axotomy and may appear as a chain of dark droplets in longitudinal sections with interposed constrictions (see Lamparter *et al.*, 1967, 1969), so that an affected fiber might not be recognized in even transverse section through a nerve or fascicle. Dark osmiophilic patches appear in lengths of axons devoid of synapses as well as in parts exhibiting "*en-passant* junctions" and in synaptic terminals (the identification of which depends on electron microscopy). It cannot be determined from light-microscopic examination whether degenerative products represent degenerated terminals, boutons, spines, fragmented axons, or phagocytotic glial inclusions.

Figure 12-2 clearly shows the characteristic participation of glial cells situated at the border of and sending their protrusions into the neuropil. This glial rim is filled with dark reaction products at a site where nerve fiber profiles rarely occur. In some cases, intact, relatively thick and translucent nerve fibers can be detected adjoining black patches, thus indicating possible functional contacts (Figs. 12-3 and 12-4). A correct interpretation, however, can only be established by electron microscopy. Since these dark patches cannot be confused with organelles or segments of normal intact nerve fibers—as seen from controls—they allow a light-microscopic mapping and a three-dimensional reconstruction from serial sections of the projective field of axons from severed neurons (Lamparter *et al.*, 1969; Fig. 12-5).

Ultrastructural Observations

Ultrastructural studies of degeneration, which trace neuronal pathways and their functional connections, must be referred to elements that have kept their spatial integrity and that allow the identification of their synap-

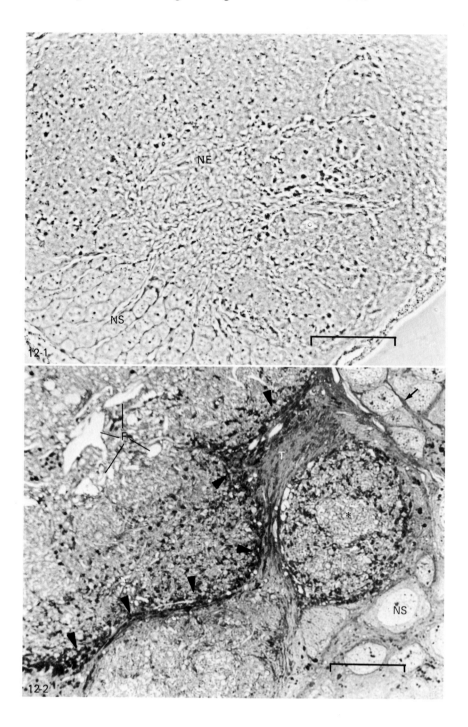

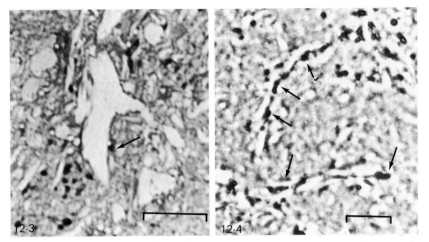

Fig. 12-3. An enlargement from Fig. 12-2. Dark spots represent degenerating fibers. A small degenerating element appears to contact the thick fiber (*arrow*). Scale 10 μm.

Fig. 12-4. Magnification from Fig. 12-3; phase contrast picture. Dark degenerative elements line white intact fibers (*arrows*). Scale 10 μm.

tic specializations despite the increase in opacity of the axoplasmic matrix. This transient state is preceded and followed by discernible structural changes, which are described below.

Axonal Degeneration

Early structural changes in axons, in either nerves, tracts, or connectives, involve a marked swelling of mitochondria (Lamparter *et al.*, 1967; Rees

---

Fig. 12-1. Phase contrast picture of anterograde fiber degeneration showing scattered black degeneration dots in the neuropil (*NE*) and the layer of nerve cell somata (*NS*) in the antennal lobe in the brain of *Locusta migratoria*. Degeneration of sensory fibers 3 days after cutting the antenna. 1-μm plastic section. Scale 100 μm.

Fig. 12-2. Toluidine blue-stained 1-μm plastic section of anterograde fiber degeneration. Dark spots (dark blue in color) mark degenerative changes; thicker fibers (*F*) appear white; smaller fibers that are not resolved by light microscopy are uniformly gray (light blue in color). Note the massive degeneration reaction at the rim of the antennal glomeruli (*arrowheads*), where glial cytoplasm is known to be prominent, and compare with the intact unaffected central core in a glomerulus (*asterisk*). A tract (*T*) of the fine sensory fibers shows degenerative changes. Nerve cell somata (*NS*) are separated by a glial sheath (*arrow*). The toluidine-stained section gives more information than does unstained material, which must be examined by phase contrast microscopy. *Calliphora erythrocephala*. Brain: antennal lobe, 10 h after ablation of the ipsilateral flagellum. Scale 20 μm.

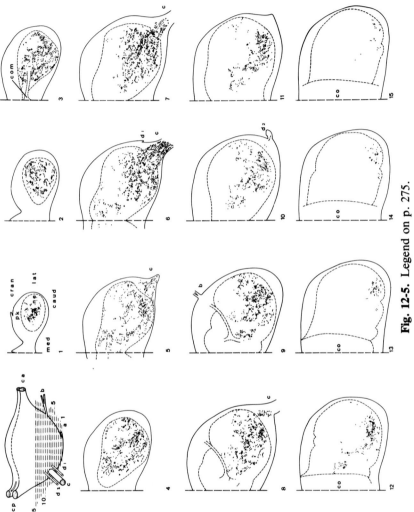

Fig. 12-5. Legend on p. 275.

and Usherwood, 1972). The enlargement of mitochondria and a constriction of the axon cylinder may result in a "string of pearls" appearance of the fiber. The mitochondria are, in addition, subject to increased electron opacity but can preserve their internal structure for several weeks. Membrane-bound vesicles develop in the axon; these have been interpreted as swollen derivatives of the components of the smooth endoplasmic reticulum (Vial, 1958; Farley and Milburn, 1968; Rees and Usherwood, 1972).

Increase of vesicular components (diameter 30–50 nm) has been reported in motor axons and interneuron axons (Melamed and Trujillo-Cenoz, 1963; Boulton, 1969; Rees and Usherwood, 1972). The neurotubuli—normally longitudinally aligned—typically become disoriented and sometimes clumped (after 3 days postoperatively in *Schistocerca* motor axons). The successive increase in electron opacity involves mitochondria and the axoplasmic matrix, and leads to a homogeneous appearance of fiber contents (Fig. 12-6), making the distinction of organelles progressively more difficult and finally impossible. This marked darkening—the main feature for the detection of a degenerating axon—is often realized after 24 h and, in general, becomes pronounced after 1–6 days but may appear much later (Fig. 12-7). Neurotubules disappear or become invisible after several days. The fibers then undergo progressive shrinkage, which is best demonstrated by giant axons (Farley and Milburn, 1969; Spira *et al.*, 1969). Extracellular spaces increase and finally become occupied by glial elements (Lamparter *et al.*, 1967; Rees and Usherwood, 1972). Individual fibers are at this stage no longer discernible and lose their integrity by virtue of the destruction and fragmentation of their axolemmas (e.g., 6 days after neurotomy in motor axons, fragmentation of sensory axons 4 days postoperatively) and an axon may disappear entirely 8 days postoperatively (Rees and Usherwood, 1972; Lamparter *et al.*, 1967).

Preterminal and Terminal Degeneration

In the insect central nervous system synapses are restricted to the neuropil. They frequently occur *en passant* in thicker parts of an axon and are

Fig. 12-5. Mapping of sensory fiber distribution in the ipsilateral prothoracic ganglion of *Formica lugubris* following leg amputation at the coxal level. Degeneration time is 10 days. 15 sections were taken at 10-μm intervals from the ventral half of the ganglion. Drawings were made from phase contrast photomicrographs of 0.8-μm-thick, unstained, plastic sections. The figure in the upper left corner indicates the position of the 15 sections. Degeneration from the sensory axon of the leg nerve (*c*) is restricted to the ventral part of the ganglion. Nerve roots: *b, c, d₁, d₂*; anterior connective = *ca*; caudal = *cau*; cranial = *cran*; connective = *co*; posterior connective = *cp*; lateral = *lat*; commissure = *com*; medial = *med*; perikaryal layer = *pk*. [From Lamparter *et al.*, 1969, p. 374.]

abundant in fine terminal arborization.   Both terminal and *en passant* synaptic junctions in sensory and motor axons are subjected to a sequence of identical structural alterations during degeneration (Lamparter *et al.*, 1967, 1969; Rees and Usherwood, 1972).   The process of darkening is identical to that in more distal parts of the axons: early alterations,

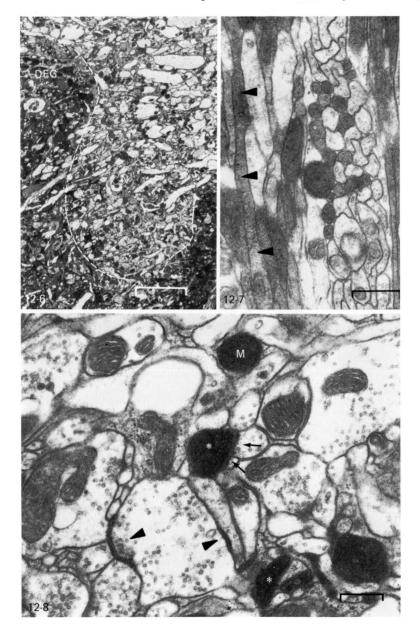

which are attributed to initial reactive events, affect the clear presynaptic vesicles, which tend to cluster and show a marked increase in number (Usherwood and Rees, 1972).

The next phase is a progressive darkening of axoplasm, enclosing the translucent synaptic vesicles, which appear slightly larger than normal and which may fill the entire fiber profile (Fig. 12-9). Vesicles remain clear in the initial phases of Wallerian degeneration and stand out clearly against a progressively darkening background until their membranes can no longer be resolved. This "honeycomb" pattern has been observed in vertebrate axons subjected to axotomy (Birks *et al.,* 1960) as well as in insect axons (Figs. 12-8,12-10,12-12) and has been interpreted as the penultimate stage of axon degeneration, prior to the breakdown of the axonal structure and to the formation of myelin-like bodies (Rees and Usherwood, 1972).

Normal presynaptic profiles of a dark type have been reported in insects (e.g., Steiger, 1967) and in vertebrates (e.g., Cohen and Pappas, 1968). Degenerating fibers should not be confused with boutons of normal axons on the basis of their honeycomb pattern. Lamparter *et al.* (1969) stated that normal boutons can be safely distinguished from their degenerating counterparts on the basis of three criteria:

1. The cytoplasmic matrix is not opaque but relatively clear.
2. Membranes of synaptic vesicles appear distinct from the axoplasmic matrix.
3. Only degenerated boutons can show ensheathment by glial processes.

In addition distended mitochondria are atypical for normal intact fibers.

Presynaptic structures are occasionally detected in truly degenerating

---

Fig. 12-6. Electron micrograph of antennal lobe neuropil, low magnification. Massive degeneration is indicated by abundant osmiophilic fiber profiles (*DEG*) at the left hand side where some scattered clear intact fibers are found. Normal neuropil (*N*) is at the right. *Calliphora erythrocephala;* degeneration time 10 h. Scale 5 μm.

Fig. 12-7. An electron micrograph of a branching tract with small sensory fibers in the antennal lobe. Some fibers appear intact while others show various degrees of electron opacity as a result of degenerative alternations. Organelles such as mitochondria and tubules can still be distinguished. A degenerating axon can show various electron densities (*arrowheads*). *Calliphora erythrocephala;* degeneration time 10 h. Scale 0.5 μm.

Fig. 12-8. Electron micrograph of dark, degenerating fiber profiles (*asterisks*) with presynaptic vesicles slightly lighter than the surrounding fiber matrix, which results in the honeycomb pattern. Subsynaptic densities are visible in two undegenerated postsynaptic fibers that synapse with a degenerating fiber (*arrows*); compare this with normal presynaptic fiber with two synaptic sites (*arrowheads*). Degenerating mitochondrion (M). *Calliphora erythrocephala.* Brain: antennal lobe; degeneration time of sensory antennal nerve fibers was 3 days. Scale 1 μm.

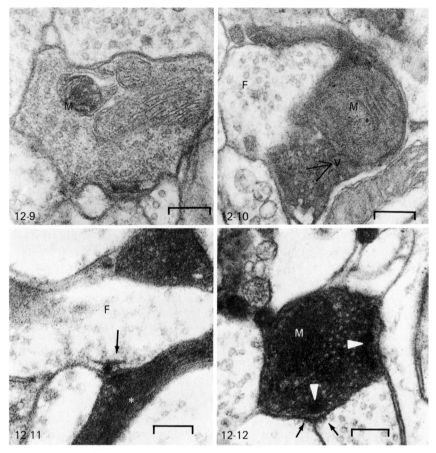

Fig. 12-9. A fiber profile, presumably in a state of early degeneration. Lysis of a mitochondrion (*M*) and a darkening of the fiber matrix (compare surrounding normal fibers) visible. The presynaptic figure (*table*) is clearly shown. *Calliphora erythrocephala*. Brain: antennal lobe; degeneration time of sensory antennal axons was 3 days. Scale 0.2 μm.

Fig. 12-10. A degenerating fiber profile (*F*) partially darkened: a mitochondrion (*M*) and synaptic vesicles (*V*) can still be detected in the electron-opaque matrix. *Calliphora erythrocephala*. Brain: antennal lobe; degeneration time of sensory antennal nerve fibers was 3 days. Scale 0.2 μm.

Fig. 12-11. A degenerating fiber (*asterisk*) connected with a normal presynaptic fiber (*F*) by a dyad-synapse with a presynaptic figure (*arrow*). *Calliphora erythrocephala*. Brain: antennal lobe; degeneration time of antennal sensory nerve fibers was 10 h. Scale 0.2 μm.

Fig. 12-12. A degenerating fiber in the honeycomb stage: a darkened mitochondrion (*M*) and synaptic vesicles are still visible. Presynaptic figures (*arrowheads*) are present as black spots. Synaptic clefts remain widened and subsynaptic densities (*arrows*) can be recognized in the normal postsynaptic nerve fibers. *Calliphora erythrocephala*. Brain: antennal lobe; degeneration time of sensory antennal nerve fibers was 3 days. Scale 0.2 μm.

profiles as slightly electron-denser spots (Fig. 12-11).   Usually they are masked by the complete, homogeneous opacity of the matrix.   However, the site of a synaptic contact is often revealed by persistent widened clefts and the specific subsynaptic densities on the postsynaptic membrane (Figs. 12-8 and 12-12) of the neighboring fiber (Lamparter *et al.*, 1967; Boeckh *et al.*, 1970; Schürmann and Wechsler, 1970).   Presynaptic fibers, with typical presynaptic apparatus, contacting postsynaptic fibers in an advanced state of degeneration, have been demonstrated in Diptera by Campos-Ortega and Strausfeld (1973).   Some synaptic sites retain their normal appearance for 13 days postoperatively in ant prothoracic ganglion and are thus characterized as remarkably stable structures (Lamparter *et al.*, 1967).

Glial Reactions

The loss of spatial integrity of axonal units, their displacement, reduction, and final abolition during degeneration are closely related to reactive changes of glial cells induced by the lesion itself and/or alterations of nerve fibers disrupted from their somata.   Comprehensive and detailed descriptions of the response of the different types of supportive cells of the insect nervous system ("Schwann" cells, perineurial cells, neuropilar glial cells; see Strausfeld, 1976) after neural surgery are given by Hess (1960), Boulton (1969), Lamparter *et al.* (1967, 1969), and Rees and Usherwood (1972).   Glial cells of the neuropil proliferate, divide, ensheath, and invade degenerating fibers to finally occupy the space donated by shrinking and vanishing nervous elements.   The nerve fibers lose their normal spatial relationships and their fragmentation is achieved by autophagocytosis, phagocytotic activity, and digestion by glial cells.   Glial cells do not seem to attack and incorporate intact fibers synapsing with degenerating elements.   Dark glial elements are sometimes indistinguishable from degenerating axons.

In summary, the anterograde degeneration process in axons of the insect nervous system shows features very similar to those of the electron-dense type of degeneration in vertebrate nervous systems (see Gray and Guillery, 1966; Guillery, 1970).

Physiologic and Chemical Correlates of Anterograde Degeneration

Investigations on physiologic changes in axons subjected to experimental degeneration are even scarcer than studies on structural alterations.   In general it can be stated that the events of functional breakdown (with respect to spike propagation and synaptic transmission) must precede gross structural changes of darkening, autophagocytosis, fragmentation, and glial phagocytosis. The reader is referred to selected studies dealing with behavioral and electrophysiologic investigations related to structural observations (Bodenstein, 1957; Drescher, 1960; Usherwood, 1963a,b; Usherwood et al., 1968; Usherwood and Rees, 1972; Guthrie, 1967; Farley and Milburn, 1969; Spira et al., 1969; Clark, 1976a,b).

Biochemical studies on anterograde degeneration in the insect nervous system are lacking, and the biochemistry of the metabolic processes that accompany and cause structural changes in degenerating fibers is not fully understood even for the vertebrate nervous systems. The assumption by Rees and Usherwood (1972) that darkening in degenerating fibers is due to the breakdown and lysis of external and internal membranes of axons and reflects bonding of spread membranous lipids to osmic acid seems plausible with respect to cyto- and biochemical studies on vertebrate nervous systems (Adams, 1965; Domonkos, 1972).

Concluding Remarks

Experimental anterograde degeneration as a neuroanatomical tool suffers from various experimental restrictions that can only be overcome by careful studies of timing of degeneration phases within a specific subsystem of the central or peripheral nervous system. A major difficulty is the adequate choice of the postoperative survival period, which is very important for the achievement and correct interpretation of results. No general statement on the response of a nerve fiber deprived of its soma can be made for the insect nervous system. Some severed brain interneurons are reported to be rapidly destroyed and not substituted by regenerative processes (Vowles, 1955; Drescher, 1960). Ventral nerve cord interneuronal fibers possibly survive for longer periods, as can be deduced from fiber counts of normal and severed connectives (Rowell and Dorey, 1967; Boulton, 1969).

Regeneration of motoraxons and sense organs—at least in larvae and sometimes in imagos—is known for several species (Bodenstein, 1957; Edwards and Palka, 1974, 1976; Young, 1973), and prolonged autonomy of the axon after separation from the soma has been described from observations of an identified motorneuron in the ventral nerve cord of the cricket Teleogryllus oceanicus (Clark, 1976a,b). Maintenance of postoperative structural and functional integrity might be a widely distributed feature for some nerve cell types in insects and is known to be the case for nerve fibers of Crustacea (see Wine, 1973; Krasne and Lee, 1977). However, as far as is known, there is no evidence from insect nerve fibers that absence of structural decline may follow soma ablation. At least the distal part of the axon degenerates quite rapidly when separated from the proximal axon segment and its arborization (collaterals, dendritic branching; Clark, 1976a).[2]

For careful and correct interpretation of the fine structure of Wallerian degeneration several possible interfering and misleading events must be considered: spontaneous degeneration, gerontologic degradation of neuropil, retrograde changes, and transneuronal degeneration. The possibility of error from spontaneous and gerontologic degeneration can be greatly reduced and mainly ruled out when young intact animals are used and a number of controls are compared. Retrograde changes in insect neurons have been used for mapping nerve cell bodies (Young, 1973), and electron-dense bodies were described by Farley and Milburn (1969) in fibers; however no honeycomb structures have been revealed in proximal segments. Transneuronal degeneration—known for different parts of the vertebrate nervous system (Cowan, 1970)—was discussed for the insect nervous system (Campos-Ortega and Strausfeld, 1972) but known structural details differ from our concept of normal anterograde degeneration events.

Although the methods for experimental anterograde degeneration suffer from the limitations described above, the technique represents an important tool for insect neuroanatomy. Many recent studies employing Wallerian degeneration have extended our knowledge of neural relationships and have yielded information that was not or could not be gained using other methods. The main advantage of the technique is that topographies and synaptic connections of selected fibers can be qualitatively explored by a method that simply combines both light and electron microscopy.

Addendum

Dagan and Sarne (1978) report a loss of the enzyme choline acetyltransferase in the ventral nerve cord of *Periplaneta americana* after induced degeneration of the giant fiber system (postoperative survival period, 30–60 days), thus showing some evidence for a cholinergic nature of the giant neurons.

---

[2] Local ablation at the outer surface of the fly lamina results in normal degenerative changes (darkening) of tangentially oriented fibers (terminal processes of wide-field centrifugal cells). However, monopolar cells whose perikarya have been destroyed give rise to other initial structural changes (1 day postoperatively). Their initial dendritic segments swell, become electron translucent and contain large vacuolar structures. Eventually, cell membrane portions and the more distal portions show darkening (J. A. Campos-Ortega and N. J. Strausfeld, unpublished).

Chapter 13

Simple Axonal Filling of Neurons with Procion Yellow

G. E. Gregory

Rothamsted Experimental Station
Harpenden, England

The first successful use of the textile dye Procion yellow to fill individual neurons was reported by Kravitz *et al.* in 1968. Stretton and Kravitz (1968) then gave a fuller description of the method and of its use to reveal neuronal geometry in abdominal ganglia of the lobster. They had screened a total of 63 Procion dyes and related compounds and of these the four best for neurobiological purposes proved to be Procion brown H3RS, brilliant red H3BNS, navy blue H3RS, and yellow M4RS. Procion yellow was appreciably more fluorescent than the other three and was finally used exclusively. A full account of its selection and the development of the technique has been given by Stretton and Kravitz in the book on intracellular staining by Kater and Nicholson (1973). The dye could be introduced intracellularly with a micropipette and would spread throughout the neuron, revealing it under ultraviolet illumination virtually in its entirety. Iontophoresis was the method originally employed to inject the dye and rapidly became the standard procedure, though pressure injection has also sometimes been used (Remler *et al.,* 1968; Milburn and Bentley, 1971). The chemistry of Procion yellow and many aspects of its use have since been written about in considerable detail in the book by Kater and Nicholson (1973), particularly in chapters by Stead (chemistry), Nicholson and Kater (scope of uses), Kater *et al.,* and Mulloney (techniques of introduction into the cell), Van Orden (fluorescence microscopy), Purves and McMahan (use as an electron-microscopic marker), and Murphey (in characterizing an insect neuron). Intracellular injection of Procion yellow is further discussed in Chapters 14 and 15 of the present volume.

Although introduction of the dye into the cell by injection has proved an extremely valuable tool and is now widely used in neurobiological studies, it has its limitations. In particular, it requires a considerable degree of technical expertise, and normally only one neuron, or at most only a very few, can be filled in any one preparation. To study whole groups of neurons demands a different approach—filling the cells with dye by making it enter the cut ends of their axons. Two basic methods of doing this were developed independently in different laboratories. One method, axonal iontophoresis, was evolved by J. F. Iles at Oxford to study motor neurons in the metathoracic ganglion of the cockroach *Periplaneta americana* (L.) (Iles and Mulloney, 1971). It involved driving the dye ions into the cut end of the selected nerve by passing this through an insulating (petroleum jelly) barrier from a pool of saline solution into a pool of dye and imposing an electric current in the appropriate direction through electrodes placed in the two solutions. Details are given by Mulloney (1973). The second method, simple immersion in a dye solution, was developed by the present author at about the same time for displaying the bundles of fibers that form the roots of the peripheral nerves in the cockroach mesothoracic ganglion (Gregory, 1974a). Payton *et al.* (1969) and Payton (1970) had noticed that when crayfish ventral nerve cord ganglia were immersed in a Procion yellow solution some of their axons stained, and they suggested that this could be due to dye entering from the cut ends of the nerves. Tests by the present author with cockroach ganglia confirmed that staining in this way was feasible and indeed delightfully simple (Gregory, 1973), and the method has since worked well with a large number of preparations. However, a two-bath method, a simplified version of axonal iontophoresis, without the use of an imposed current, has also been tried and found to be effective. Both this and the simple immersion technique are described below and their relative merits assessed.

The Immersion Method

Practical Procedure

The Procion yellow dye was originally provided by Imperial Chemical Industries Ltd., but is now available through the usual histologic supply houses. For this method, it is used as a saturated solution in an appropriate insect saline solution diluted to maintain isotonicity. For the cockroach the saline solution of Yamasaki and Narahashi (1959) has been used routinely and requires about 2.5 g of dye per 100 ml saline solution. The addition of 11.4 ml distilled water yields a suitably diluted solution, though the effects of small variations in dilution have not been extensively tested. Oily orange-colored droplets form in the solution but do not seem to affect its staining ability. It appears to keep well in a refrigerator. The final dye concentration in the solution is less than 2% at 2°C, the temperature normally used for staining. Saline solutions used for other insect species would no doubt need to be diluted differently to avoid shrinkage or swelling of their nerve tissue.

Insects can be narcotized with carbon dioxide if necessary and are then dissected under a suitable saline solution, often most easily from the ventral side. The chosen ganglion is removed, together with the ganglion in front of and behind it, keeping its peripheral nerves as long and undamaged as possible. The nerve to be filled with dye is then cut short, close to the ganglion. More than one nerve can be treated in this way and filled, if desired. The chain of three ganglia is then immersed in the staining solution in a covered dish kept at around 2°C. The time needed for staining depends upon the length and thickness of the axons being filled, but a good starting point for trials is usually about 1–2 h. Dye is prevented from entering the ganglion along the interganglionic connectives by the ganglia anterior and posterior to it, while keeping the other nerves long helps to prevent entry of dye along them. The fairly low temperature slows the rate of movement of the dye and so makes it easier to adjust the staining time so that the chosen neurons are sufficiently filled but staining of other neurons by entry of dye along other nerves is minimized. Individual nerve branches can similarly be cut and filled if they arise close to the ganglion, but if they lie more distally, other, shorter nerves may fill preferentially and perhaps confuse the final result.

After removal from the stain the three ganglia (it is helpful to leave the two unwanted ganglia attached for the time being, to facilitate handling) are washed in three changes of saline solution, 10 min each, at room temperature (20–25°C), to remove dye from the surface, and are then placed in fixative. "Aged" alcoholic Bouin (Duboscq-Brasil) or its synthetic substitute, as used in the Bodian silver technique (see Chap. 6), gives good results, and the ganglia should be fixed overnight at room temperature and then washed out in 70% ethanol. The addition of 10% ammonium acetate to the ethanol removes the yellow color due to picric acid more speedily and completely but is hardly necessary as the color seems to have no discernible effect on the fluorescence of the tissue. Raising the temperature also speeds processing, as described in Chapter 6. The ganglia are then dehydrated in ascending ethanol grades, cleared, embedded, and sectioned, the unwanted ganglia being cut away at whatever stage is most convenient. The processing regime given in Chapter 6 can again be used, except that it is usually better not to stain temporarily with eosin to aid orientation in the block, as eosin itself fluoresces and, if used, must later be completely removed by thoroughly washing the sections in 90% ethanol. Orientation of unstained ganglia in the block can prove difficult, however, as they are generally much the same color as the wax, so it is usually best to trim one side of the block very close to the specimen and orient it from this view. Sections can be cut at whatever thickness is most appropriate to the material, but thicker (20 μm) sections are usually better for tracing and mapping fiber pathways, while thinner (10 μm) sections tend to be better for photography. The sections are mounted on slides with albumen and dried in the usual way. The warning given in Chapter 6 against smearing the albumen on slides with the finger again applies here, as skin cells are autofluorescent. The sections are dewaxed— and taken down to 90% ethanol to wash out eosin if it was used—and mounted in a nonfluorescent mountant. They are viewed by fluorescence microscopy using ultraviolet illumination. Suitable exciter filters are the Schott BG12 + BG38 (red-absorbing) and as barrier filter the Zeiss 50, or other manufacturers' equivalents. The stain is subject to photodecomposition, and to limit fading, sections should not be exposed to ultraviolet light for longer than necessary (for more detail, see Van Orden, 1973; Chap. 15, this volume).

Results

During staining, dye enters all the cut ends of nerves and also the tracheae and the outermost layer (neural lamella) and sometimes also the underlying perineurium of the ganglion, but it reaches the central fibrous core (neuropil *sensu lato*) at first only along the axons of the shortened nerve or nerves (Fig. 13-1; for other illustrations of results given by this technique, see Gregory, 1974a, Figs. 16–22). When observed using the filter combination given above, the axons and their branches thus appear brilliant yellow against the green background of the autofluorescent unstained tissue. Longer staining yields a brighter color and more complete filling of the finer, branch-fibers, but selectivity gradually becomes impaired as axons of other nerves fill. Attempts to prevent this by sealing their cut ends with petroleum jelly, liquid paraffin, or cautery have proved unsuccessful. Contrast between the stained fibers and the background increases until (in ganglia the size of those of cockroach) after 4–6 h the background begins to stain pale greenish yellow, apparently because dye travels inward from the nerves and the ganglion surface along the glial cell processes that ensheathe the fibers. After about 12–24 h, the entire ganglion is stained bright yellow and is of no further use. Successful staining

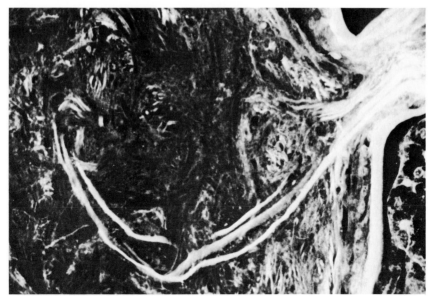

Fig. 13-1. Cockroach mesothoracic ganglion, laterally oblique frontal section, 10 μm thick, showing axons of nerve 3, root 4, filled with Procion yellow; anterior at *top*. Immersion method, nerve 3 cut short, stained for 2 h at 2°C.   × 309.

thus depends upon selecting an appropriate, short, staining time, and on reducing the length of the axons to be filled to a minimum.

The Two-Bath Method

Practical Procedure

This, as already noted, is essentially a simplification of J. F. Iles's axonal iontophoretic technique, omitting the use of electrodes or current. Iles and Mulloney (1971) reported that without the aid of an imposed current dye was prevented from entering the cut ends of motor axons by injury currents, but no problems of this kind have so far been encountered by the present author using cockroach material. The procedure closely parallels that described by Pearson and Fourtner (1973) for the axonal infusion of cobalt (see Chap.19 of this volume) and now in general use for that purpose.

A single ganglion is dissected out as in the immersion method but this time only the nerve to be filled need be kept long. The ganglion is placed in a bath of suitable saline solution and the nerve led across a sealing barrier into a second bath containing the dye. A saturated solution (about 6% w/v) of Procion yellow in distilled water suffices, and there seems to be no problem with osmotic effects. Two shallow cavities almost touching in a piece of Plexiglas (Lucite) form convenient baths. They may be connected by a small groove but this is unnecessary for small nerves and not essential even for larger ones. A narrow band of sealant is applied between the cavities, conveniently by means of a bent needle. Petroleum jelly is the sealant commonly used, but a stickier, more water-repellent material such as Apiezon-N high-vacuum grease (Apiezon Products Ltd., distributed by Shell Chemicals U.K. Ltd., 41 Strand, London, England) offers some advantages, particularly if staining is carried out at higher temperatures (see below). A good seal is obtained more easily if the baths are only partly filled with liquid at this stage. Also, sealing around larger nerves and connectives, which are sometimes difficult to seal adequately, is improved if their surface is blotted briefly with a sliver of filter paper and lightly smeared with grease before they are laid in position. Once in position the nerve, of whatever size, is covered with another band of grease, which is pressed carefully around it. The baths are then topped up to give a convex meniscus each side, which touches the grease barrier and ensures that all parts of the ganglion and nerve are immersed. Finally, it is helpful to draw the end of the nerve gently further into the dye bath until the ganglion is close against the grease barrier and then to cut through the nerve under the dye solution; dye seems to penetrate more readily into the freshly cut axons. An alternative method that sometimes helps is to cut the nerve under saline solution or distilled water in the second bath, and then replace this with dye solution. The nerve must not be unduly stretched or cut too close to the barrier or it may later draw back and become buried in the grease. The preparation should then be covered to restrict evaporation, for instance in a petri dish, and if staining is likely to be prolonged the surrounding atmosphere should be kept moist by, say, lining the dish with filter paper dampened with saline solution. The temperature used for staining can be varied according to the needs of the material. For large nerves and connectives, which usually fill rapidly, it is often advisable to lower the temperature, as in the immersion method, while for finer nerves, room temperature or higher, up to 37°C or so, may be needed to speed up staining. The time required will again depend on the length and thickness of the axons being filled, and also on the temperature used, but can range from a half hour or less for large, short nerves to several hours or overnight for small nerves. The color of satisfactorily filled axons is not usually visible through the ganglion, even though this is mostly unstained, and so cannot normally be used as a safe indicator of when staining should be stopped.

At the end of staining the ganglion is removed from its bath, drawing the nerve gently from the grease barrier, and rinsed briefly in saline solution. It is then fixed, dehydrated, and sectioned as described for the immersion method. Whole mounts of entire ganglia have not proved really satisfactory with either staining method, because the depth of field available for observation with ultraviolet microscope illumination is generally insufficient and detail is usually further obscured by staining of the outer layers of the ganglion, even using the two-bath method (see below).

Results

Filled axons and their branches have the same general appearance as with the immersion method. However, they can often be stained for longer periods and hence more intensely and completely without selectivity being diminished by entry of dye from the ganglion surface, although it still enters through the glial tissue of the nerve being filled, so that the background eventually begins to stain from the base of the nerve inward. Staining should therefore not be too prolonged. Dye also travels from the nerve through the neural lamella of the ganglion, so that this again stains brightly (Fig. 13-2). Successful staining with this method therefore depends less on selecting the optimum staining time than on manipulative skill in making an efficient seal between the nerve and the grease barrier.

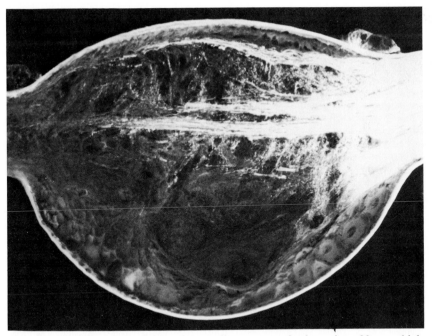

Fig. 13-2. Cockroach metathoracic ganglion, parasagittal section, 20 μm thick, showing longitudinal fiber tracts filled with Procion yellow from right anterior connective; dorsal at *top*. Two-bath method, stained for 2 h at 37°C. Note staining of neural lamella around ganglion. × 129.

The Two Methods Compared

The results given by the two methods are very similar, the main difference being that in the immersion method staining of the background tends to begin peripherally around the whole of the ganglion, whereas in the two-bath technique it appears at first only in the vicinity of the base of the nerve being filled. However, each method has its own advantages for particular purposes. The immersion method is better for filling nerves that are very short, e.g., the ventral thoracic nerve 7 of the cockroach, or too delicate easily to survive the operations called for in the two-bath technique, while the latter method often gives a cleaner background and better staining of fine branch-fibers. On the debit side, problems may be encountered with the immersion method in making the dye solution adequately isotonic with the tissue, while the two-bath technique requires considerably more time and practice to set up. A further difference is that it is difficult to fill more than one nerve or connective at a time using the two-bath technique, though it can be done with care, whereas it is quite feasible and indeed extremely simple with the immersion method.

Chapter 14

Intracellular Staining with Fluorescent Dyes

L. G. Bishop

University of Southern California
Los Angeles

Classically, neuroanatomy has been studied using a variety of tissue-applied stains. Such stains vary in their ability to highlight cells. At one end of the spectrum most of the neurons present are stained to reveal the gross architecture of the brain. Speculations can then be made about the connectivity between those cells that appear to come into close association. At the other end of the spectrum individual cells are stained selectively, revealing the shapes of single cells as though they were taken from their environs.

Combining the intracellular electrode with simultaneous intracellular staining created a powerful probe of the nervous system (Kater and Nicholson, 1973). With this technique it is possible to record electrical activity (spikes, synaptic potentials) of a single cell and then subsequently study the detailed anatomy of the same cell with the light or the electron microscope. One can refer to the electrical activity and structure of an *identified neuron*. Hopefully its structure can be related to the cells that surround it, and its connectivity can be surmised by an analysis of synaptic structures in the electron microscope. The essential feature of this technique is that the cell has lost its anonymity. One knows where it is and what it looks like.

291

Procion Dyes

Ideally one would prefer to record and inject dye from the same electrode. The best possible circumstance is that the dye dissociate readily in water into a charged ion. In those situations in which penetration of the cell cannot be maintained for a long time, it is important that the dye should migrate readily out of the electrode. It should not react with the glass or precipitate to clog the electrode tip. It should not react readily with the intracellular environment in a way that will prevent it from migrating throughout the cell. The dye should fill even the finest cell processes and preferably be electron dense. The dye should survive fixation and histologic processing. It should outline the boundaries of the cell clearly, e.g., it should not leak out of the cell and present a "fuzzy" definition of cell boundaries. The dye should be clearly visible for viewing and photography, and should be stable with time (for storage and reexamination) and with the exposure to light necessary for photography. It would be convenient if the dye came in a variety of recognizably different colors so that more than one cell could be stained in the same preparation. It is not essential but certainly preferable that the dye be nontoxic to the cell. The dye should be readily available commercially and should not be expensive. It should have a long shelf life. The dye and the chemicals used to process it (if necessary) should not be toxic to the experimenter.

This would seem too much to ask from a single dye but, thanks to a search by Kater and Nicholson (1973), a useful series of dyes, the Procion dyes, was found. Procion dyes are derived by substitution of a chromagen on cyanuric chloride. They have a molecular weight of the order of 3000 and a negative charge in aqueous solution. At physiologic intracellular pH, Procion dyes react primarily with amino groups. Procion dyes are available from Poly Sciences, Inc. (Paul Valley Industrial Park, Warrington, Pennsylvania 18976) and Inolex Corp. (4221 Western Blvd., Chicago, Illinois 60609) in powder form at a nominal cost for the purified powder. This powder can be used without further treatment and can be stored indefinitely in the freezing compartment of a refrigerator (4°C).

In my experience a saturated solution of Procion dyes in glass double-distilled water makes an excellent recording medium. My experience is that the concentration of the dye solution is not critical, i.e., a solution of Procion yellow that looks dark maroon to the eye is adequate. Fresh stock solution should be made approximately once a week and should be refrigerated. In my experience Procion-filled electrodes are usable for a few hours, but cannot be stored overnight. The resistance (see below) of a Procion-filled electrode is 3 to 4 times that of the same electrode filled with KCl or K-citrate. Migration of the dye out of the electrode, while not 100% dependable, is acceptably high. These dyes fill even the finest cell processes (Fig. 14-1, Plate 10).

Procion as a Recording Electrode

For this technique the recording electrode should be small enough to penetrate and stay in a cell, of low enough resistance for acceptable signal-to-noise ratio, and capable of passing sufficient current to iontophorese enough dye into the cell to mark it.   Theoretically, one would expect the electrode resistance to be the controlling factor relevant to recording noise and current-carrying capacity.   This should be a function of the resistivity of the fluid in the electrode and of the geometric configuration of the electrode.   The shape of the electrode will determine its value for penetrating, and to some extent, its ability to remain inside cells.

Many problems were moderated by the introduction of the glass fiber technique.   In this technique a fine strand of glass is made by pulling a piece of capillary tubing rapidly in a flame, or one or several strands of glass wool are inserted into a piece of capillary tubing prior to the pulling of the capillary into a micropipette.   In the micropipette puller, the glass fiber melts along with the capillary tubing and fuses to its side.   The micropipette may then be filled by injection from the blank end, for example, with a syringe and No. 30 hypodermic needle.   The electrode fills by capillary action and can be used within a few minutes.

This technique is generally adequate, but suffers from two problems: Insertion of the glass fibers is tedious and time consuming and the glass fibers sometimes do not fuse to the side wall of the capillary tubing and are not drawn down along with the capillary tube.   This is especially problematic when a vertical pipette puller is employed.

The technique I prefer is to manufacture the capillary tubing so that a glass fiber adheres to the inner side wall of the tubing.   A device to do this is pictured on page 161 of Kater and Nicholson (1973).   Capillary tubing (1 mm o.d.) is placed inside large tubing (9.5 mm o.d., Cat. No. 237650 Corning) and the combination is drawn through an oven by rollers emerging at an o.d. of 1 to 2 mm.   With such an apparatus enough tubing can be made in an afternoon to supply a busy laboratory with four people for a month.   Such an apparatus was used in this laboratory.   It was obtained by special order from A. Broers (Information Science Booth Computing Center, California Institute of Technology, Pasadena, California 91109) or it can be obtained from the Narashige Instrument Company (1754-b Uarasuyama-Cho, Setagaya-Ku, Tokyo, Japan).   A more convenient, but more expensive, way to get such tubing is to purchase it cut into 4-inch or 6-inch lengths from A-M Systems, Inc. (P. O. Box 7332, Toledo, Ohio 43615), W-P Instruments, Inc. (10 Marietta Street, Hamden, Connecticut 06514), or David Kopf Instruments (P. O. Box 636, Tujunga, California 91042).   Multibarrel tubing is also available.

For work on insects typically four or six electrodes are pulled on a vertical pipette puller (David Kopf Model 700 C).   They are filled by injection from a syringe with a No. 30 hypodermic needle (keep these in distilled water when they are not in use), and placed tip down in a holder (e.g., wax placed on a plastic ring with three legs) and placed in a humid chamber, e.g., a jar with water in the bottom. Such storage chambers may be obtained from the Kopf Instrument Company. The electrode tips fill immediately, and the shank is filled by applying constant pressure upon the syringe while withdrawing the hypodermic needle from the electrode. This may leave air gaps in the electrode barrel, but these will disappear within a minute. Stored in this manner, the electrodes are usable for up to 4 h.

What is a good electrode? The answer is one that works. In my experience the best recordings and cell fillings were obtained from broken electrodes. Unfortunately electrode breakage is an uncontrollable variable. So, we are left with making electrodes that are judged a priori to be good. What criteria are to be used to tell a good electrode from a bad one? Initially I screened electrodes by measuring the DC resistance in distilled water, and/or inspecting them in a compound microscope at $450\times$. Electrode resistance was measured with a platinum foil indifferent electrode of about 1 cm $\times3$ cm and with a Keithley model 360 electrometer. Electrodes with resistances between 100 and 200 megohms were selected. In some cases the electrode tips were ground (bevelled) on a Narashige Microelectrode Beveler (Narashige Instrument Co. Model No. 5). This instrument measures the electrode resistance in an AC wheatstone bridge at a frequency of 130 Hz. Procion-filled electrodes, measured in excess of 300 megohms before bevelling, were bevelled to resistances of 30–200 megohms. In my experience bevelled electrodes penetrated tissue more readily than electrodes that were not bevelled, but the success rate of good fills was unaffected. At the present time I screen unbevelled electrodes only by the noise level when the electrode is in solution in the preparation. Bevelled electrodes are viewed first at $450\times$ in a compound microscope to check that the tips are clean and are not broken.

Injection of Procion

The Procion molecule is negatively charged; hence, it may be ejected from the electrode tip by passage of hyperpolarizing current (i.e., the Procion electrode is made negative relative to the indifferent electrode). There are many ways to eject dye, and each experimenter has his own favorite. One method is to place the recording electrode in momentary contact with a large, negative DC potential, e.g., -180 V. Good, single-cell fills can be achieved with this technique. However, this technique can lead to extracellular spilling of the dye and/or injection of dye into more than one cell. For example, the use of this technique led to confusion in the identification of movement-detecting cells in the optic lobe of the fly (see Dvorak *et al.*, 1975a,b).

More effective isolation of single cells is achieved by controlling the magnitude of the injected current. Currents up to 10 nA can be achieved with most commercially available preamplifiers by means of an externally controlled applied voltage. The amount of current passed is then a function of the resistance of the electrode. The preferred way to inject dye is to use a constant-current source such as the WPI (10 Marietta Street, Hamden, Connecticut 06514) Model 160 Microiontophoresis Programmer. When employing such a device one must disconnect the recording electrode from the preamplifier and connect it to the constant current source. This may be done remotely, i.e., away from the site of the electrode, with some commercially available preamplifiers, for example, the WPI Model 701 preamplifier with a model BB-1 breakaway box. A problem is that in the current-passing mode there may exist current-carrying pathways to ground other than through the preparation. One can determine the current passing through the preparation only by measuring it directly.

Substances have been injected by applying pressure to the recording micropipette. Pressure from a tank of nitrogen was applied to a glass micropipette through a reduction valve and gated electronically by a General Valve Company (No. 9-82-902) solenoid valve. To date, experience with this method of injection is that the electrode tips must be larger than those used for iontophoresis so that the electrode tips do not plug up under pressure. Procion yellow, Procion red, and Procion yellow/red mixtures have been injected into fly movement-detector cells in the optic lobe by the application of a 30-psi 0.5 Hz pressure signal for a minute or longer using electrodes with a DC resistance of 30 megohms.

The presence of air bubbles anywhere in the micropipette will interfere with dye injection by pressure. When micropipettes made from filament tubing are filled with a hypodermic needle, some bubbles are usually formed in the taper, but away from the tip. Sometimes all of these bubbles will leave the taper and rise through the barrel to be released to the atmosphere. Usually this is not the case. Such bubbles will not interfere with electrical recording or iontophoretic dye injection. For pressure injection, or actually for any purpose, the tip and taper may be filled quickly (within 30 s) by capillary action, without bubbles, by placing the untapered end of the pipette into a pool of the dye.

Some difference of opinion exists with regard to how long to allow for the Procion dye to migrate throughout the cell. Initial reports in the literature suggested that 6–12 h should elapse prior to fixation. This rule was followed faithfully at first but, as a test, flies were fixed immediately after injecting the cell with Procion yellow. This means the fly brain was in fixative within 5 min after the cell was injected. Cells in brains treated this way were filled as completely as in those brains allowed to sit 12–14 h prior to fixation. A visual observation confirmed this finding: The motion-detecting cells in the fly optic lobe are large enough that, under 250× in the (Wild M-5) dissection microscope, Procion red (which can be seen readily in white light) can be seen as it enters the cell. With the application of a constant current of 20–40 nA, these cells are filled within 5 s. Hence, the preparation is allowed to sit for 30 min or so before it is placed in fixative.

Fixation

Since the fluorescence of a filled cell must be viewed against the fluorescence of the tissue as background, the fixative must not impart a high fluorescence to the unstained tissue. We have found Susa's fixative (mercuric chloride sat. in 6% NaCl 50 ml, trichloroacetic acid 2.0 g, glacial acetic acid 4.0 ml, conc. formalin 20 ml, dist. H_2O 30 ml) and FAA (100 ml 70% ETOH, conc. formalin 5 ml, glacial acetic acid 5 ml) to be suitable for use with Procion yellow or red. Note that with these fixatives the tissue does fluoresce to a degree sufficient to outline the unstained tissue and even to show unstained cells as "shadows." This shadowing can be enhanced and used (at the loss of contrast for filled cells) by fixation in glutaraldehyde. Figure 14-5 is an example of a cell filled with Procion yellow in juxtaposition with an unstained cell. Synaptic contact of these two cells is presumed.

Viewing

The Procion molecule absorbs maximally at approximately 475 nm and emits maximally at approximately 600 nm (Kater and Nicholson, 1973, p. 67). Light for excitation may be incident upon the section, i.e., from the same direction as the viewing direction, or may be transmitted through the material toward the direction from which the material is viewed. Our experience has been exclusively with transillumination. In our work we have used a high-pressure mercury burner (Osram, HBO 200 w/4), which has a high emission in the near ultraviolet and blue spectral regions. This light source in combination with exciter filters BG-38 and BG-12 and barrier filter 530 (53) (Fig. 14-2) produces a specimen coloration like that shown in Fig. 14-3 (Plate 11). A standard Zeiss RA microscope was used. This microscope has a trinocular head and chromatic objective lenses (3.2×, 10×, 20×, 100×). Dark-field illumination was used.

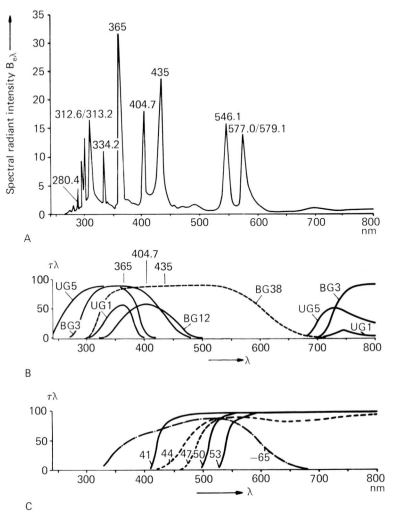

Fig. 14-2. Data reproduced from Leitz publication G 40-215/l-e. Cells shown in the figures in this chapter were photographed using transmission from a super-pressure mercury lamp HBI 200W, at BG 12 exciter filter, and a No. 53 barrier filter. **A** Spectral emission of the super-pressure mercury lamp HBO 200 W. **B** Spectral transmission of the exciter filters and filter BG 38. **C** Spectral transmission of the barrier filters.

Photography

The results of intracellular staining with the Procion dyes are best observed in the microscope. Second in effectiveness is photoreproduction in color. However, color reproduction is too expensive for routine laboratory use and very expensive for publication. The majority of the photographic recording for cell reconstruction is done on black and white film. Both 35-mm format (Leitz 35-mm camera back and parfocal viewing tube) and 2½ × 2½ inch format (Hasselbladt 500C/M) have been used. Eastman Kodak Tri-X film (ASA 400) is satisfactory when developed in Eastman Kodak D-76, Acufine, or Edwal No. -7. Exposure times of 2–3 min produce negatives with good contrast, and enough background to reveal the tissue surrounding the stained cell.

In the case of smaller cell processes, focusing may be difficult and an estimated focus must be made. With complex cells, such as the large cells of the fly optic lobe, each section may be photographed at several levels of focus. Thus, as many as 36 photographs may be taken to reproduce a cell. Reconstruction of the cell is achieved by constructing a montage from 8 ×10 inch prints made from these negatives. Eastman Kodak polycontrast paper (preferably not single weight) used without filters and KP-5 graphic paper have been used and produced satisfactory results. Montages made in this way are then rephotographed in 4 ×5 format. The montages can be retouched with white ink to improve the visibility of the smaller cell processes. The final product of this process is a two-dimensional projection of the stained cell. Color photographs of individual sections have been made with high-speed Ectachrome transparency film (Eastman Kodak). Prints can then be made by the Cibachrome process (see Figs. 14-1 and 14-3, Plates 10 and 11).

Example: Procion-Filled Cells in Insect Optic Lobe

We used flies and bees in our studies. The experimental procedures used were as follows:

For experiments on the fly the back of the head capsule was removed, i.e., access to the optic lobe was achieved from the rear of the head. After the experiment the head was further dissected until only the compound eyes connected by a ''bridge'' of exoskeleton in the front between the two compound eyes remained. The head was then severed from the body leaving the optic lobes and midbrain untouched in a cradle formed by the corneas of the compound eyes and the exoskeleton. This head is immediately placed in fixative. Most of the time the head will sink in the fixative. If it does not, it can be made to sink by placing the head in fixative in a vacuum for a few minutes (e.g., in a vacuum oven connected to a fore pump).

The duration of fixation does not appear to be critical; a fixation time of 2 h at room temperature seems to be satisfactory. Longer fixation times, e.g., overnight or over a weekend, appear to make the tissue brittle and to increase background fluorescence.

Dehydration was achieved in ethyl alcohol. From the fixative the head was placed successively in 70%, 80%, 95%, and pure ethyl alcohol. The duration for each of these solutions was judged from the time the tissue takes to "clear" (become translucent) in the next solution, xylene. This is of the order of 3 min. Hence, 10 min at each stage is considered sufficient. After 10 min in the xylene the head was placed in an oven at 56°F in a 50–50 mixture (by volume) of xylene and paraplast (Sherwood Medical Industries, Brunswick Corp., Brunswick Ctr., 1 Brunswick Plaza, Skokie, Illinois 60077), and then in pure paraplast. Ten minutes is sufficient in each of these stages; longer times are not detrimental to the final results. The heads were then embedded in blocks and placed in the refrigerator.

Sections were cut on a Spencer rotary microtome (American Optical Co. Scientific Instrument Division, Buffalo, New York 14215). Razor blades (Valet) were used in an adaptor made by American Optical Co. (Scientific Instrument Division, Buffalo, New York 14215). We routinely cut 12-μm-thick sections. Prior to use the microscope slides were coated (wiped by finger) with albumin (J. T. Baker Chemical Co., Phillipsburg, New Jersey 08865), which was allowed to dry. The ribbons of serial sections were floated on distilled water, picked up on slides, and the slides were placed upon a slide warmer until the water evaporated. A very dilute solution of methylene blue or toluidine blue may be used instead of distilled water. This will stain the tissue lightly so that it can be viewed in white light.

The paraffin was removed by placing the slides into xylene (usually two baths) for approximately 10 min. Coverslips were secured with Histoclad (Clay Adams, Division of Becton, Dickinson and Company, Parsippany, New Jersey 07054) diluted with xylene such that the Histoclad solution ran freely from a glass stirring rod. One must be careful to use enough mounting medium and to place the coverslip on so that no bubbles remain under the coverslip. Air spaces that do remain once the coverslip is in place can be removed by gentle application of pressure on the coverslip with forceps. The slides are then placed upon the slide warmer. They may be viewed in a few minutes if handled carefully. Stained cells prepared in this manner have remained unchanged for over 3 years.

For experimentation with the honey bee, *Apis mellifera,* the procedure was slightly different. Access to the optic lobe was achieved by removing a section of the front of the head capsule. The exoskeleton of the honey bee is too hard for sectioning in paraffin. The heads of honey bees were dissected to remove the front part of the head, except for the compound eyes. Fixation in FAA was followed by dehydration in alcohol as with flies but the preparation was embedded in Epon. Sections 12 μm thick were made using a sliding microtome (American Optical Co., Scientific Instrument Division, Buffalo, New York 14215, Model 860).

In experiments with tethered walking or flying flies, Goetz (1968, 1972) found that the fly's visual nervous system separates the horizontal component of a moving stimulus from the vertical component. That is, in the fly nervous system the visual input for the control of turning is processed separately from the visual input for the control of lift and thrust. Electrophysiologic experiments showed the presence in the lobula plate of movement detectors selective for horizontal or vertical movement (Bishop and Keehn, 1967; Bishop et al., 1968).

From histologic studies Pierantoni (1973, 1976) and Strausfeld (1976) described two systems of large cells in the lobula plate: One system of cells was located in the anterior region of the lobula plate. They named these cells the horizontal (H) cells (Fig. 14-4). They also described a system of cells located in the posterior region of the lobula plate, and named these the vertical (V) cells (Fig. 14-5). Intracellular recording combined with iontophoretic injection of the fluorescent dye Procion yellow (Dvorak et al., 1975) identified cells of the H system as horizontal, directionally selective (see below) movement detectors (HDSMDs), and some smaller cells as VDSMDs. Eckert and Bishop (1978) have shown that the large V cells are also VDSMDs. Another type of HDSMD is shown in Fig. 14-6; this cell type was not described in the work of Pierantoni or Strausfeld.

Each of these cell types shows the property of directional selectivity as follows: The cells will respond to movement of an object in one particular

Fig. 14-4. Photomontage of 12-μm frontal sections of a "north" horizontal cell in the lobula plate of the fly optic lobe. Marker 100 μm.

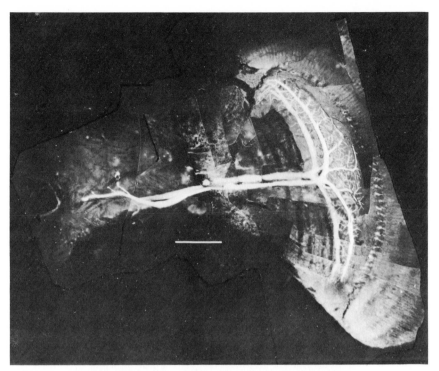

Fig. 14-5. Photomontage of 12-μm frontal sections of two individually stained vertical cells in the lobula plate of the fly optic lobe.   Marker 100 μm.

direction and little or not at all to movement at right angles to that direction.   For example a "horizontal" cell will respond to vertically oriented stripes moved in the horizontal direction, but not at all to the same pattern if it is rotated 90° and then moved.   Along the "preferred-null axis," the cell is excited by movement in one direction (preferred direction) and inhibited by movement in the opposite direction (null direction).   The three large horizontal cells described by Pierantoni receive visual information from both compound eyes, while the large vertical cells receive information only from the ipsilateral eye.

A detailed discussion of the electrical responses of these cells is not appropriate here (see Dvorak *et al.,* 1975; Hausen, 1976a; SooHoo and Bishop, 1979; Eckert and Bishop, 1978).   These cells do, however, provide an excellent example in which behavioral, anatomical, and electrophysiologic data can be brought into congruence by the powerful technique of intracellular staining.

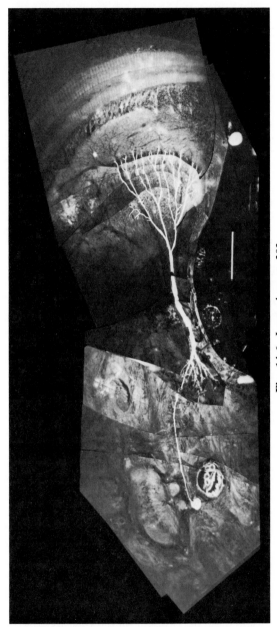

Fig. 14-6. Legend on p. 303.

Lucifer Yellow

Lucifer yellow is a superfluorescent dye, recently developed at the National Institutes of Health (Stewart, 1978). The synthesis of the dye and experiments concerned with its use are described in detail by Stewart, so only its use in insect tissue is described here. Lucifer yellow CH, compound 5 (most frequently used), has a molecular weight of 457.3. Like the Procion dyes, Lucifer yellow can be injected iontophoretically into a cell by the passage of hyperpolarizing current. The same light source, exciter, and barrier filters may be used with Lucifer yellow as are used with Procion dyes. At this writing Lucifer yellow is available only from Dr. Stewart; its commercial production is being arranged. A 5% (approximately 0.1 M) solution of this dye is used. The useful characteristic of Lucifer yellow is that the quantum yields (ratio of photons emitted to photons absorbed) in water of the Lucifer dyes are approximately 500 times that of the Procion dyes. This means that Lucifer yellow is about 100 times more fluorescent (brighter) than Procion yellow. Thus, injection of a small amount of Lucifer yellow into a cell will produce a brightly stained cell.

Lucifer yellow has been injected into neurons in the nervous system of the fly. Since the dye is available in limited quantities, pipette electrodes are filled only at the tip. This is achieved by pulling the electrode, then placing the blank end of the pipette into a drop of the dye solution. Within 30 s the electrode tip will fill well up into the shank (without bubbles). The rest of the electrode can then be filled with a different solution, with a syringe and No. 30 hypodermic needle. Procion red has been used for this in this laboratory. The red dye allows one to observe the degree of mixing (i.e., the dilution of the Lucifer solution) of the two dyes, which is not appreciable for at least 2 h.

Some photographs of cells stained with Lucifer yellow are shown in Fig. 14-7. We have found this dye to be excellent in the fly nervous system (Bishop and Bishop, 1979). The anatomical detail seen in a cell stained with Lucifer yellow is about the same as that seen in a cell stained with Procion yellow. The real advantage of the use of Lucifer yellow as an intracellular stain is that since it is more highly fluorescent than Procion, much less of it is required than Procion and consequently brightly

---

Fig. 14-6. Photomontage of 12-μm frontal sections of a horizontal directionally selective movement detector in the lobula plate of the optic lobe of the fly (slightly retouched). Note the position of the cell body in the contralateral optic lobe, and the length of fine neurite over which the dye traveled. The horizontal cells (north, equitorial, south) can be seen in outline lying anteriorly to the plane of this cell. Marker 100 μm.

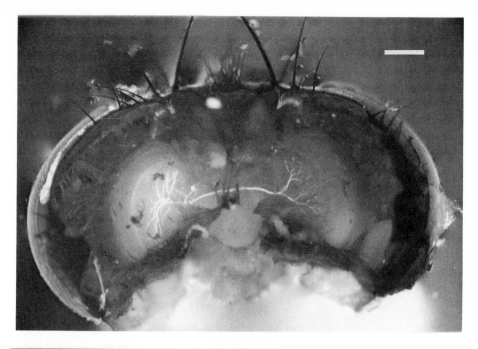

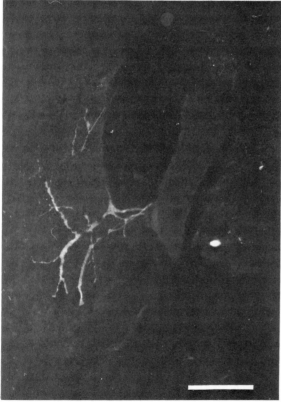

Fig. 14-7.
A Photograph of a fresh whole mount of a fly's head. After injection with Lucifer yellow the head was prepared as described in the text, placed in Ringer solution with the back of the head down on a microscope slide, and thus viewed frontally. This cell responded with action potentials to objects moved in the horizontal direction in the left eye; it did not respond to object movement in the vertical direction in the left eye or to movement of any kind in view of the right eye. Photographed with Kodak Ektachrome 400 film. Marker 300 μm.
B A 12-μm cross section of the cell shown in part **A** of this figure. The head was fixed for 24 h in 2% paraformaldehyde, dehydrated in a series of ethanol solutions, and embedded in paraffin. Marker 100 μm.

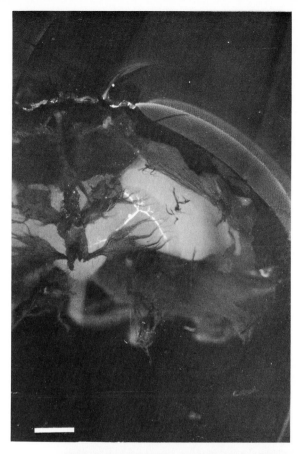

Fig. 14-7.
C Frontal view of a fixed whole mount of a fly's head. This cell responded to movement in the vertical direction in the ipsilateral eye only (Eckert and Bishop, 1978; SooHoo and Bishop, 1978). Marker 300 μm.
D Reconstruction of another vertical cell from 30-μm frozen sections. Marker 100 μm.

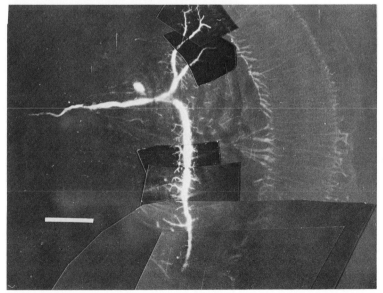

stained cells can be produced even in cell penetrations of short duration. Thus, cells that are rarely filled with Procion are much more readily filled with Lucifer (e.g., Fig. 14-7A). The overall result is that our access to the fly nervous system with intracellular staining is greatly increased.

In our experience Lucifer yellow does not survive fixation in FAA or Carnoy, i.e., the dye leaks from the cell. However, the dye will remain inside the cell during fixation in aldehyde fixatives. We fix the whole head (exoskeleton partially removed as described above) for 24 h. Lucifer-stained cells may be viewed in fresh or fixed whole amount under epifluorescence, and in frozen sections of fresh material or in paraffin sections (Fig. 14-7) under epifluorescence or transmission fluorescence.

Chapter 15

Intracellular Staining of Insect Neurons with Procion Yellow

Roland Hengstenberg
Bärbel Hengstenberg

Max-Planck-Institut für biologische Kybernetik
Tübingen, F.R.G.

The principal advantages of simultaneously recording from, and staining of single neurons, buried deep in tissue, is obvious and need not be stressed. Particularly in insects, where most cell somata are electrically silent, there is so far no way to correlate neuronal activity with a particular cell other than to penetrate and to label small fibers simultaneously (Hengstenberg, 1971). Great efforts have been undertaken during the past years to develop methods that would meet the many and often conflicting demands that an ideal intracellular marking procedure has to satisfy (see Nicholson and Kater, 1973). The injection of Procion dyes (Stretton and Kravitz, 1968; Christensen, 1973), cobalt(II) ions (Pitman *et al.*, 1972; Tyrer and Bell, 1974), and horseradish peroxidase (Muller and McMahan, 1976), followed by appropriate histochemical and histologic procedures, seem at present to approach most closely the requirements of an ideal marking method. Each of the three has distinct advantages to recommend it for a particular problem and drawbacks that may prohibit its application in other instances.

We chose Procion yellow when searching the blowfly's brain for visual interneurons that might be involved in movement perception, because the dye does not seem to interfere with neuronal activity in acute experiments. It can easily be injected, and the necessary histologic processing is fast, simple, and reliable. Our procedure is based upon the experience of other investigators (Kater and Nicholson, 1973). It has been empirically modified to yield satisfactory results with the insect (blowfly) nervous system, but no systematic studies have been undertaken to optimize the Procion yellow technique per se. Consequently, this chapter is addressed to those who wish to apply a reliable staining procedure straightaway. Negative experience is included to save unnecessary effort, and tentative "explanations" serve to communicate our limited understanding of the basic cellular processes. Visual interneurons of diptera, which have been stained by this or closely related techniques, were published earlier (Autrum *et al.*, 1970; Dvorak *et al.*, 1975a; DeVoe and Ockleford, 1976; Hausen, 1976a,b; Hengstenberg, 1977).

Recording

Satisfactory intracellular staining requires a satisfactory and stable penetration in the first place. Unfortunately insect nervous tissue does not facilitate this task. Therefore it seems useful to outline briefly our recording procedure and some observations about the penetration of neurons.

Microelectrodes are drawn by a two-stage solenoid puller from 1.0 mm o.d. × 0.5 mm i.d. borosilicate glass tubing with a fused-in filament of 0.1 mm (Hilgenberg, Malsfeld, FRG), inspected at a magnification of 1260× (Zeiss: Plan 63/0.90 oD) and back-filled with a 30-gauge steel capillary (Small Parts, Miami, Florida), connected to a 0.01-μm filter cartridge (Sartorius, Göttingen, F.R.G.), which is attached to a 2-ml syringe, containing 4% Procion Yellow M4-RAN (Serva, Heidelberg, FRG) in distilled water. Broken pipettes filled with saline solution serve as reference electrodes. Both pipettes are fixed in holders, contact is made by Ag/AgCl coils in saline solution. This electrochemical chain is essentially symmetric, yielding small and stable bias potentials (≈ 1 mV/h).

The recording amplifier (H. Wenking, unpublished) consists of a single-ended, FET-follower (gain, 1×; $R_i > 10^{12}$ Ω; $I_i < 3 \times 10^{-12}$ A), a capacity compensating circuit (0–30 pF), a bandwidth limiting amplifier (10 kHz/50 kHz), which summates the signal, and a variable bias voltage (± 100 mV) to compensate tip potentials, and it increases the overall gain tenfold. The bandwidth of the recording section, with the capacity compensation adjusted to a flat frequency characteristic, is 5 kHz at 10^8 Ω source impedance. Polarizing and staining currents are injected through the recording electrode by a feedback-controlled constant current source of virtually infinite resistance to ground. The signal source resistance is measured by injecting a square wave of 10 Hz, $\pm 10^{-10}$ A, and the I.R.-

drop voltage across this resistance is compensated by balancing an acitve bridge circuit. A monitor signal indicates the source resistance ($<10^9$ Ω).

From many hundreds of penetrations it became evident that cell somata of interneurons usually do not participate in the nervous signal traffic: They are electrically silent. The property of many motor neurons (e.g., Hoyle and Burrows, 1973) and some large interneurons (e.g., Murphey, 1973) that signals spread electrotonically back into the cell soma seems to be the exception rather than the rule. It is therefore necessary to penetrate nerve fibers, either in the neuropil, in tracts, or in peripheral nerves.

Profile diameters in visual neuropils of the fly range from less than 0.1 μm to about 10 μm, where the mass of small profiles represents terminal branches of dendritic and axonal arborizations, and the largest profiles are axons of a few "giant" interneurons. Golgi stains show that most cell types are about 2–5 μm thick (Strausfeld, 1976). This indicates the fiber size to be penetrated, if a representative part of the neurons constituting the fly's nervous system is to be investigated. Similar figures apply to most other insects. Accordingly, microelectrodes must have fine tips, and the two types described in Table 15-1 proved to be usable: The "long" type is necessary for fibers below 5 μm; the "short" type is acceptable for larger fibers.

The basal lamina of the perineural sheath seems to be the most serious obstacle for penetration. A large proportion of microelectrodes breaks during the initial penetration, and they have to be replaced. This is particularly so with the "long" type of pipettes. The lamina can be torn off, thereby facilitating penetration, but the risk of injuring fibers just below the sheath is high, and the preparations do not survive equally well. Therefore desheathing does not seem to be advisable, if the signal flow through the central nervous system is not to be disturbed.

The second serious obstacle to good penetration is given by the tracheal system. Microelectrodes very often pick up a small trachea, which firmly sticks to the tip, even if the pipette is advanced or withdrawn. This results in excessive injury when cells are penetrated and is recognized by comparatively small membrane potential values and/or a rapid potential decay. A high electrode impedance, usually considered to indicate an in-

Table 15-1. Microelectrode Properties

Borosilicate glass 1.0 × 0.5 mm, fused-in fiber 0.1 mm	Long type	Short type
Tip length from shoulder	35 mm	15 mm
Tip outer diameter	0.07–0.15 μm	0.10–0.20 μm
Tip outer cone angle	4°	10°
Resistance 2 m KAc/saline	250 MΩ	100 MΩ
Resistance 5% Procion yellow/saline	500 MΩ	200 MΩ
Current carrying capacity	2 nA	10 nA

tact small tip diameter is badly misleading in this case, and the micro-electrode must be replaced.

If a microelectrode is smoothly advanced through nervous tissue, it will generally not yield satisfactory recordings, probably because of pressure buildup in front of the tip. The strain in the tissue can be minimized by advancing very slowly and relieved by mechanical vibration (Chowdhury, 1969) or by high-frequency current (brief overcompensation of input capacity).

Bevelled electrodes seem to pass more readily through sheath and tissue, but we could not obtain stable recordings in fly neurons. Siliconizing pipettes should facilitate their sliding through tissue and prevent cells from sticking to the glass, but we did not observe a noteworthy improvement in the ease and stability of penetration.

The distribution of membrane potential values observed in insect neuropil ranges from less than -10 mV to about -80 mV. There is no clear-cut grouping of potential values that would allow the separation of acceptable penetrations from unacceptable ones. Potential differences probably do exist in intact animals, e.g., between dendritic, somatic, and axonal areas of a given cell, and possibly also between different cells, depending upon their momentary activity, metabolic load, etc. This distribution is not accessible to measurement in small nerve cells because the perforation of the cell membrane generates a significant injury conductance at the site of penetration, which shunts an unknown fraction of the membrane potential. A simplifying generalization is: The larger the membrane potential, the better is the penetration. We usually start with the functional characterization of interneurons and/or dye injection if the membrane potential is at least -40 mV and stable for 5 min.

Dye Injection

Once a cell is penetrated, one would like to spend most of the time performing physiologic measurements and practically none on dye injection. This could be achieved either by injection of a brief strong pulse or by a continuous low rate of infusion that does not interfere with neuronal activity and recording. Diffusion of dye from the electrode tip would be most convenient, but even when we have recorded for more than an hour from one cell, no noticeable fluorescence was found.

Pressure injection has so far not been tried in this laboratory because manageable pressures are thought not to eject sufficient dye from such fine electrodes. Very high pressure injection (Bader *et al.*, 1974) seems to work in bee photoreceptors without noticeably disturbing the cells.

High-voltage single-shot iontophoresis (Llinas and Nicholson, 1971) is in our experience not to be recommended because usually several cells in the vicinity of the exploding electrode tip seem to be hurt and to incorporate dye from a common pool: We have often found multiple and incomplete stainings, which leave the suspicion of artifact, even if only one cell has been stained (DeVoe and Ockleford, 1976).

Square pulse regimes of iontophoresis were reported to reduce the probability of electrode "blocking" (Stretton and Kravitz, 1973). We could not see this effect distinctly and therefore used mainly DC-iontophoresis. With low injection currents, and with the bridge circuit balanced, this procedure has several advantages: (1) Neuronal activity can be continuously observed during the injection, and dye extrusion can be stopped immediately when the cell is being lost from the electrode. (2) The injection current can be optimized for each cell, again by observing its membrane potential and electrical activity. (3) The amount of dye can readily be estimated by integration of the control voltage of the constant current source. (4) Procion dyes are injected by hyperpolarizing currents, which tend to compensate depolarizing injury currents in less than perfect penetrations.

If the injection current is slowly increased from zero, rather than abruptly switching it on, the electrode resistance first decreases and reaches a minimum at about 0.8, the resting value. With further increasing current it rises again, and finally the electrode "blocks" with a distinct jump to very high resistance values ($> 10^9\ \Omega$). The background noise increases steadily with current, but briefly before "blocking" occurs, it increases dramatically. We have the impression that electrode blocking can be avoided by quickly reducing the current for some time, whereas it tends to be irreversible once it has happened.

Current values at which blocking occurs depend largely upon the shape of the electrode tip and upon the condition that it is not plugged, e.g., by a trachea. Electrodes whose resistance increases even at very low currents will usually block soon. The long, slender type of electrode (Table 15-1) will allow about -2 nA to be injected, whereas the short, blunt type yields more than -10 nA. These currents seem small, but if one estimates the DC input impedance of a fiber of 5 μm diameter to be 20–40 MΩ (Katz, 1966), -10 nA would locally hyperpolarize the cell membrane by 200–400 mV. The membrane will of course break down under such a strong field, probably at the penetration leak around the electrode, and an appreciable fraction of the injected dye may leak out of cell.

In many cases, particularly in the beginning of this work, when we tried to inject as much current as possible, we often observed local swelling at the site of penetration (see Fig. 15-1E, Plate 12) and a halo of diffuse stain in the vicinity of the injected fiber (see Fig. 15-1E). With small injection currents (<1 nA) swelling has never been observed, the site of penetration cannot be recognized, and a halo is usually absent (see Figs. 15-6 and 15-7). We think that among others the "transport capacity" of the cell, i.e., the rate at which it can remove dye from the site of injection, is a rate-limiting factor for intracellular staining.

The quantity of dye to be injected in order to obtain an acceptably stained neuron is difficult to judge, because the transport number, i.e., the proportion of current carried by the dye is not known under the prevailing circumstances. Procion yellow M4–RAN is a triply charged molecule with a molecular weight of about 650 (Stead, 1973). One can therefore expect that most of the injection current is carried by potassium ions migrating into the electrode. The *de*crease of electrode resistance observed under current injection may be due to this potassium entry, whereas in an *in*crease of electrode resistance is observed with pressure injection (Kater et al., 1973). In wide-field tangential neurons of the fly's lobula plate (see Fig. 15-7), 1 μC is about the minimum charge to be displaced for an acceptable stain, where the fine terminal branches are just recognizable. Eight microcoulombs will give superb stains even in the most extensively branched cells so far known in the fly's brain (VCH-cell, Fig. 15-7). Smaller and less branched cells, e.g., photoreceptors or monopolar cells, need much less dye (see Fig. 15-6). It follows from these observations that perfect stainings cannot be obtained without investment of recording time: 1 μC/1 nA $\approx$ 15 min. The compromise to be chosen depends upon the particular intention of the experiment.

Procion dyes are alkylating substances that form a covalent bond by nucleophilic substitution (Stead, 1973). It is remarkable how little it affects neuronal activity in acute experiments. Figure 15-2 shows action potentials from one of the large vertical cells in the lobula plate after the injection of 23 μC during 2.5 h. No significant changes in the response properties of this cell to light intensity changes or pattern movements in the receptive field were observed. It may be that the rate constant of the foregoing reaction at room temperature and physiologic pH is sufficiently small.

Acceptable stains can however be obtained even when the tissue is fixed only 1 h after injection with an acid fixative, which should wash out all dye that has not yet been bound. This point has not been followed because our flies routinely stay overnight in a moist chamber at 4°C in the refrigerator. All of them survive. The intention is that the dye spread homogeneously throughout the cell, that the binding reaction be complete, and that the schedule of histologic processing be maintained. Excessive diffusion times or death of the animal lead to autolytic tissue damage, e.g., vacuolization. The stained cell appears "grainy" and the cytoplasm seems to clump close to the cell membrane (Fig. 15-1G, Plate 12).

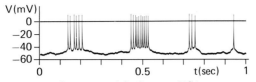

Fig. 15-2. Spontaneous action potentials from a VS1 neuron, after injection of Procion yellow for 2.5 h.   With 23 μC of displaced charge, the cell received about ten times as much dye as is necessary for a brilliant staining of this cell type, resulting in a prominent halo (see Fig. 15-1F Plate 12).   The electrical activity did not noticeably change during this time; neither the action potentials nor the synaptic activity are overtly impaired.

Histology

We use a simple, conventional paraffin technique.   Fixation: ethanol-formalin-acetic acid (85 ml 100%, 10 ml 40%, 5 ml 100%; ≥4 h, room temperature).   Dehydration at room temperature in graded ethanol (15 min 85%, 15 min 90%, 15 min 95%, 3× 30 min 100%).   Intermedia: terpineol (3 h), xylene (2 × 30 min).   Infusion: xylene-paraffin (1 : 1, 1 h, 65°C), pure paraffin (Fisher tissue prep. m.p. 56.5°C, 3 changes).   Blocks are cast in forms of aluminum foil.   Serial sections are cut at 7–12 μm with steel C-knifes on a rotary microtome (Jung, Autocut 1140), deparaffinized in xylene, and mounted (Fisher Permount).

The preparations can be viewed as whole mounts, if cleared in methylsalicylate after dehydration, but we skip this step because tracheae tan during fixation, and the autofluorescence of solid tissue outshines the Procion yellow fluorescence, especially from small fibers.   The structural preservation of this procedure is sufficient to reveal characteristic details in the fine branching pattern of dendritic or axonal arborizations (Fig. 15-1A,B, Plate 12; Figs. 15-6 and 15-7), dendritic spines (Fig. 15-1C, Plate 12), or varicose axon terminals (Fig. 15-1D, Plate 12).   Preparations of this kind are perfectly stable when stored at room temperature in darkness.   No fading of fluorescence or leakage of dye from stained fibers has been observed after more than three years.

Improved structural preservation can be obtained by fixation in buffered aldehydes (Karnovsky, 1965), plastic embedding, and semithin sectioning.   When applied to whole insect heads, however, this procedure yields a bright, yellow tissue fluorescence, probably because buffered aldehydes dissolve some of the retinal screening pigment.

Fluorescence Microscopy

The identification of stained profiles in a whole mount or section is a detection problem, where the locally emitted fluorescence radiation is considered as signal and the spatial distribution of all other light as noise. Detection is reliable when the signal exceeds the noise level by some factor that is specified by the spatial fluctuation of noise. To maximize the probability of detection, all steps of the procedure and the instrumentation involved must be optimized.

We use a Zeiss "Universal" microscope, equipped with a xenon-arc lamp (XBO-150 W/1), an incident light illuminator (Zeiss III RS, system Ploem), and modified FITC filtering. The xenon lamp was chosen because its spectral emission in the excitation range of Procion yellow (400–500 nm) is about equal to the output of the mercury lamp HBO 200; at shorter wavelengths (300–400 nm), however, where tissue autofluorescence is preferentially stimulated, the output of the HBO-200 is about 100 times higher than that of the XBO-150. The same holds for the emission range of Procion yellow (500–700 nm), where the HBO-200 has two prominent emission lines (546 and 577 nm), whereas the xenon spectrum is flat and at about the same level as the mercury spectrum between these lines. The excitation spectrum is shaped by the following sequence of filters (Zeiss, Schott): 3 mm KG 1, 3 mm BG 38, KP 490, 3 mm BG 38, BP 455–490, and the dichromatic mirror FT 510. The excitation spectrum, as calculated from the spectral emission of the XBO-150 W/1 and the filter characteristics, is shown in Fig. 15-3. It is centered upon the excitation maximum of Procion yellow-stained neurons (Van Orden, 1973), and has very little energy in adjacent spectral regions. Rigid suppression of incident energy at λR 500 nm is essential, because the emission from a small nerve fiber is many orders of magnitude smaller than the lamp output in this range. Barrier filters are: FT 510 and 2 mm OG 520, if the full emis-

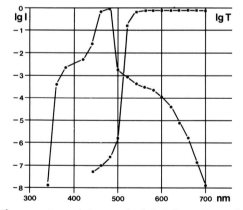

Fig. 15-3. Excitation spectrum (●), as calculated from the emission spectrum of the xenon lamp XBO 150 W/1, and transmission spectra of excitation filters. Beam intensities are roughly in milliwatts per nanometer (■). Transmission spectrum of barrier filters. For details, see text.

sion spectrum is to be viewed; additional longpass filters (e.g., OG 570) can be inserted if the green part of the autofluorescence is to be suppressed in order to increase the intensity contrast for black and white photography. A sensitive test for the spectral separation of the microscope is to replace the specimen by a diffuse metallic reflector.

Recently developed, high-aperture fluorite lenses (e.g., Zeiss Plan Neofluar 25 × 0.8 W-Oel, Leitz Fl 25 × 0.75 Oel) are very useful because they combine low magnification with high resolution and high-incident intensity. They are designed for fluorescence microscopy and have therefore very little intrinsic fluorescence. The Plan-Neofluar is particularly well suited for photography because of its smaller curvature of the field. With a 63-mm projection eye piece we obtain 65× magnification on 35-mm film, which corresponds to a field of 370 × 560 μm in the object. At an optical resolution of 0.4 μm a film resolution of $\leq$40 mm^{-1} is required to store the 2×10^7 data points given by field size and resolution. Modern black and white films (e.g., Agfapan 25, Agfapan 100) have a resolution limit of 100–350 line pairs per millimeter (lpm). This limit corresponds to the spatial frequency, where the contrast transfer function has decreased to 10%. At 40 lpm the transmitted contrast is 60–80%, which is sufficient to resolve aperiodic structures without significant loss. Most color films resolve 70–100 lpm, such that slightly higher magnification may be necessary. We generally use High Speed Ektachrome (Kodak EHB-135) with satisfactory results; other brands may serve equally well. Color slides are especially helpful, if faintly stained neurons have to be traced. They appear distinctly red against the green surroundings, and can therefore be discriminated, even if their intensity contrast would not allow reliable detection (e.g., Fig. 15-1G, Plate 12). Black and white halftone prints can be obtained directly from color slides with panchromatic paper (Kodak, Panalure); stained cells appear dark on light background (see Fig. 15-6; Rl). Additional red filtering during the printing process suppresses the green autofluorescence, and thereby increases the contrast of prints.

Fluorescence of tissue and of stained profiles fades with strong and prolonged irradiation. Fading occurs most rapidly with high-power lenses, where the full beam intensity is focused onto a small object area; it is almost neglectable with low-power lenses ($\leq$25×). Figure 15-4 shows the normalized time course of fluorescence decay, when 10-μm sections are illuminated via the lens Fl 63/1.30 Oel. Both microfluorometer traces show an approximately exponential decay, but the autofluorescence fades much faster ($\tau \approx 15$ min) than the Procion yellow fluorescence ($\tau \approx 60$ min). The autofluorescence decay includes a second, much larger time constant ($\tau \approx 2$ h), which dominates in later parts of the bleaching process. Possibly one of the two represents the decay of aldehyde-induced fluorescence (Elofsson and Klemm, 1972), and the other

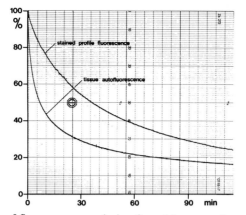

Fig. 15-4. Fading of fluorescence emission from 10-μm sections (lens, F1 63/1.30 Oel; recording aperture, 4.4-μm diameter). Tissue autofluorescence and Procion yellow fluorescence decay with time constants of 15 and 60 min, respectively. Therefore, color contrast decreases and intensity contrast increases with bleaching, especially during the first few minutes.

true tissue autofluorescence, which is known from freeze-dried tissue (Zettler and Järvilehto, 1971). After very long irradiation (>6 h), the autofluorescence intensity is less than 10% of its initial value, without yet having reached a true steady state. This shows that diffuse reflection of exciting light from the tissue sections is insignificantly small.

The difference in bleaching time constants shows that the relative color contrast between tissue and stained profiles decreases with bleaching time, whereas the relative intensity contrast increases, especially in the early phase of bleaching. This difference can be utilized in color or black-and-white photography.

Reconstruction

We tried several ways to reconstruct cells from sections or slides: (1) Direct reconstruction with the drawing tube exposes the preparation unnecessarily long. It is also quite tedious and spoils contrast of the microscope image. (2) Stacking of enlarged transparencies (X-ray film or drawings on foil) suffer from multiple reflections between foils. (3) Back projection on the ground-glass screen of a microfilm projector suffered from graininess and parallax due to the glass plate thickness between the ground-glass surface and the drawing foil. Replacing ground glass with clear glass still leaves the graininess of the transparent paper (4) Photographic enlargers are usually too dim at sufficient magnification. (5) Most convenient, simple, and satisfactory is a slide projector (LEITZ Prado

Universal), equipped with 50-mm optics, to which is attached a front surface mirror (200 × 200 mm, Spindler & Hoyer, Göttingen) at 45° to the
optical axis. When placed on a sideboard, 70–80 cm above the table top,
and accurately leveled, it will give a bright and virtually distortion free
picture in a convenient drawing position. With the 25 × microscope
lens the total magnification of this arrangement is about 1000 ×, such that
fibers of about 0.5 μm can still be drawn.

Alignment of subsequent sections is quickly and precisely achieved by
translation and rotation of the drawing paper, particularly when the section to be drawn contains several stained profiles. Otherwise alignment
is achieved by matching prominent structures of the surrounding tissue.
Large unstained profiles appear as autofluorescence "shadows" (Fig.
15-1F, Plate 12) and may serve as a reference framework for the position of
a stained cell in the neuropil. Outlines of ganglia, neuropil areas, and
fiber tracts can be added to yield further structural detail. Cell reconstructions obtained this way are shadow casts in a plane parallel to the
plane of sectioning. This has two obvious disadvantages: (1) the loss of
the third dimension and (2) the fact that the plane of sectioning determines
the direction of view in the shadow cast. This direction may not be optimal for the representation of a particular neuron: Planar arborizations like
Purkinje cells of the vertebrate cerebellum, horizontal cells of the vertebrate retina, or large-field tangential cells of the fly's lobula plate ought to
be represented parallel to the major plane of their arborization. In most
other cases, such as vertebrate cortical neurons, interneurons and motor
neurons in aperiodic invertebrate ganglia, no single plane can be defined
that would represent the whole branching pattern of such cells unambiguously.

A general solution to these kinds of problems is to replace the drawing
paper by a graphics digitizer of sufficient size and resolution (Lindsay,
1977). Spatial coordinates of stained profiles can then be stored section
by section on magnetic disc or tape, and a suitably programmed computer
can perform arbitrary operations upon these data, such as coordinate
transformations (to generate shadow casts from any direction), computation of stereograms, anisotropic normalization, etc. The very same system can be used to process data from serial electron micrographs. Combining a digitizer with the foregoing alignment procedure utilizes the
speed and accuracy with which an experienced operator can match subsequent slides and can recognize errors due to dirt particles, scratches,
folds, and other sources of "noise" in the histologic specimen. This pre-
selection and preprocessing of data to be acquired saves considerable
data storage space, computing time, and programming effort. Pertaining
to the issue of this chapter, computer-generated stereograms of reconstructed cells have the following principal advantages: (1) three dimen-
sional representation with selectable direction of view, (2) independence

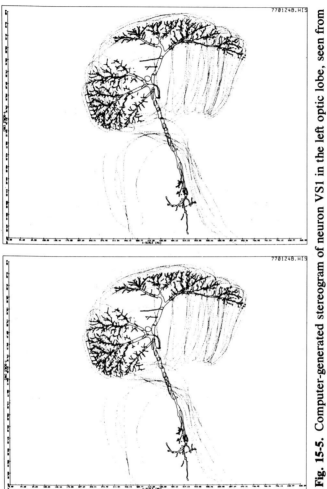

Fig. 15-5. Computer-generated stereogram of neuron VS1 in the left optic lobe, seen from anterior. Cell profiles (*solid lines*) were digitized from 7-μm paraffin sections after Procion yellow injection. Notice the cup shape of the lobula plate neuropil (*dotted*), the location of the cell body and major dendritic branches in the posterior (deep) surface, and the S-shaped course of the axon. Axon terminals bend anteriorly in the protocerebrum, and close inspection of the dendritic arborizations reveals that the dorsal fan nestles against the anterior surface of the lobula plate, whereas the distal brush (right) predominantly resides in the curved posterior surface. Disparity angle for stereo computation, 2.5°; basis 65 mm for commercial lens stereoscopes.

of the plane of sectioning, (3) infinite "depth of focus," and (4) multiple choice of supplementary structures to be displayed. A principal disadvantage of this method is the anisotropy of resolution, which is determined by the optical resolution (<0.4 μm) within the plane of sectioning and by the thickness of sections (paraffin 7 μm, plastic 2 μm) normal to the plane of sectioning. First experiments with this procedure (H. Bülthoff and R. Hengstenberg, unpublished) show, however, that even very crude stereograms reveal a variety of significant features of cell structure (Fig. 15-5) that are not recognizable in shadow casts.

Fly Interneurons

A few interneurons from the blowfly's central nervous system shall serve to demonstrate the range of applicability of Procion yellow, and the overall cell structure as reconstructed from serial sections. Some cells were chosen because they have been stained before with totally different staining procedures. References are given in such cases, to promote critical comparison. Other cells have been selected because their existence has not been demonstrated to date by any other procedure. All examples come from studies about visual movement perception, with particular emphasis on such cells that might be involved in locomotor control. The cell drawings are shadow casts, as described before. All cells except DN3 in Fig. 15-6 are essentially fan shaped and are sectioned parallel to the plane of their major arborization, such that shadow casts are sufficiently representative of cell structure. Very simplified indications of the physiologic properties of these cells are given; the largest part of the functional specification procedure, namely, stimulus conditions to which cells are indifferent, and controls are omitted.

   1. *R 1 (Fig. 15-6)*: one of the six peripheral photoreceptors in the open-rhabdom ommatidium of Diptera. Responds with graded depolarizing potentials when the receptive field is illuminated.
   2. *L 1 (Fig. 15-6)*: one of five monopolar neurons in a "cartridge" of the first visual neuropil (lamina). Receives input from six peripheral photoreceptors (Braitenberg, 1967; Boschek, 1971), which share the same direction of view (Kirschfeld, 1967), and responds with graded hyperpolarizing potentials to illumination of the receptive field (Autrum, et al., 1970). Fiber size, <2 μm (compare Strausfeld, 1976, frontispiece).
   3. *OCl (Fig. 15-6)*: spiking interneuron of the ocellar system, running from the lateral ocellus, via the protocerebrun (soma and small-blebbed aborization) through the cervical connective to the contralateral mesothoracic neuromere. It is silent in the dark, tonically active in the bright, and responds vigorously to flickering light. Fiber size, <2 μm.

4. *OCH* (*Fig. 15-6*): Heterolateral interneuron of the ocellar system (Strausfeld, 1976, plate 7.29).   No responses could be recorded, presumably because the ocellar nerve was transsected.   Fiber size, <1.5 μm.

5. *DN 3* (*Fig. 15-6*): complex descending neuron, silent in the resting state, fires 1 or 2 spikes with fixed delay in response to bright flashes, but

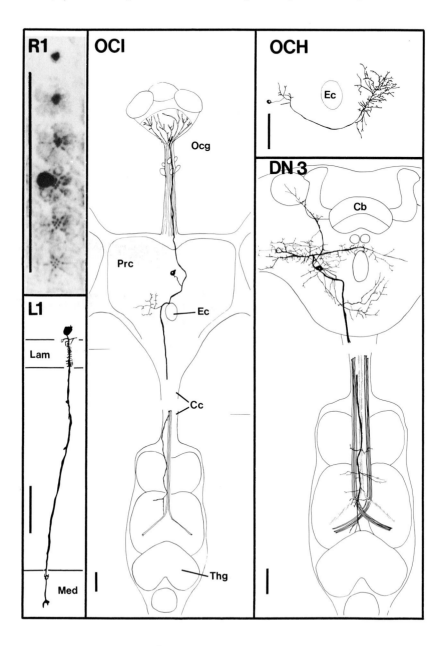

habituates rapidly with repetitive flashes. Moving striped patterns elicit no response. Axon diameter, ≈ 4.5 μm (compare Strausfeld, 1976, plate 7.26B, cell 3).

6. *H2* (*Fig. 15-7*): Heterolateral tangential neuron whose dendritic arborization covers more than the dorsal half of the lobula plate, and whose axon terminates in a bush that wraps around the endings of contralateral HS-neurons. This cell is monocularly excited if patterns move from back to front, in the visual field of the right eye. Spontaneous activity is inhibited by movement in the reverse direction. The exclusive input from the right eye yields the direction of signal flow in the cell and justifies the notation of axon and dendrite. Axon diameter, <5 μm; details on Fig. 15-1B (Plate 12).

7. *VCH* (*Fig. 15-7*): Tangential neuron that covers primarily the ventral half of the lobula plate. In contrast to the former cell, its second arborization is ipsilateral. It is again predominantly excited by horizontal movements, but inputs from both eyes combine such that the cell responds best if the world rotates clockwise around the fly. The physiologic polarity of this cell type is not quite clear (Hausen, 1976b), but the small medial arborization is presumably dendritic, and the large fan in the lobula plate is probably telodendritic, the main direction of signal flow then being centrifugal. Axon diameter, <6 μm; details on Fig. 15-1C (Plate 12).

8. *VS1* (*Fig. 15-7*): homolateral neuron with a bipartite arborization in the lobula plate. It is associated with the set of large "vertical cells" (Pierantoni, 1976; Hausen, 1976a,b) and, like these, is excited by ipsilateral downward movements, and inhibited by upward movements. This may be true only for movements in the anterior, ventral part of the visual field, because the corresponding part of the dendritic arborization is in the posteriorly directed surface layer of the lobula plate, where neurons that process vertical movements are found (Hausen, 1976b). The second arborization, corresponding to a dorsal and lateral part of the visual field, resides entirely in the anteriorly directed layer of the lobula plate, where

---

Fig. 15-6. Reconstructed visual interneurons of the blowfly *Calliphora*. *R1*: cross-sectioned ommatidia with stained peripheral photoreceptor No. 1. Negative print from color slide on panchromatic paper. *L1*: lamina monopolar cell. *OC1*: small ocellar interneuron, descending directly to the contralateral mesothoracic neuromere. About 1 mm of the cervical connective is omitted. Fiber, <2 μm. *OCH*: Heterolateral interneuron of the ocellar system. Fiber size, <1.5 μm. *DN3*: Strongly habituating descending neuron with extensive arborizations in different protocerebral areas and in the pro- and mesothoracic neuromeres. See text for details about these cells. *Cb* central complex; *Cc* cervical connective; *Ch* chiasma externum; *Ec* esophageal canal; *Lam* lamina ganglionaris; *Med* medulla; *Ocg* ocellar ganglion; *Prc* protocerebrum; *Thg* thoracic compound ganglion. Scales, 100 μm.

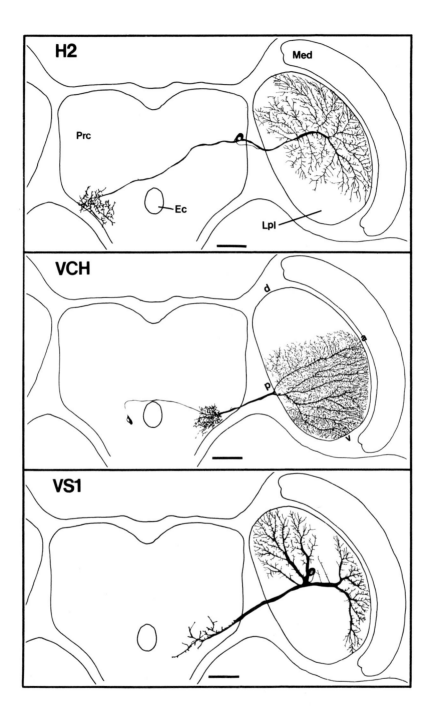

horizontal movements in the visual field are processed preferentially. It is not known at present whether these two areas of dendritic arborization have different directional specificities, and what their functional significance would be (compare Fig. 15-5).

The latter three cells all respond maximally at low pattern speeds, which in the intact animal elicit the strongest velocity-dependent optomotor reactions (Goetz, 1964; Goetz and Wenking, 1973; Reichardt and Poggio, 1976). A distinct second group of cells, which respond best to high pattern speed, presumably serve to elicit the landing response (Hengstenberg, 1973 and unpublished).

Concluding Remarks

The use of Procion yellow for simultaneous determination of functional characteristics and the detailed structure of insect central neurons has proved to be extremely satisfactory. The remarkable property of Procion yellow, not to disturb significantly the electrical activity of neurons in acute experiments, and the outstanding reliability of the staining procedure recommend it highly for such purposes.

Very small fibers, which constitute most of any insect brain, can be penetrated and successfully stained. The limiting factor is still the achievement of acceptable and stable penetrations; perfect staining is then almost guaranteed. When fibers are penetrated, the dye spreads consistently in anterograde and retrograde directions, and fills extensive dendritic and telodendritic arborizations uniformly, even if they are millimeters away. Injecting dye into the soma usually does not yield acceptably stained cells, presumably because too little dye can be transported

---

Fig. 15-7. Movement-sensitive tangential neurons of the lobula plate. *H2*: monocularly driven heterolateral cell responding to horizontal movements in the right eye's visual field; converges upon contralateral neurons with horizontal preferred direction of movement sensitivity. The dendritic arborization occupies predominantly those parts of the lobula plate that correspond to dorsal (*d*) and anterior (*a*) parts of the visual field (retinotopic projection). Details on Fig. 15-1B. *VCH*: One of two binocularly driven tangential cells with predominantly, though not exclusively, horizontal movement sensitivity. The homologous cell occupies the dorsal part of the lobula plate. In both cells the small proximal arborization is presumably dendritic, and the large fan-shaped arborization is telodendritic. Details on Fig. 15-1D. *VS1*: monocularly driven homolateral neuron with preferred vertical movement sensitivity in the anterior and ventral parts of the right visual field. The role of the fan-shaped second dendritic arborization, which resides in the anteriorly directed surface of the lobula plate, is not yet clear. Presumably it conveys to this neuron an additional horizontal movement sensitivity in the corresponding dorsal and lateral parts of the visual field. Details on Fig. 15-1F.

through the very thin soma fibers of most interneurons.   Any anatomical type of neuron will be completely stained if dye is injected and the cell survives.   We have never seen a case where the dye got stuck at the injection site, nor have we observed that dye moved only in one direction. The kind of electric activity that a neuron generates has no recognizable effect whatsoever on the success of staining.

The "completeness" of any staining procedure cannot be ascertained by itself.   However, comparing the same unique neurons, when stained by totally different procedures, allows a critical comparison (Strausfeld, 1976; Strausfeld and Obermayer, 1976; Strausfeld and Hausen, 1977). Wherever this comparison could be made, all significant features were present in Procion-stained preparations, and sometimes even details are found that are missing in Golgi preparations (soma of DN3; cf. Strausfeld, 1976, plate 7.26 B).   However, quantitative branch-by-branch comparisons can not be carried out, because of the natural variation of ramification patterns (K. Hausen, 1978, in preparation).

The resolution of this staining procedure is apparently limited by the resolving power of light microscopy (Berek, 1950) and not by uneven distribution of dye inside stained cells.   Details of cell structure like the "spiny" or "blebbed" appearance of arborizations can be recognized reproducibly, even with the crude histologic procedure that we use presently.   The thickness of paraffin sections causes an unduly bright autofluorescence, which tends to mask thin and faintly stained profiles.   Both improved structural preservation and reduction of autofluorescence can readily be achieved when maximum resolution is required by application of advanced histologic techniques.

Procion yellow has three shortcomings: (1) Some of the recording time is at present required for dye injection.   This may be reduced if dyes with stronger specific fluorescence (Stewart, 1978; see Chapter 14) become available.   (2) Whole mounts, which would save time, effort, and money, cannot be recommended for thin profiles, because of the tissue autofluorescence.   (3) The lack of electron opacity complicates ultrastructural studies to be carried out in singly stained cells.   For many applications, however, these drawbacks are by far outweighed by the simplicity, reliability, and nontoxicity of the procedure.

Note and Acknowledgments

This article was submitted for publication in June, 1976.   We are deeply grateful to Drs. R. D. DeVoe and K. Hausen for their expert critique, advice, and help throughout this work; to Professor K. G. Goetz for stimulating suggestions and steadfast encouragement; to Dipl. Phys. H. Wenking for designing the electronic equipment; to Miss C. Hanser for typing the manuscript; and to all members of this institute, who let us participate in their understanding of the fly's brain and behavior.

Chapter 16

The Use of Horseradish Peroxidase as a Neuronal Marker in the Arthropod Central Nervous System

Hendrik E. Eckert

Ruhr University Bochum
Bochum–Querenburg, F.R.G.

C. Bruce Boschek

University of Giessen
Giessen, F.R.G.

Electrophysiologic methods provide us with a means of studying functional pathways within nervous systems. By themselves, however, such studies cannot tell us about the morphologic pathway underlying the flow of information. On the other hand, morphologic studies can tell us about the shape and distribution of nerve cells, their synaptic contacts, and can thus provide us with information about the structural contiguity in functional pathways. Obviously, it would be a great advantage to bring both experimental approaches together for the study of neural networks.

Such experimental analyses have recently become possible: Cells could be characterized functionally by intracellular recording techniques and simultaneously, they were identified anatomically by intracellular injection of dyes like Procion yellow (Stretton and Kravitz, 1968; Iles and Mulloney, 1971) and cobalt ions, later precipitated as sulfide (Pitman *et al.*, 1972), allowing visualization by light-microscopic techniques. At the electron-microscopic level, however, neither Procion yellow nor cobalt proved to be very useful: Procion yellow is not electron opaque, and cobalt sulfide, though electron dense and thus easily recognized, does not allow the visualization of cell organelles because all structural details are obscured by the black, torn appearance of the cobalt precipitate.

Another dye studied for its potential use in ultrastructural investigations was Procion brown (Christensen, 1973). This dye is difficult to trace in light-microscopic sections and thus stained cells are difficult to detect for further processing at the electron-microscopic level. Another difficulty arises from its capricious diffusing properties.

A new marker substance, horseradish peroxidase (HRP), originally used for tracing the early stages of protein absorption in the renal tubules by histochemical methods (Straus, 1959; Graham and Karnovsky, 1966), proved to be particularly useful as a neuronal marker (Kristensson and Olsson, 1971; LaVail and LaVail, 1972). The reaction product of this enzyme can be seen both in the light and electron microscope and has been used extensively in vertebrate preparations for the study of afferent and efferent neuronal pathways, as well as for ultrastructural investigations at the electron-microscopic level (e.g., LaVail and LaVail, 1972, 1974; Nauta et al., 1974; Bunt et al., 1975; Colman et al., 1976; Geisert, 1976; Keefer et al., 1976; Cullheim et al., 1977; Herkenham and Nauta, 1977; Hunt et al., 1977; Mesulam et al., 1977; Rastad et al., 1977; Cullheim and Kellerth, 1978; Keefer, 1978; Magalhaes-Castro et al., 1978; Vanegas et al., 1978).

Only recently has this enzyme been put to use on invertebrate preparations (Hirudineae: Carbonetto and Muller, 1976; Muller and McMahan, 1976; Yau, 1976a,b; Mason, 1978; Diptera: Coggshall, 1978; Ghysen and Deak, 1978; H. E. Eckert and S. O. E. Ebbesson, unpublished results; Orthoptera: A. Eichendorf and K. Kalmring, unpublished results; Crustacea: W. Schröder, H. Stieve, and I. Claassen-Linke, unpublished results; Xiphosura: W. Schröder, H. Stieve, and I. Claassen-Linke, unpublished results). Thus, the use of HRP in invertebrate preparations is in a very preliminary stage. However, the investigations carried out to date already prove its potential use for the study of fiber projections and synaptic connections at the light- and electron-microscopic levels.

History and Chemical Nature of Horseradish Peroxidase

The name "peroxidase" was first given by Linossier (1898) to a class of enzymes that catalyze the oxidation of the following substrates by using hydrogen peroxide as oxidant: phenols, aromatic primary, secondary, and tertiary amines, certain leuko-dyes and heterocyclic compounds (e.g., ascorbic acid, indoles), and iodide (review by Saunders et al., 1964). Until that time, these peroxidases were not distinguished from oxidases and catalases. Benzidine, now widely used in neuronal labeling techniques, was first used by Joslyn and Marsh (1933) as substrate for the enzyme. Today, several benzidine compounds are applied in neuronal marking techniques including 3,3′-diaminobenzidine (DAB), benzidine dihydrochloride (BDHC), and o-dimethoxybenzidine (o-dianisidine).

Peroxidase was first isolated from human pus (Loew, 1901) and was subsequently found in the tissue of many plants (cf. Saunders *et al.*, 1964). It occurs in rather high concentrations in the roots of horseradish (Bach and Chodat, 1903). This "horseradish peroxidase" could be purified and a brown crystalline product was obtained (Thorell, 1942). Subsequently, several isolation procedures were published (Schwimmer, 1944; Paul, 1958; Mann and Saunders, 1960).

Peroxidase is a hemoprotein with a molecular weight of 40,000, consisting of a protohematin IX and a glucoprotein. The glucoprotein is composed of approximately 160 amino acids and uronic acid; the protohematin IX is composed of a protoporphyrin IX with a Fe^{3+} center.

Recently, a new peroxidase, microperoxidase, was introduced; it has a molecular weight of only 2000 (Brightman *et al.*, 1970, quoted from Kuffler and Nicholls, 1976). Because of its low molecular weight, this enzyme can be injected iontophoretically much more easily than horseradish peroxidase. This is a distinct advantage if microelectrodes with a small tip diameter are to be used.

Materials and Methods

Histologic Procedures for Light-Microscopic Studies

As in vertebrate preparations, histologic procedures applied to date follow the treatment given by Graham and Karnovsky (1966), which is based on the histochemical procedures by Straus (1959, 1964a,b). Modifications of this basic procedure consist mainly of differing buffer solutions, fixatives and varying concentrations of the substances used. Furthermore, the duration of treatment in the different solutions and the temperatures of the solutions may vary to a great extent in the different preparations. The following section describes the different procedures applied to invertebrate preparations after introduction of HRP to the tissue.

Method 1: Central Visual System of Diptera
(H. E. Eckert and S. O. E. Ebbesson, unpublished)
After dissecting out the brain, it was placed in fresh egg yolk and fixed overnight at 4°C in formaldehyde vapors until the egg yolk had an elastic composition (Ebbesson, 1970). The fixed brains were then frozen and 33-μm-thick sections were cut and were collected in 1 M cacodylate buffer (pH 7.4). After incubating the tissue for 5 min in Trizma buffer solution (0.05 M; pH 7.4 at 25°C) containing 50 mg 3,3′-diaminobenzidine (DAB) per 100 ml buffer, the reaction product was formed by adding 66 μl conc. hydrogen peroxide per 100 ml of the DAB solution and leaving the sec-

tions in this solution for 30 min. Subsequently, the sections were floated in alcoholic gelatin (7.5 g gelatin; 500 ml distilled water; 500 ml 80% ethanol) for 30 min, rinsed in distilled water, mounted and dried for 2 h. After counterstaining with a 0.1% solution of cresyl violet acetate, the sections were embedded.

Method 2: Intracellular Application of HRP
(Muller and McMahan, 1976)

Intracellular injection of HRP was first used in studies of sensory and motor neurons in the leech (Carbonetto and Muller, 1976; Muller and McMahan, 1976; Yau, 1976a,b). Injection of HRP was achieved by applying a pressure of 0.3–1 atm for 5–30 s to the electrode, which was filled with a solution of 2% HRP in 0.1 M KCl. After 30–90 min to permit spread of the enzyme, the ganglion was fixed (30 min) by replacing the Ringer's solution with 0.8% glutaraldehyde and 1% formaldehyde in 0.1 M phosphate buffer (pH 7.3). The aldehyde-fixed ganglion was rinsed in 8% sucrose (15 min) and incubated in the sucrose solution containing 0.5% benzidine dihydrochloride (10 min). Then about 1 drop 3% H_2O_2 was added per milliliter of benzidine solution (1–5 min). The reaction was stopped by briefly rinsing the ganglion in 8% sucrose and then bathing it in cold 15% sodium nitroferricyanide (10 min). The ganglion was dehydrated by transferring it directly to absolute ethanol (1–2 min), cleared in xylene (1–2 min), and mounted.

In principle, this same procedure has also been applied to the visual system of dipterans (H. E. Eckert and C. B. Boschek, unpublished results), to the acoustic system of orthopterans (K. Kalmring, unpublished results) and to the receptor cells on the ventral nerve of the horseshoe crab, *Limulus* (W. Schröder, H. Stieve, and G. Maaz, unpublished results). For the *Limulus* photoreceptor cells, this procedure was modified: DAB and hydrogen peroxide were applied simultaneously and the preparation left for 2 h at room temperature. The subsequent treatment follows method 3.

Method 3: Diffusion of HRP into the Extracellular Space
(W. Schröder, H. Stieve, and I. Claassen-Linke, unpublished)

This method was employed to demonstrate the extracellular space in the rhabdom of the crayfish, *Astacus*. A 4% aqueous solution of HRP was injected into the retina and allowed to diffuse for 3 h at room temperature. The eye stalks were dissected out and placed in fixative (1.5% formaldehyde and 1.5% glutaraldehyde in van Harreveld buffer, pH 7.4) for 2 h at room temperature. After briefly rinsing in the buffer solution, the eye stalks were placed in a buffered 0.05% solution of DAB (pH 7.4) for 2 h. One to three drops of a 1% solution of hydrogen peroxide were added for each milliliter of DAB solution, and the preparation was stored overnight at 4°C. Postfixation in 1.5% osmium tetroxide in 0.1 M phosphate buffer (pH 7.4) for 2 h was followed by dehydration and embedding in Epon.

Method 4: Diffusion of HRP into Cut Nerve Ends
(A. Eichendorf and K. Kalmring, unpublished)

The tympanal nerve of a locust was carefully dissected out and the cut end placed into a little glass bowl made from glass capillaries. The glass bowl was filled with a 40% solution of HRP in 0.1 M KCl and placed into the animal. The bowl was sealed with Vaseline to prevent diffusion of HRP into the hemolymph. The preparation was then placed into a moist chamber and left overnight in the refrigerator. After dissecting out the ganglion with the attached nerve, the tissue was placed in 3.5% glutaraldehyde for 2 h, washed in phosphate buffer (pH 7.6), and subsequently processed according to method 2.

Method 5: Intracellular Injection of HRP into Muscle Fibers
(Coggshall, 1978)

This method is listed separately because by injection of HRP into muscle fibers the corresponding motor neurons were stained possibly due to uptake of the enzyme through the neuromuscular junction.

Approximately 5 nl of a 10% aqueous solution of HRP were injected into the dorsal longitudinal muscle fibers of intact living fruit flies (*Drosophila*, Diptera). The animals were allowed to survive for 16–24 h. Fixation took place in cold glutaraldehyde (pH 7.2, Sorensen phosphate buffer, 24 h). Subsequently, a procedure modified from Graham and Karnovsky (1966) was followed: The preparation was placed in 0.1 M Tris buffer (pH 7.5; 2 × 15 min); the remaining aldehydes were reduced in a solution of 0.1 M NaBH$_4$ in Tris buffer (30 min), washed in Tris buffer and were placed in a Tris-buffered 0.05% solution of DAB for 1 h. Subsequently, the preparation was immersed in a DAB solution containing in addition 2 drops of 3% hydrogen peroxide per milliliter of DAB solution (3 h, 0°C); it was washed in Tris buffer (15 min) and 0.2 M Sorensen phosphate buffer (pH 7.2; 15 min) and postfixed in a 2% solution of OsO$_4$ in 0.1 M Sorensen phosphate buffer (pH 7.2; 1 h). After washing in phosphate buffer, the tissue was dehydrated in an ethanol series and embedded in Epon via propylene oxide.

Histologic Procedure for Electron Microscopy
(Modified after Muller and McMahan, 1976)

The histologic procedure for preparations processed for electron-microscopic studies follows closely the one given by Muller and McMahan (1976). Stained cells are shown in Figures 16-11–16-16.

Two hours after injection of HRP into cells of the visual neuropil in flies, *Phaenicia* (*Calliphoridae*), the head capsule was immersed in fixative (0.8% glutaraldehyde + 1% formaldehyde in 0.1 M phosphate buffer at pH 7.3) for 12 h. The fixation period was later reduced to 2 h. After rinsing in phosphate buffer (10 min), the head was placed in a solution of 5 mg DAB per 10 ml of 0.1 M phosphate buffer at pH 7.4 (30 min). Two drops of 1% hydrogen peroxide per milliliter of DAB solution were added to form the reaction product (20 min). After rinsing twice in 0.1 M phosphate buffer (pH 7.4; 2 × 10 min), the head was placed in a phosphate-buffered concentrated osmium tetroxide solution (pH 7.4; 2 h), dehydrated for 15 min each in 30%, 50%, 70% methanol and placed in a 10% solution of uranyl acetate in 70% methanol for 2 h at 4°C. After further dehydration in 90% methanol (15 min) and three steps of 10 min each in 100% methanol, the head was processed through two baths of propylene oxide (PO) for 15 min each, changed into a 1:1 mixture of PO and Araldite (Durcupan ACM, Fluka, Neu-Ulm, W. Germany) for 30 min, and then placed into a 1:2 mixture of PO and Araldite overnight (8–12 h) in a closed container. Subsequently, the heads were placed in an open petri dish with pure Araldite (8 h at room temperature) and afterward polymerized for at least 36 h at 60°C. The complete histologic treatment except for the last two steps was done at 4°C.

Sections with golden interference color were cut using a diamond knife on a Reichert Ultracut microtome. Sections were either left unstained to obtain maximal contrast of the marked fibers or were double-stained with uranyl acetate and lead citrate. Micrographs were made using a Siemens Elmiskop 101 at 60 kV and 80 kV, respectively.

Preparation of Microelectrodes

A 50% solution of HRP (Type VI, Sigma, St. Louis, Missouri) was back-injected into microelectrodes drawn from fused fiber capillaries. The electrodes were used for microiontophoretic injection of HRP into the third visual neuropil of blowflies, *Calliphora*. For the intracellular injections of HRP, 5% solution of HRP in 0.15 M potassium acetate was used in order to reduce the impedance of the electrodes. The electrodes were used 10 min after filling.

Application of HRP

In the light-microscopic investigation (see Figs. 16-1–16-3 and Fig. 16-5, Plates 14–16, and Figs. 16-6–16-8), HRP was injected extracellularly and taken up by cells of the third visual neuropil. After positioning the electrodes in the lobula plate region, depolarizing current pulses of 20 to 100 nA were applied by means of a microiontophoresis unit (Model 160, WPI Instruments, New Haven, Connecticut). Pulse duration was 0.5 to

30 s, intervals varied between 0.5 and 5 s. Typically, 1-s-long pulses of 20 nA were applied 10 times at 1-s intervals. Thus, the time integral yielded between 120 and 1000 nA · s. It did not seem to make a difference whether the same amount of charges was injected by current pulses or by a single DC current step.

For the electron-microscopic investigations (Figs. 16-11–16-16) HRP was injected intracellularly by applying a pressure of 2 atm for 10–30 s. The cell's response could be monitored during pressure injection.

Results and Discussion

Light Microscopic Application of Horseradish Peroxidase

The following section describes results obtained by extra- and intracellular application of horseradish peroxidase (HRP). Results of the following experimental approaches are described and illustrated:

1. Microiontophoretic injection of HRP into the third visual neuropil of flies (Figs. 16-1–16-3, Plates 14–16, and Figs. 16-6–16-9; method 1)
2. Extracellular application of HRP into the retina of the crayfish (Fig. 16-4, Plate 16; method 3)
3. Diffusion of HRP into cut nerve ends (Fig. 16-10; method 4)
4. Intracellular injection of HRP into photoreceptor cells of the horseshoe crab (Fig. 16-5, Plate 16; method 2) and cells of the third visual neuropil of flies (Figs. 16-11–16-16)

Microiontophoretic Injection of HRP into the third
Visual Neuropil of the Blowfly, *Calliphora*
The experiments were carried out on blowflies, *Calliphora erythrocephala* Meig., wild type and mutant chalky, taken from the strain bred in the institute (6–14 days after eclosion). The animals were prepared according to the procedure described in detail elsewhere (Dvorak *et al.*, 1975). After injection of the HRP, the animals were left for 3–9 h following the injection to allow the enzyme to be distributed throughout the nerve cells. Thereafter, the retina and visual neuropils of the heads were dissected out completely, all air sacs and tracheal tubes removed, and the animal was then decapitated and the head placed in fresh egg yolk in which fixation took place (Ebbesson, 1970). The subsequent histochemical treatment was described previously (method 1).

In the dipterous central nervous system, extracellularly applied HRP is taken up readily by nerve cells and transported retrogradely and anterogradely throughout entire cells (H. E. Eckert and S. O. E. Ebbesson, unpublished experiments). Using DAB as a substrate for the enzyme, stained cells can be visualized as light to dark brown profiles (Figs. 16-1–16-3, Plates 14–15, and Figs. 16-6–16-9). The contrast of stained cells

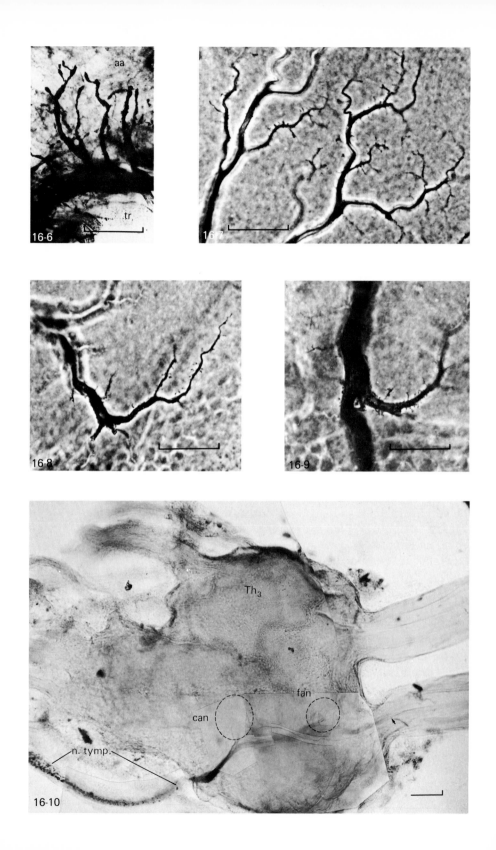

16-6

16-7

16-8

16-9

16-10

can be enhanced by using either benzidine dihydrochloride (Straus, 1972) or o-dimethoxybenzidine (Colman *et al.*, 1976), which produces a dark brown to dark bluish-brown color (H. E. Eckert, unpublished results).

Gross morphologic features (Figs. 16-1–16-3, Plates 14–16) as well as fine cytologic details (Figs. 16-6–16-9) can be resolved in the light microscope, thus providing information about neuronal architecture and topographic relationships between marked cells. Second, fine details such as dendritic spines (Figs. 16-7–16-9) or blebs can be resolved, providing clues about the pre- and postsynaptic nature of dendritic arborizations.

The retro- and anterograde (orthograde) transport of HRP in nerve cells is demonstrated in Figs. 16-1 and 16-2 (Plates 14–15). Fig. 16-1 shows a single section cut in the frontal plane in which the injection site in the lobula plate (arrowed) and parts of two types of motion-sensitive neurons can be recognized, a horizontal (EH) and a vertical cell (V). These neurons can be identified according to previous intracellular Procion yellow (Dvorak *et al.*, 1975; Hausen, 1976) and cobalt (H. E. Eckert, unpublished experiments) injection and recording techniques. Part of a horizontal cell with its dendritic field in the lobula plate, its axon (a_H), its cell body (cb), and the axonal arborization (aa) is depicted, demonstrating the retrograde transport of the enzyme; the orthograde transport is demonstrated in Fig. 16-2 (Plate 15), which depicts the axonal arborization (aa) and axon terminals (at) of the same cell type, an H-cell. Fig. 16-5 (Plate 16) shows the axonal arborization (aa) at higher magnification.

HRP stains fibers of all diameters (Fig. 16-3, Plate 15, and Fig. 16-15).

Figures 16-6 through 16-9 demonstrate fine cytologic details of HRP-marked fibers at high magnification.

Fig. 16-6. Bulbous endings of the axonal arborization (*aa*) of a horizontal cell in the posterior slope of the ventrolateral protocerebrum. Note the absence of dendritic spines in this suspected input/output region of the cell. *tr* trachea. *Calliphora*. Marker: 20 μm.

Fig. 16-7. Part of the dendrite of a horizontal cell (EH of Fig. 16-1) in the lobula plate. Note the presence of dendritic spines. Phase contrast microscopy. *Calliphora*. Marker: 20 μm.

Fig. 16-8. Part of the ventral dendrite of a vertical cell (V of Fig. 16-1) from a consecutive section shown at higher magnification. Note dendritic spines along the side branches. Phase contrast microscopy. *Calliphora*. Marker: 20 μm.

Fig. 16-9. Part of the dendrite of the same vertical cell as depicted in Fig. 16-8, but slightly more dorsally. Note the difference in the appearance of dendritic spines along this side branch. Phase contrast microscopy. *Calliphora*. Marker: 20 μm.

Fig. 16-10. Diffusion of HRP into cut nerve ends. HRP-marked fibers from the tympanal nerve (*n.tymp.*) enter the metathoracic ganglion (*Th₃*) and fan out into the frontal (*fan*) and caudal (*can*) acoustic neuropil. Note marked fibers entering the connective to the mesothoracic ganglion (*arrow*). *Locusta* (Orthoptera). Marker: 100 μm. [Courtesy of A. Eichendorf and Dr. K. Kalmring.]

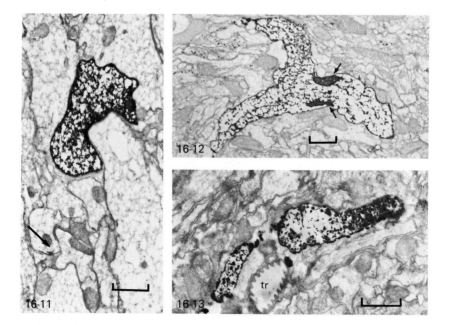

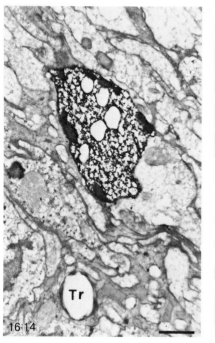

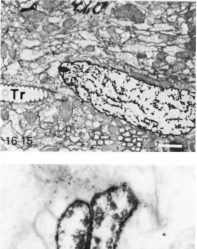

This might prove to be an important tool in the identification of cells that are too small for electrophysiologic recording techniques. Such cells are, e.g., those that are suspected to be presynaptic to the horizontal and vertical cells. They can be stained, however, by injection of HRP into the inner chiasma (H. E. Eckert and S. O. E. Ebbesson, unpublished results). If in such a preparation, the large horizontal and vertical cells are injected with HRP by intracellular application, then the functional contiguity between the presumed presynaptic elements and these giant fibers can be investigated at the ultrastructural level.

Figures 16-6–16-9 demonstrate fine cytologic details of dendritic branches belonging to horizontal and vertical cells at their synaptic input sites in the lobula plate (Figs. 16-7–16-9) and a possible input and simultaneous output site (Hausen, 1976) of a horizontal cell, its axonal arborization in the posterior slope of the ventrolateral protocerebrum (Fig. 16-6). Such stainings can provide clues or confirmation about the polarity of cells and can thus provide an important tool for the identification of functional pathways on the fine structural (light-microscopic) level.

Figures 16-11 through 16-16 show electron micrographs of HRP-marked fibers. All magnification markers indicate 1 μm except for Fig. 16-16.

Fig. 16-11. Cross section of the axon within the lobula plate, 15 μm from the site of injection. Interdigitation of the various neurons is apparent. The *arrow* indicates a presynaptic ribbon in an unmarked fiber as a demonstration of the quality of tissue preservation following the required histochemical treatment. Section contrast enhanced for 5 min with uranyl acetate and for 1 min with lead citrate. Accelerating voltage 80 kV. $\times$ 12,000.

Fig. 16-12. A bifurcation of the axon depicted in Fig. 16-11 is seen in the medulla. The *arrows* point to mitochondria visible within the stained profile. Unstained section. Accelerating voltage 60 kV. *Calliphora*. $\times$ 7000.

Fig. 16-13. The upward pointing branch on Fig. 16-12 again bifurcates at a point about 10 μm more distal and passes close to a small trachea (*tr*). Unstained section. Accelerating voltage 60 kV. *Calliphora*. $\times$ 12,000.

Fig. 16-14. A large-diameter profile of an axon within the lobula plate near the site of injection. Note the vacuolarization and the darkly stained mitochondria located around the periphery of the axonal membrane. Section contrast enhanced with uranyl acetate for 30 min and lead citrate for 5 min. Accelerating voltage 80 kV. *Calliphora*. $\times$ 10,900.

Fig. 16-15. The same cell as depicted in Fig. 16-14 approximately 100 μm more proximal; the axon can still be readily traced in a semitangential section. $\times$ 6400.

Fig. 16-16. The axon arborizes extensively in the periesophageal region resulting in small-diameter profiles such as the one on the *right*. This profile is seen to be in intimate contact with a second, dark, small-diameter profile of a different neuron. Note the absence of synaptic vesicles in both marked fibers as well as in surrounding unmarked cell processes. Unstained section. Accelerating voltage 60 kV. Magnification marker: 200 nm.

Diffusion of HRP into the Extracellular Space in the Retina

The use of HRP as a marker for the extracellular space is demonstrated in Fig. 16-4 (Plate 16), which shows a longitudinal section through the eye of a crustacean (*Astacus, Decapoda,* method 3; W. Schröder, H. Stieve, and I. Claassen-Linke, unpublished results). Alternate layers of microvilli are arranged such that the long axes of the microvillar tubules are perpendicular to one another from one layer to the next. The stain can be most easily detected in the layer in which the microvilli are cut perpendicularly to their long axes.

Diffusion of HRP into Cut Nerve Ends

The method of applying a dye to cut nerve ends has been used extensively for the demonstration of fiber projections using, e.g., Procion yellow (Gregory, 1974; Strausfeld, 1976), cobalt (Tyrer and Bell, 1974), and, recently, HRP (A. Eichendorf and K. Kalmring, unpublished results). Figure 16-10 shows an example of such a preparation: The tympanal nerve of a locust (*Locusta, Orthoptera*) was cut and the end placed into a 4% solution of HRP. The subsequent histochemical treatment is given on p. 329 (method 4).

The tympanal nerve projects into two areas in the metathoracic ganglion, the frontal and caudal acoustic neuropils, and contains in addition ascending fibers to the mesothoracic ganglion. The resolution of stained fibers as shown in this whole-mount preparation is not as good as when using the cobaltous chloride diffusion technique (e.g., Tyrer and Bell, 1974), especially when combined with the silver intensification technique (Bacon and Altman, 1977). However, the advantage of the HRP diffusion technique lies in its potential use for electron-microscopic investigations.

Intracellular Application of HRP

An example of a cell marked by intracellular injection of a 12% solution of HRP is shown in Fig. 16-5 (Plate 16), which depicts photoreceptor cells that are closely attached to the ventral nerve of the horseshoe crab, *Limulus* (W. Schröder, H. Stieve, and G. Maaz, unpublished results; histochemical treatment according to method 1, p. 327).

Electron Microscopic Observations of HRP-Injected Neurons

Fibers that have been injected with HRP and stained accordingly are readily visible in the electron microscope (see also Muller and McMahan, 1976). What appears as a diffusely stained cell in the light microscope (H. E. Eckert and S. O. E. Ebbesson, unpublished results) shows up as a granular, electron-opaque substance that fills the axon to a greater or lesser extent. If the concentration of the reaction product is not excessive, it can be seen to be selectively bound to organelles within the cell such as microtubules and microfilaments, endoplasmic reticulum, mito-chondria, and various vesicular structures. It seems that, if the amount of injected HRP is too great, the fiber becomes stained so heavily that internal structures are almost indistinguishable. The density of the individual grains of the osmiophilic reaction product within the cell appears to be independent of the total HRP concentration within the cell; therefore, it is obviously advantageous to limit the concentration of HRP for electron-microscopic investigations.

Near the injection site, the axoplasm is characteristically dense and vacuolarized. These vacuoles are devoid of internal structures and are of relatively uniform diameter, about 500 nm (Fig. 16-14). The general quality of tissue preservation is adequate for most morphologic investigations and the marked fiber is itself preserved well enough to allow visualization of the various ultrastructural components.

Identification of the injected neuron is rarely difficult due to the great electron density of the HRP reaction product. Even in sections in which the contrast has been enhanced with uranium and lead salts, the marked neuron can be readily located even at low magnification. In very thin sections (silver to gray), it is advantageous to reduce the accelerating voltage of the electron microscope in order to provide additional background contrast (Figs. 16-12, 16-13, 16-16).

By following individual injected neurons through numerous sections it has been possible to observe bifurcations and extensive arborizations until the remaining marked fibers are 100 nm in diameter and even smaller. Within the neuropil, these small fibers are occasionally seen to make contact with other small-diameter elements (Fig. 16-16). It has not yet been possible in these pilot studies to characterize these contacts or to identify the origin of the participating fibers. Synaptic vesicles have been observed in these small-diameter elements (Fig. 16-16). Both fibers were stained individually as shown by serial sectioning.

Nonspecific diffusion from one intact neuron to another was never observed under the conditions used in this study. Multiple marking only takes place when more than one cell is damaged by the electrode tip. This can hardly be avoided in such preparations when recording intracellularly, since the electrode tip has to be advanced through the tissue and is bound to injure cells during this passage. Even small amounts of HRP leaking out of the electrode seem to be sufficient to be taken up by injured cells, which subsequently will be visible in the electron microscope.

Uptake and Transport of HRP by Nerve Cells

The processes underlying the uptake and transport of the enzyme by nerve cells are still not understood. The motor neurons labeled by intracellular injection into muscle fibers in *Drosophila* (Coggshall, 1978) might have taken up the enzyme by pinocytosis via the neuromuscular junction. However, as already argued by Coggshall (1978), it cannot be excluded that damaged axonal endings of the motor neuron deep within the muscle fiber took up the HRP. Intra-axonally, the enzyme is probably transported actively since its reaction product can be found within minutes after the injection in processes far away from the injection site. Literature on this subject has recently been reviewed by Herkenham and Nauta (1977) and by Vanegas *et al.* (1978).

Technical Considerations

No general histochemical procedure is applicable to any system under study. A bulk of literature on this subject has been accumulated, devoted to the study of influences exerted by modifications of the procedure by Graham and Karnovsky (1966). These modifications include, e.g., the use of different fixatives, their subsequent reduction (Kraehenbuhl *et al.*, 1974), the influence of counterstains, the use of different substrates, and the additional treatment with cobalt and silver staining procedures. One of the most reliable methods appears to be the procedure published by Malmgren and Olsson (1978). This literature was recently reviewed by Adams (1977).

Concluding Remarks

Extracellular application of HRP can be used to characterize populations of neurons, or, under certain conditions, the extracellular space around cells. Fiber tracts and projection fields may be identified by diffusion of HRP into cut nerve ends. The intracellular application of HRP offers a high-contrast, high-resolution marking technique for neurons that have been characterized by electrophysiologic recording techniques. Since the reaction product is readily visible at the light as well as the low-electron-optic magnification, it is unnecessary to prepare tedious and cumbersome high-magnification montages of electron-microscopic sections. This is a distinct advantage in the morphologic characterization of neurons in neuropil regions in which the structural organization is unknown or has been inadequately studied. Since HRP seems to diffuse into even the smallest diameter endings of neurons, it appears to be within the grasp of the investigator to identify, map, and reconstruct marked elements and identify synaptic contacts with other fibers as well. Since the internal ultrastructural organization of the marked fiber remains visible, it appears to be possible to identify the polarity of synapses when recognizable synaptic specializations are present.

It would be very useful to be able to visualize the marked neuron in the light microscope in thin sections. Thick sectioning techniques, as they are applied to Golgi–electron-microscopic studies, might provide a possibility, although visualization in semithin sections would be much less cumbersome. This technique would allow very rapid mapping of neurons over long distances by alternately cutting semithin and ultrathin sections. Enhanced contrast for the light-microscopic observation of thin sections may be achieved by using a combination of benzidine compounds or by coinjection of a fluorescent or visible dye along with the HRP.

A very promising method seems to be the characterization of motorneurons by intracellular injection of HRP into corresponding muscle fibers (Coggshall, 1978). A similar technique was described by Kristensson *et al.* (1971). The main advantage lies in the fact that the muscle fibers are usually much larger than their corresponding motor neurons and can thus be penetrated much more easily. In addition, motorneurons that are too small for intracellular recording techniques might become identifiable.

Acknowledgments

This study was supported by the Deutsche Forschungsgemeinschaft (Ec 56/1a + b) and a grant provided by the National Science Foundation (NSF-BMS 74-21712 awarded to H.E.E. and L. G. Bishop). We wish to express our gratitude to Dr. S.O.E. Ebbesson and his co-worker Lolyn Lopez, in whose laboratory the first of our HRP stainings were obtained and who introduced H.E.E. to this technique (supported by National Eye Institute Grant EY 00154 and NASA Grant NGR-47-005-186 to Dr. Ebbesson). We are most grateful to our friends and colleagues Drs. I. Claassen-Linke, A. Eichendorf, K. Kalmring, G. Maaz, W. Schröder, and H. Sieve, who provided unpublished material (Figs. 16-4 and 16-5, Plate 16, and Fig. 16-10).

Chapter 17

Cobalt Staining of Neurons by Microelectrodes

Michael O'Shea

University of Chicago
Chicago, Illinois

The purpose of this chapter is to provide the basic technical information required to trace the course of individual neurons from which recordings of electrophysiologic activity have been made. Two basic methods, intracellular and extracellular, have been established for achieving this goal. Both methods depend upon the introduction of a dye or staining substance through a recording microelectrode into the recorded neuron. The geometry of the neuron is then revealed after suitable histologic treatment.

The history of microelectrode staining of neurons includes the use of many substances including various metal ions, radioactive amino acids, horseradish peroxidase (HRP), and a number of dyes. In the early 1960s steel microelectrodes were first used to deposit ferric ions in goldfish Mauthner cells (Furshpan and Furukawa, 1962). It was not, however, until the late 1960s that the first really successful technique for revealing the anatomy of extensive regions of neurons *beyond* the site of electrode penetration was described. The history and literature pertaining to the techniques available prior to 1968, when Kravitz *et al.* described their intracellular Procion yellow injection technique (Kravitz *et al.*, 1968), was reviewed extensively by Kater and Nicholson (1973) and will not be reiterated here.

The impact of the Procion yellow technique was enormous; for the first time it was possible to explore the relationship between structure and function at the level of the individual neuron. This led to an explosion in the literature of papers applying the new technique in vertebrate and invertebrate preparations. In spite of this, however, and partly because the Procion yellow technique is highly labor-intensive, requiring sectioning and fluorescent microscopy, it was superseded as a routine intracellular method by cobalt injection (Pitman *et al.*, 1972).

The development of the intracellular cobalt method led to a second explosion in the literature because it proved to be less technically demanding than the Procion method and resulted in stained neurons that were visible, without sectioning, with conventional light-microscopic techniques. Interest in the cobalt technique developed rapidly and, probably because of the immediacy and striking results obtained by its application, many were encouraged to develop modifications of the basic technique in a way which did not happen after the introduction of Procion yellow.

It was found, for example, that intracellular penetrations were not necessary for staining individual neurons (references cited in General Methods section) and that neurons recorded extracellularly with cobalt-filled glass microelectrodes could be extensively stained. Previous extracellular methods served only to indicate the location of the recording tip of the electrode. Now, however, cobalt provided a means of revealing details of the structure of units recorded extracellularly.

In addition to this, the pleasing results obtained with intracellular cobalt injection reawakened interest in the technique of axonal iontophoresis (Iles and Mulloney, 1971). This method was first described as an extension of the microelectrode Procion yellow technique, but the use of cobalt as a dye substance promoted its widespread adoption. The axonal iontophoretic technique of intracellular staining without the use of microelectrodes is the subject of another chapter.

Here both intracellular and extracellular microelectrode techniques will be described that combine single neuron recording with subsequent staining with cobalt. Such techniques and their use over the past few years have led to the functional anatomy of the central nervous system (CNS) being discussed at the level of the individual neuron. Combined recording and staining techniques have been employed to answer questions concerning the extent of individual anatomical variability, development of identified neurons, connectivity of neurons, and neuroanatomical correlates of specific neural integrative functions.

A crucial assumption that underlies these applications is that both electrophysiologic data and anatomical data are gathered from the same neuron. In other words, the neuron recorded from is the one stained. Techniques differ from one another with respect to our confidence of the validity of the basic assumption. Some space will be devoted to this issue after the basic methods have been described.

General Methods

Certain basic techniques are required for both intracellular and extracellular cobalt staining. These include making and filling of glass capillary microelectrodes, recording and passing of current through a high impedance, preparing and filtering of solutions, preparation of tissue after cobalt injection, and methods of photography, drawing and viewing of stained neurons. The differences between intracellular and extracellular techniques will be discussed as they arise. It will be assumed that readers are familiar with standard techniques of recording and of current passing through high-resistance micropipettes. No special equipment is required in cobalt injection other than that normally found in a laboratory equipped for microelectrode recording.

It should be stressed that cobalt injection is a capricious technique. This is especially true for extracellular filling in which neurons can be stained without actively passing current through the electrode. Uncertainty leads to superstition, and many superstitions surround microelectrode staining techniques. Here I will outline only the basic parameters; adaptation is encouraged to suit the technique to the different conditions that arise in different preparations.

This section is divided into four subsections. The first describes the preparation of microelectrodes. Next the methods employed in recording and staining are described, followed by a description of the various methods available for preparing the stained neuron for viewing. Finally, the methods of viewing and recording the anatomical data are explained.

Preparation of Microelectrodes

The shape and resistance of the microelectrode made for cobalt injection will depend on the type of staining to be attempted (intracellular or extracellular) and in the case of intracellular staining, the size of the neuron that is to be penetrated. Sharp electrodes with a high impedance ($\sim$80 MΩ) will facilitate the penetration of fine neural processes but will pass current (cobalt ions) rather poorly and therefore may not be suitable for cobalt injection.

Electrode resistance can be reduced without affecting the sharpness of the electrode by increasing the concentration of the cobalt electrolyte. This strategy, however, leads to an increased rectification in the current-passing characteristics of the electrode. Unfortunately, cobalt-filled electrodes pass depolorizing current rather poorly; as the injection progresses the electrode resistance increases and reduces the flow of cobalt ions into the neuron. This problem is more pronounced when the concentration of the cobalt electrolyte is high. A 1–50 mM solution of cobaltous chloride (CoCl$_2 \cdot$6H$_2$O) was originally prescribed by Pitman *et al.* (1972), but except for rather blunt electrodes this low concentration produces a high electrode resistance. I have successfully used solutions from 250–500 mM cobaltous chloride.

It has been claimed that cobalt nitrate and cobalt acetate give better results and reduce the tendency of electrodes to block when passing depolarizing current. Storage of electrodes filled with cobalt nitrate ($\sim$1 M) has been claimed to reduce subsequent "electrode noise" (Kien, 1976). I have tried cobalt nitrate and cobalt acetate at concentrations between 50 mM and 1 M, and in my hands the difficulty of passing depolarizing current remains one of the major problems with the technique.

Since the introduction of the fiberglass-threaded tubing method of making micropipettes (Tasaki *et al.,* 1968), the filling of microelectrodes is a rapid and simple procedure. Today microelectrode filling has been made even more routine by the availability of ready-made microfiberglass capillaries (Kwik-Fill Glass Capillaries from W-P Instruments, Inc., Box 3110, New Haven, Conn. 06515, for example). Electrodes made with this type of glass can be filled by first dipping the tips into the electrolyte solution for about 30 s and then injecting the electrolyte into the electrode stem with a syringe fitted with a 31-gauge needle. Electrodes prepared in this way are ready for use immediately. The glass fiber in the electrode not only aids filling by internal capillarity but also imparts strength and rigidity to the electrode tip (Pearson and Fourtner, 1975), thus preserving the sharpness of the electrode as it passes through the ganglion sheath.

The cobalt salt solution should be micropore filtered in order to reduce possible problems of electrode blockage. If microfiberglass tubing is used to make the microelectrode, the most convenient time to filter the electrolyte is while syringe-injecting the electrode. This is achieved by attaching to the syringe a Swinny filter apparatus (Millipore Corp., Bedford, Mass.) fitted with a 0.22 μm or less Millipore filter. I have not found it necessary to clean precut and fiber-filled glass tubing.

In relation to finding a solution to the problem of passing cobalt ions through the microelectrode by electrophoretic technique, pressure injection is a technique that has been applied successfully in mollusks and the leech for Procion yellow and horseradish peroxidase (see Kaneko and Kater, 1973). Success with pressure injection, however, depends upon rather large electrode tip sizes and therefore may not be suitable for intracellular injection of cobalt into small insect neurons.

Grinding or beveling of cobalt-filled glass microelectrodes so that the tip resembles the tip of a hypodermic needle is another technique worth

considering. Several investigators have found that beveled microelectrodes have reduced impedance, cause less damage on neuron penetration, and have more favorable dye and current-passing properties. A
simple apparatus that has been employed in the production of beveled microelectrodes is shown in Fig. 17-1. It consists of a small turntable onto
which an adhesive abrasive film is stuck (Alumina Abrasive film, 0.3 μm,
Arthur Thomas Co., Philadelphia, Penn. 19105). The turntable is rotated
at 1–5 Hz by a DC motor and the filled electrode is slowly lowered onto
the grinding surface at about 30°, pointing in the direction of rotation by
means of a micromanipulator. If the abrasive surface is wetted with
physiologic saline solution, the electrode resistance can be monitored
during grinding by measuring between a wire in the microelectrode and a
wick electrode that touches the turntable surface. A binocular microscope should be used to observe the microelectrode while it is lowered
onto the turntable surface.

No matter how careful one might be in the construction of this piece of
equipment, it will be almost impossible to achieve a perfectly flat rotation
of the grinding surface. When the edge of the grinding surface is viewed
horizontally through a microscope it will appear to rise and fall slightly as
it rotates. This apparent defect can be turned to advantage. With the
electrode appropriately poised, fine control over the amount of grinding
can be achieved by allowing the rising grinding surface to brush the electrode tip once per revolution. Each contact will remove a small amount
of the tip and thereby reduce the electrode resistance. Using this technique, cobalt-filled microelectrodes (250 mM CoCl$_2$·6H$_2$O) have been reduced in resistance from 80 MΩ to 12–20 MΩ with six or seven contacts
with the abrasive surface.

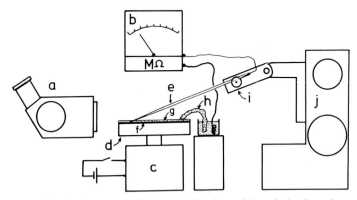

Fig. 17-1. A simple apparatus for the production of beveled microelectrodes: *a*
binocular microscope; *b* megohm meter; *c* slow DC motor; *d* turntable; *e* cobalt
chloride-filled microelectrode; *f* adhesive abrasive film; *g* film of saline solution, *h*
cotton wick between surface of turntable and small container of saline solution; *i*
microelectrode holder; *j* micromanipulator.

Such beveled cobalt microelectrodes have been used by Palka *et al.* (1977) for cobalt staining of primary afferent fibers in the cercal nerve of the cricket. Palka *et al.* claim that beveled microelectrodes could pass much higher currents for longer times than unbeveled microelectrodes. This is an encouraging result, and beveling seems to be a useful technique but one that in insects has been insufficiently tested to determine whether it is a solution to one of the major problems of cobalt iontophoresis, i.e., electrode blockage.

Recording and Cobalt Injection

Difficulties encountered in employing cobalt-filled microelectrodes rather than potassium acetate or some other conventional electrolyte are those associated with the generally higher resistance produced by low concentrations of cobalt solutions and the marked tendency of the electrode impedance to increase during iontophoretic injection of cobalt. Despite this difficulty, however, neurons that can be impaled for 10 min or so with conventional microelectrodes can be stained by intracellular injection using cobalt electrodes. Long penetrations are not required for good cobalt staining, and in my experience microelectrode penetrations in the neuropil produce better and more rapid results than cell body penetrations.

In the original description of the cobalt technique (Pitman *et al.*, 1972), two methods were described. The first was pressure injection in which microelectrodes were filled with a 1 *M* solution of cobaltous chloride containing 0.4% Procion navy blue. The dye was added so that the filling of the neuron could be observed. Neurons injected iontophoretically were impaled with electrodes filled with 1–50 m*M* cobaltous chloride and positive current pulses of 5×10^{-8} A, 0.5 s in duration, were delivered at 1 Hz for 1 h.

Techniques for pressure injection were discussed above. Difficulties encountered with iontophoretic injection are related to rectification. As discussed above, the beginning electrode impedance can be reduced by filling the pipette with more concentrated cobalt solutions than recommended by Pitman *et al.* (1972). When positive current is passed, i.e., cobalt ions carry current out of the electrode tip by making the electrode positive with respect to the preparation ground, the beginning electrode resistance will rapidly be exceeded. This effect, which is confined to depolarizing current, becomes more marked in later stages of filling, and this, combined with the fact that cobalt cannot be seen leaving the electrode, makes it essential to monitor current flow throughout the fill and compensate for rising electrode impedance.

The simplest and most direct method of monitoring current is to measure the voltage drop across a known impedance to ground. Currents used in filling neurons range between 10^{-9} and 10^{-7} A, and a convenient impedance for monitoring these currents is 100 kΩ. This will not produce excessive noise due to the preparation being above ground, and it will generate 1 mV per 10^{-8} A (Fig. 17-2).

Intracellularly impaled neurons are filled by passing pulses of depolarizing current through the microelectrode. Long pulses or DC lead to rectification problems, since most of the current is passed early in the pulse before the electrode resistance rises. For this reason pulses of about 0.5 s duration delivered on a 50% duty cycle seem to be about optimal— avoiding serious blockage problems during the pulse and allowing sufficient recovery time between pulses for the electrode resistance to fall to its original level. Often the problem of electrode blockage becomes more severe late in the filling process; the amount of current being passed per pulse falls. An effective but temporary remedy is to reverse briefly the polarity of the current for a few cycles. When the polarity is returned to the depolarizing direction the electrode will have become unblocked and will continue to pass current before the resistance increases again. This reverse-current remedy progressively decreases in effectiveness, and it must be applied more and more frequently as the fill proceeds.

If the impaled neuron is a spiking neuron and if the electrode is placed either in or close to a spiking region, then the depolarizing current pulse

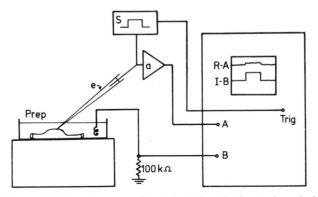

Fig. 17-2. Determination of current flow (I) during the iontophoretic injection of cobalt through a microelectrode. The preparation (*Prep*) is in saline solution and impaled with the microelectrode (*e*). The microelectrode amplifier (*a*) is connected to the upper (*A*) beam of a dual beam oscilloscope. A positive rectangular voltage is applied to the electrode from the stimulator (*S*). A trigger pulse that precedes the stimulus pulse is connected to the external trigger input of the oscillosope (*Trig*). The preparation is connected to ground through a 100 kΩ resistor. Input B of the oscilloscope monitors the voltage across the resistor and is connected to the Prep side of the resistor. This arrangement will provide 1 mV per 10^{-8} A, which flows through the microelectrode. A fall in the current flow without a change in the applied stimulus indicates an increase in electrode resistance.

will probably initiate action potentials. The amplitude of the action potentials is a rough guide to the progress of the fill. The cobalt will gradually poison the neuron and the action potential amplitude will fall. This poisoning presumably spreads from the site of penetration, and the action potential is presumably being recorded more and more remotely from the recording site. It will become necessary to increase the amplitude of the depolarizing pulse in order to excite spikes as the fill progresses.

The duration of the fill will generally be determined by how long a good penetration of the neuron can be maintained. In any case, it should not be necessary with insect neurons to continue to fill for more than 1 h, and often excellent results can be obtained in very short (10 min) filling times; this is especially true for neuropil penetrations. Any indication that the penetration has been lost or that the neuron has been badly damaged should terminate the fill. The continuation of such a fill often results in the marking of more than one neuron.

On termination of the fill the current should be turned off and the electrode removed from the neuron. The preparation can now be treated immediately with ammonium sulfide to precipitate the cobalt or it can be left for an hour or so at room temperature or at 4°C in order to allow for diffusion of cobalt ions throughout the neuron. Allowing an incubation period for diffusion is probably a good strategy when the filling time has been brief, when the penetration was made in a large cell body of a neuron with extensive ramifications, or when long-distance filling of axons is required. It has been found to be a particularly useful technique when combined with the Timm's intensification technique of Tyrer and Bell (1974).

Movement of cobalt out of insect neural somata tends to be rather slow: The cell body apparently "holds" cobalt ions. This problem is especially true for neurons with large cell bodies in which the primary neurite is very narrow—as in dorsal unpaired medial (DUM) neurons of the locust, for example. For such neurons, diffusion incubation combined with intensification may be necessary for revealing the full extent of the neurons' ramifications. This approach appears to be required in obtaining good "fills" from the cell bodies of giant interneurons in the terminal ganglion of the cricket (R. K. Murphey, personal communication).

The techniques described above assume that the filled neuron is intracellularly impaled by the cobalt-filled microelectrode. A technique of revealing the anatomy of relatively large neurons with extracellularly placed cobalt-filled microelectrodes has been described (Rehbein et al., 1974). Such a technique may even be suitable for marking neurons from which extracellular recordings have been made from very fine processes in the neuropil (Kien, 1976). This is potentially an important technique because it may allow us to see neurons that cannot be impaled and filled by intracellular means.

An extracellular microelectrode technique was first described by Rehbein *et al.* (1974). These authors used glass electrodes filled with 3 *M* CoCl₂·6H₂O, which has resistances between 2 and 8 MΩ. Such blunt microelectrodes are unsuitable for intracellular recording. The electrodes were used to reveal the structure and function of large acoustic neurons in the thoracic nerve cord of the locust, *Locusta migratoria*. Rehbein *et al.* claim that the recorded neurons are stained with cobalt without the application of a depolarizing voltage to the electrode. It is thought that the spiking activity of the neuron draws cobalt ions from the blunt electrode, but the precise mechanism of cobalt migration is unknown. Extracellular recordings with glass microelectrodes often record activity from more than one neuron, and under these circumstances more than one neuron is stained. Rehbein *et al.* claim that when single units are recorded only one neuron is stained, and when two are recorded, two neurons stain. The best staining is achieved when the extracellularly recorded spike amplitude is high.

Extracellular cobalt staining of insect neurons has been achieved using sharp microelectrodes recording activity from fine processes in the neuropil (Kien, 1976). The electrodes employed by Kien were filled with 30% cobalt nitrate and had resistances ranging from 15 to 25 MΩ. The use of cobalt nitrate and the incubation of the electrodes at 40°C overnight prior to use appears to be crucial to the recording of fine neural processes. The extracellularly recorded neurons were stained by passing current (about 2×10^{-8} A) pulses at 1 Hz on a 50% duty cycle. The potential advantages and problems of the extracellular cobalt staining technique will be discussed in the final section of this chapter.

Preparation for Viewing

Neurons filled with cobalt ions are not made visible until the cobalt is precipitated as cobalt sulfide. This is done by the application of a dilute solution of ammonium sulfide to the preparation. Before doing this the ganglion containing the filled neuron should be washed briefly in saline solution in order to remove any cobalt chloride solution that might have contaminated the ganglion surface. The ganglion is then immersed in a 1% solution of 44% ammonium sulfide in saline solution for about 5 min. Higher concentrations of ammonium sulfide should be avoided since they will cause severe tissue damage and, if the tissue is to be prepared for sectioning, lower concentrations should be used. After precipitation, the tissue should be washed in saline solution to remove the ammonium sulfide prior to fixation.

At this stage it is often possible to see parts of the stained profile of the neuron in the whole unfixed, and uncleared ganglion. Generally the cell body of a well-stained neuron will be seen since it lies superficially, but usually little can be seen of the neuropil branches that will lie in the core of the ganglion.

The fixative used will be determined by whether or not the tissue is to be prepared for sectioning. If the neuron is to be viewed as a "whole mount" in the unsectioned ganglion, the most convenient and rapid fixative is Carnoy's (60% absolute alcohol, 30% chloroform, 10% acetic acid). The ganglion should be transferred from saline solution to the Carnoy's solution supported on a small piece of paper on which the nerve roots can be arranged in their natural positions. If the ganglion is placed unsupported into Carnoy's, the rather severe reaction of the tissue, saline solution, and fixative will often cause a deformation of its normal outline and appearance. Tissue as large as the brain or metathoracic ganglion of the locust can be fixed in this way in about 10 min. As much as possible of adhering fat, air sacs, and trachea should be removed prior to fixation and what remains should be removed after fixation. Normally the ganglion will become detached from the paper support during fixation and will be "frozen" in the way it was arranged on the paper. Dehydration is a rapid process beginning in 96% alcohol (two changes of 10 min each), then 100% (two changes of 10 min each). The whole ganglion can now be cleared in methylbenzoate, a process that takes 10–15 min. Any drawing, photograph, or analysis of the results should be carried out as soon as possible because the intensity of the filled neuron will fade gradually. Fading is, however, very slow in methylbenzoate, and I have stored preparations in it for up to 4 years with significant but tolerable deterioration. Whole-mount preparations can be Timm's intensified in order to enhance the contrast of the filled neuron (see Chap. 20), and this can be carried out on material that has faded as a result of long-term storage in methylbenzoate.

Tissue fixed in Carnoy's can be sectioned, but more gentle fixatives that cause less damage and shrinkage are preferable. Bouin's fixative is probably the most convenient and suitable for insect material, and full instructions for its use and the preparation of sections can be found in Pantin's (1964) book on microscopic technique. A method for Timm's intensification of sectioned material can be found elsewhere in this volume (Chap. 21).

Recording Anatomical Data

Standard techniques of drawing, photography, and serial section reconstruction can be used on sectioned cobalt preparations. Recording of anatomical information from whole-mount preparations of microelectrode injected neurons, however, deserves special consideration.

There are several reasons why photography or drawing of whole-mount preparations may present special problems. The filled neuron may, for example, have extensive arborizations within the ganglion, and some of the arborizations may be very fine or may present rather poor contrast with the background. In this situation one is presented with the problem of photographing the whole neuron with sufficient depth and extent of field and with sufficient acuity to record adequately its morphology. The striking appearance of a whole-mount cobalt fill and the advantage over the sectioned preparation that the entire morphology of a neuron can be visualized make the technical difficulties involved in successful recording of the image worthwhile. It should be cautioned, however, that the whole-mount method of recording data cannot place the neuron in context with other neuroanatomical structures of the ganglion. Accurate appreciation of anatomical relationships between the inject neuron and other neurons and neuropil regions can only be gained by examination of serially sectioned material.

Most of the problems of photographing whole-mount preparations can be overcome by the use of large-format high contrast film, direct rather than trans-illumination, and a Zeiss Luminar objective lens.

The technique employed to photograph the lobula giant movement detector (LGMD) interneuron in the brain of the locust (O'Shea and Williams, 1974) (Fig. 17-3) was as follows. The brain containing the injected neuron was fixed in Carnoy's and cleared in methylbenzoate. It was placed in a depression slide in methylbenzoate, and the slide was placed on a white background on a stage. Two microscope lamps were focused on the white background and arranged so that minimal shadows were cast by the brain. The preparation was viewed with a 16-mm luminar objective lens that focused the image on a 5 × 7-in. ground-glass screen supported about 3 ft above the lens on the open end of a long rectangular box. The photograph was taken by exposing 5 × 7-in. Kodak Graphic Arts film, Kodalith Ortho Film Type 3 to the focused image. The luminar lens was stopped-down to provide the required depth of focus and exposure times as long as 4 min were used. The film was developed immediately in DK 50 developer under "safe light" control, and further photographs were taken if the exposure or lighting required adjustment. Stereo pair pictures were taken without a tilting stage (see O'Shea and William, 1974; O'Shea *et al.,* 1974) by placing the neuron's image first to the far left of the field and then to the far right. This provided sufficient angular displacement to allow for visualization of a good stereo image from the pair of photographs so produced. The apparatus described here was designed by Professor Hugh Rowell at the University of California at Berkeley, and the procedures were subsequently developed by him, Dr. Carol Mason, Dr. Corey Goodman, and myself. They have been used by us successfully to produce excellent photographs with good acuity and depth of focus of whole insect neurons (see, e.g., Mason, 1973; Goodman, 1974; O'Shea and Williams, 1974; O'Shea and Rowell, 1975).

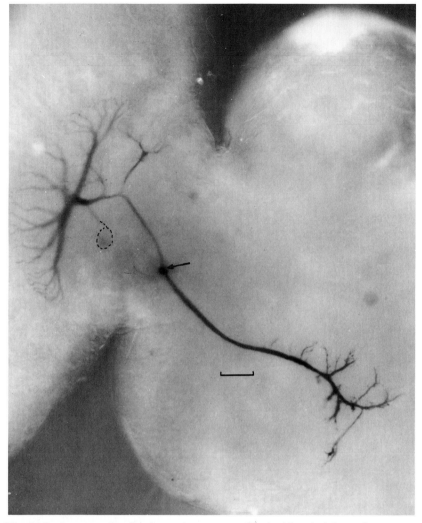

Fig. 17-3. An example of an insect interneuron filled with cobalt by an intracellular microelectrode (see O'Shea and Williams, 1974). The *arrow* indicates the site of microelectrode penetration in the axon. The cell body of the neuron is out of focus and is outlined by a *dashed line*. Scale bar is 100 μm.

Applications and Interpretations

It is clear now that techniques that simultaneously allow for physiologic and anatomical analysis of single neuron function have been developed. These are the techniques in which physiologic recording and dye injection are combined by the use of dye-filled microelectrodes. The impact of these techniques on the development of neuroscience in recent years has been enormous. Even if I confine myself to insect neurobiology—since this book will presumably be consulted primarily by entomologists—the contrast between pre- and postintracellular dye techniques is striking. Beginning with Procion yellow and more recently with cobalt, intracellular microelectrode injection techniques have enabled several groups of workers to elevate a small number of insect preparations to a level where they currently yield valuable information of fundamental neurophysiologic importance. This has been possible because individual neurons have been identified physiologically and anatomically.

The third thoracic ganglion of the locust is a good example of an advanced physiologic preparation, the development of which was made possible only by the application of intracellular (microelectrode and axonal back-filling) dye staining techniques. A significant number of motor neurons, a few interneurons, a sensory neuron, and a few special aminergic non–motor-neuron efferent neurons have been individually identified in this ganglion (Burrows and Hoyle, 1973; O'Shea *et al.,* 1974; Rehbein *et al.,* 1974; Burrows, 1975; Altman and Tyrer, 1977a; Hoyle *et al.,* 1974; Evans and O'Shea, 1977). Another example of an advanced insect preparation that was developed by the use of intracellular dye techniques is the terminal ganglion of the cricket, *Acheta domesticus.* In this animal, Murphey (1973) for the first time identified an insect interneuron by neuropil penetration and injection. A similarly advanced preparation is represented by the cockroach (see Fourtner and Pearson, 1977, for review).

The identification of neurons as individuals by intracellular injection has facilitated neurophysiologic study in insects and other animals in a variety of ways. Identification, for example, allows for repeated recording from the "same" neuron in different individuals of the same species. This allows for the gradual accumulation of physiologic data and makes it possible to extend the investigation to the establishment of "identified" synaptic connection between identified neurons (O'Shea and Rowell, 1975). The identification of single neurons makes possible the identification of neurotransmitters associated with identified and excised neuronal somata. This is one step in the process of showing that a given substance is the neurotransmitter of a *particular* known neuron (Emson *et al.,* 1974; Evans and O'Shea, 1977).

Intimate knowledge of the anatomy and physiology of single identified neurons is the only way in which we will come to understand the subtle relationship between structure and function in the nervous system at the cellular level. In relation to this, insects have provided suitable preparations in which structure/function correlation has been attempted (see, e.g., Rowell *et al.,* 1977), made possible by prior identification of an interneuron by intracellular cobalt injection (O'Shea and Williams, 1974). Recently, intracellular recording and dye staining (using Lucifer yellow;

Corey Goodman and Nick Spitzer, personal communication) have been used to trace the development of individually identified neurons in the locust embryo.

In summary, the identification of individual neurons by intracellular dye staining has been powerful and influential in our understanding of the role of individual neurons in the functioning of the CNS. To a large extent its power is derived from the near certainty that the neuron recorded and the neuron stained are one-and-the-same. It is worth considering the extent to which the methods described here provide for such confidence.

Clearly, intracellular methods are to be preferred. They provide the assurance that at least part of what is stained represents at least some of the morphology of the recorded neuron. It should be cautioned, however, that intracellular techniques do not guarantee either that portions of other neurons are not also filled trans-synaptically or via nonsynaptic dye-permeable pathways or that the entire arborization of the impaled neuron will be stained. Both multineuronal filling and incomplete filling present serious problems of interpretation. Trans-synaptic filling or filling via dye-permeable pathways of unimpaled neurons presents two main problems depending on the extent of staining in the unimpaled neuron.

First, if the unimpaled neurons are filled well enough to be able to see additional cell bodies, the problem is to resolve the boundaries of the filled neurons. Clearly, the question of which neuron provided the electrophysiologic data is raised. Fortunately, unambiguous cases, i.e., more than one cell body visible, of multineuronal filling with intracellular microelectrode technique are rare. Uncommon though it may be, the fact that it happens is reason to consider the second problem presented by multineuron filling. This is that it is not possible in a single fill of a neuron to determine whether or not partial multineuronal filling has occurred. Partial trans-synaptic filling here means filling of the branches but not the cell body of the unimpaled neuron. The presence, therefore, of a single filled cell body and associated arborizations is not absolute proof that the arborizations belong only to a single neuron. My experience with intracellular cobalt filling of visual interneurons in the locust brain suggests that this form of transneuronal filling does occur and should therefore be considered as a possible interpretation of intracellular fills of what appear to be single neurons.

The descending contralateral movement detector (DCMD) and the lobula giant movement detector (LGMD) are connected in the locust brain by a spike transmitting electrical synapse. Intracellular cobalt injections of either neuron in the vicinity of the synapse normally fill only the impaled neuron. On one occasion, however, an electrode placed in the DCMD filled both the DCMD and a portion of the LGMD's arborizations and axon, but not its cell body. If the DCMD and LGMD were not well known and much studied interneurons, the anatomy of the DCMD might have been described including a piece of its main presynaptic neuron, the LGMD. This is an obvious and fortunately easily avoided example of the type of contamination of anatomical data that can occur even with intracellular injection. It remains unknown to what extent low levels of this type of contamination might occur in every intracellular cobalt fill. It is likely, in addition, that if it does exist it is exaggerated by the Timm's intensification technique, which generally reveals far more arborization of a neuron than is seen in the unintensified preparation. The possibility that some fraction of the new branches so revealed belongs to other neurons that are weakly cobalt-coupled to the injected neuron should not be overlooked.

One might expect the occurrence of multineuronal filling discussed above and the partial filling of impaled neurons to be determined by a complex of difficult-to-control variables. Given this, the problems are to some extent solved by building up an anatomical description from many penetrations and fills of the same neurons and by the use of different techniques and dyes. Descriptions based on single fills are likely to be unreliable.

In relation to extracellular techniques of microelectrode cobalt staining, in my hands the probability of both multineuronal and partial filling is increased. The extracellular techniques must be considered less reliable than the intracellular techniques. Problems apply with additional relevance, therefore, to extracellular cobalt staining, particularly when staining is combined with Timm's intensification.

Probably the best way to avoid misinterpretation of anatomical data derived from cobalt injection is to avoid dependence on a single technique and to avoid reliance on single fills.

Chapter 18

Nonrandom Resolution of Neuron Arrangements

J. Bacon

Max-Planck-Institut für Verhaltensphysiologie
Seewiesen über Starnberg, F. R. G.

N. J. Strausfeld

European Molecular Biology Laboratory
Heidelberg, F. R. G.

Passage of cobalt or nickel ions through axons to nerve cell branches can be achieved by introducing the salt solutions (such as acetate, chloride, nitrate, or bromide) into cut ganglia or their connectives (see Chap. 19). Distal portions of nerve cells are resolved after silver intensification. Appropriate conditions, such as low temperature, low salt concentrations (in Diptera), or high concentrations (in Orthoptera), reveal contiguous interneurons that impinge on primary-filled elements. In Dipterous insects, this method has been used to resolve the entire population of horizontal and vertical motion-sensitive neurons whose terminals are presynaptic onto a set of descending interneurons that happen to be particularly receptive to cobalt ion influx (Strausfeld and Obermayer, 1976). Comparisons between the shapes of such "trans-synaptically" filled elements, and identical neurons resolved by Procion yellow iontophoresis (Hausen, 1976a) have shown that cobalt ions pass to all parts of the neuron. Extended periods of Co^{2+} uptake may even resolve up to three consecutive sets of nerve cells (Fig. 18-1D).

The major drawback of this technique, as with the majority of filling procedures, is that neural tissue has to be dissected free and then placed in a reservoir of the solution. This demands perfect isolation of the target neuropil from the rest of the animal in order to avoid leakage. This opera-

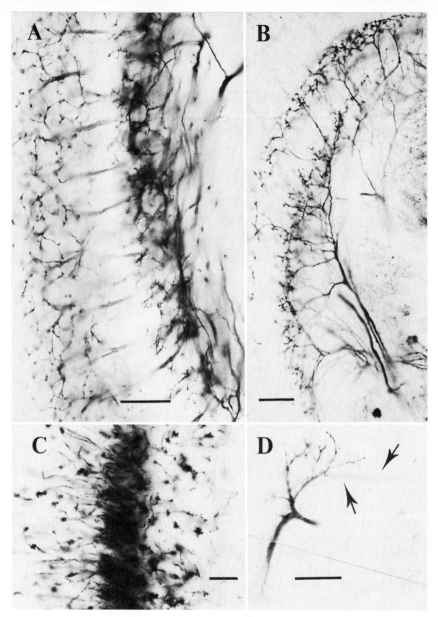

Fig. 18-1. Neuronal assemblies resolved after cobalt injection from glass capillaries. **A.** Assemblies of columnar neurons in the lobula of *C. erythrocephala*. Note different shades of silver intensification to give pale and dark staining. **B.** Tangential neurons in the lobula of *C. erythrocephala*. **C.** Mossy fiber terminals in the granule cell layer of the mouse cerebellum (courtesy of M. Obermayer). **D.** Dendrites of the giant descending neuron of *Drosophila*, after injection into the mesothoracic ganglion. Dendrites are in the brain, and are shown with "transsynaptic" uptake by contralateral afferents from the antennae (arrow). Note that secondarily filled "trans-synaptic" fibers appear pale, probably due to the low concentration of Co^{2+} in them. Scales represent 20 μm for this and other figures, unless otherwise stated.

tion may function well on larger insects such as locusts, and even on large Diptera, but can be applied only with difficulty to smaller animals, such as *Drosophila* or early instar Orthoptera. The method is further limited in that it can be used only to resolve peripherally derived (or destined) elements, as well as a few species of interneurons within the brain and thoracic ganglia.

A recent modification of the method was devised to introduce the reservoir directly into the tissue, rather than vice versa. The technique (originally described by Strausfeld and Hausen, 1977) has been used to reveal assemblages of interneurons within the brain of *Calliphora*, by means of long-term implantation of a fused glass capillary micropipette. This is filled with cobalt chloride and Co^{2+} ions spread by passive diffusion into the tissue. The present account summarizes this very simple method, and describes how it may be generally applied to resolve assemblages within brains and thoracic ganglia of *Drosophila* and *Locusta* in addition to larger Diptera, Odonata, and Hymenoptera. Two recent accounts demonstrate the method's use for resolving consistent neuron arrangements (Hausen and Strausfeld, 1980; Strausfeld, 1980: see Fig. 18-1 A,B).

The Method

Single or double-barreled fused glass capillary micropipettes (1 mm o.d., 0.58-mm i. d., containing a 0.1 mm glass filament) are pulled to an obtuse tip with a diameter of about 0.5 μm.

The tip diameter is one parameter that defines the rate of outward Co^{2+} diffusion. Because the effective concentration of Co^{2+} is not identical in different species, the dimensions of the electrode have to be modified accordingly. For *Calliphora* the tip is broken off to give a diameter of between 5 and 20 μm.

The electrode is filled with an aqueous solution of cobalt chloride (acetate, bromide, or nitrate) immediately before use. Filled micropipettes may be stored with their tips immersed in distilled water for about half an hour.

The insect is glued onto an appropriate platform that is mounted on a Plasticine gimbal so that it can be tilted or rotated. The micropipette is held in a simple manipulator: This is most conveniently built of Plasticine such that the electrode can be raised, lowered, moved laterally, and advanced (Fig. 18-2). Partial adhesion between the glass shaft and the Plasticine allows increments of advance of as little as 4 μm. The depth of penetration can be quite well gauged using a double-fused fiberglass micropipette where one shaft is broken slightly shorter than the other. The filled tip is advanced far enough so that the broken end of the shorter shaft

Fig. 18-2. Only rudimentary equipment is needed for cobalt injection into the neuropil, and several dozen fillings can be carried out simultaneously. The fly is stuck to a glass plate (*1*) mounted on a Plasticine gimbal (*2*), and a glass electrode is arranged in a Plasticine micromanipulator. (K. Hausen must be credited for the invention of this most accurate piece of equipment.) One shaft of the double-barreled electrode is broken slightly shorter than the other, which contains the cobalt solution. The longer tip is advanced a measurable distance into the neuropil.

rests against the tissue (or head capsule) surface (Strausfeld and Hausen, 1977).

Usually only a small area of the neuropil surface need be exposed prior to injection. The animal is placed in a damp chamber at between 4° and 20°C during injection and during the subsequent period of diffusion after the source of cobalt is removed. Desiccation can be prevented by sealing any openings with nontoxic (highly purified white) Vaseline. *Drosophila* cannot be treated this way because they would desiccate. Instead, the micropipette is advanced through the intact cuticle of the head into the brain. This operation usually breaks the electrode tip from its original 0.5μm diameter to 3 μm. Likewise, thoracic ganglia of *Drosophila* can be impaled without opening the cuticle. Both operations require good marksmanship.

Injection and Diffusion Parameters

It is desirable to achieve uptake of cobalt by neurons so that they can be resolved as intact elements within tissue that is not obscured by excess cobalt ions. If there is excess cobalt, this is shown up after sulfide precipitation and silver intensification.

If tissue is treated with sulfide immediately after withdrawing the micropipette, a black spot of extracellular cobalt is resolved. The extent of this precipitation is proportional to the amount of cobalt released. The quantity will depend upon the concentration of the original cobalt solution and the dynamics of its passive efflux from the micropipette. The diffusion dynamics are a function of the diameter of the micropipette tip and of the ambient temperature during which the injection takes place, all other factors being equal. Hypothetically, ideal parameters for Co^{2+} incorporation by neurons would be the maintenance of a cobalt pool over a short period of time and its subsequent total depletion by neuronal uptake.

Experiments on Diptera have shown that judicious timing of injection followed by a period of diffusion, after withdrawing the micropipette, gives rise to diminution of the cobalt pool accompanied by an increase in the number of resolved neurons. These comprise one or more entire populations of one or more species of nerve cells. Thus, the longer the diffusion with respect to a constant injection period (between 0.5 and 1 h), the greater the number of cell populations resolved. The cobalt pool diminishes, and the concentration of cobalt ions is progressively diluted among increasing numbers of neurons: It is lost from some initially filled populations and taken up by others. This phenomenon is analogous to the trans-synaptic resolution of interneurons after back-filling from the ventral cord and ganglia. Thus, after 5 h diffusion, an abundance of nerve cells can be visualized as granular rather than solid outlines after silver intensification. After prolonged diffusion periods, say 8–10 h, neither neurons nor the location of the original cobalt pool at the injection site can be resolved. The neuropil is merely darker than it should otherwise be, betraying the presence of cobalt sulfide distributed throughout the ganglia.

Species-specific Variations of the Method

It was presumed that injection and diffusion criteria that were optimal for Diptera would also apply to a range of insects. However, this has proved otherwise. The orthopterous neuropil requires carefully controlled and slow rates of cobalt ion influx in order for it to be taken up by interneurons. Electrodes exceeding a diameter of 5 μm give rise to extensive Co^{2+} pools. The size of the pool inhibits the resolution of the few neurons that have incorporated cobalt from it. This is because after sulfide precipitation the pool functions as the most powerful catalyst for silver intensification, thereby inhibiting nucleation of nascent silver elsewhere. Also, damage to the perineurium seems to cause such drastic ionic distruption within the tissue that most neurons are killed and are never properly resolved.

It was found that optimal conditions for Orthoptera required tip diameters of less than 1 μm, containing 5–20% cobalt chloride or cobalt bromide. In practice, the tip is advanced directly through the neuropil sheath, causing it to break to an effective diameter of about 1–3 μm. An injection time of 30 min, followed by a 30 min diffusion period at room temperature,

Table 18-1. Injection Diffusion Parameters for Neuropil Uptake in Several Species of Insect

Target	Species	Injection	Diffusion	Approximate tip diameter (μm)	$CoCl_2$ (%)	Temperature (°C)
Brain	*Locusta* *Schistocerca*	30 min	30 min	1–3	5–20	4–20
Brain or thoracic ganglia	*Calliphora* *Sarcophaga* *Musca*	30 min–1 h	30 min–2 h	5–20	6	4–20
Brain	*Eristalis*	30 min	30 min	5–10	4	20
Brain or thoracic ganglia	*Drosophila*	2–5 min	10–15 min	1–3	6	20
Brain and thoracic ganglia connections	Into antenna of *Drosophila*, *Calliphora*	5–10 min / 30 min	30–45 min / 1 h	Antenna into pipette tip	6	4 or 20
Thoracic ganglia	Via wing or leg of *Drosphila*, *Calliphora*	20–40 min / 40–60 min	20–40 min / 40–80 min	Appendage into pipette tip	6 / 6	4 / 4
Thoracic ganglia	Via muscle of *Calliphora*	1–2 h	40–80 min	5–20	10	4
Brain	Backfill from thoracic ganglion of *Calliphora*	3 h / 12 h	— / —	Cut ganglia into pool or pipette	6	20
Lamina and medulla	Retina of *Notonecta*	40 min	1 h	5–10	6	4

proved optimal. In comparison, injection times for *Drosophila* must necessarily be cursory because of their size. If they were not, the entire brain would be flooded by extracellular Co^{2+} and no interneurons could be resolved. Two minutes injection followed by 15 min diffusion is optimal for short-axoned cells within the brain. Five minutes injection followed by between 30 and 45 min diffusion is adequate for uptake by descending interneurons to thoracic ganglia. Sensory axons and motor neuron axons can be seen after prolonged injection periods (30 min) followed by diffusion for about 60 min. The motor neuron terminals in muscle and sensory cell bodies and afferent fibers are revealed after silver intensification of the intact animal. The neuropil is, however, pitch black throughout. Injection and diffusion parameters are summarized in Table 18-1.

Peripheral Fillings via Micropipettes

Dissection of axon bundles in appendages presents no real problem in large insects, and the cut bundle can be isolated within a reservoir of cobalt chloride solution (see Chap. 19). However, a simpler approach which achieves the same end, is to sever the appendage and introduce its stump into a micropipette; this is then back-filled with an appropriate cobalt solution. The method has been particularly useful for *Drosophila* and *Calliphora* and is now used extensively for screening patterns of projections and their contiguous interneurons in wild-type and mutant animals. For example, wings that are cut at their tips take up cobalt into only those axons that have been severed. It is thus possible to approximate the origin of most fiber bundles by filling appendages that are cut at various distances from the body. Motor neurons can be filled separately by injection directly into muscle whereby Co^{2+} is incorporated into the motor axon by retrograde leakage, presumably across the neuromuscular junction.

Cobalt Precipitation

After the diffusion period the animal is cleaned of Vaseline and, if necessary, additional cuticle is removed. The animal is immersed in a weak solution of sodium or ammonium sulfide for 3 to 10 min. The concentration of sodium sulfide is 2.0 ml of saturated $Na_2S.9H_2O$ (approx. 22 g/100 ml H_2O) in 100 ml distilled water adjusted with glacial acetic acid to between pH 7.5 and 8.3. Solutions of $(NH_4)_2S$ contain about 20% of the sulfide per volume and are purchased ready-made; 0.2 ml of $(NH_4)_2S$ in 10 ml distilled water is sufficient to precipitate cobalt sulfide, and the pH of the solution can be adjusted to between 8.3 and 9.0 with drops of glacial acetic acid, or between pH 7.4 and 8.3 with PIPES buffer (see Appendix II, Chap. 9). Phosphate buffers should not be used. The osmolarity of both solutions is adjusted by the addition of between 1 and 5 g sucrose/100 ml.

Following CoS precipitation the preparation is washed well in buffer and fixed in 80 ml 100% ethanol, 8 ml acetic acid, 12 ml formalin (AAF) (Lillie, 1965) or in Duboscq-Brazil (Gregory, 1970), Carnoy's, or Karnovsky's (1965) buffered paraformaldehyde. Neuropil is dissected out under 70% alcohol or in buffer after aldehyde fixation. *Drosophila* and early instar Orthoptera can be intensified intact using methods described in Chapters 20 and 21.

Resolution of the Cobalt Reservoir and Nerve Cells

Neurons that have taken up cobalt, as well as the extracellular reservoir, are resolved by intensification with silver nitrate and hydroquinone (see Chap. 20). If the volume of the reservoir is large, its catalyzing effect suppresses intensification at other sites. In *Calliphora*, the reservoir is resolved as an intense black deposit at the location of the micropipette tip after 2 h injection and 1 h diffusion, using a 10 μm tip-diameter micropipette containing 6% $CoCl_2$. Characteristically, the relative dimensions of this pool are greatest after an equivalent concentration of cobalt acetate, and least using cobalt bromide, other factors being equal.

Two hours injection, followed by 2 h diffusion, gives rise to a brick-brown discoloration at the vicinity of the erstwhile tip that extends nearly twice as far into the surroundings as after 1 h diffusion. After 2 h injection and 4 h diffusion, there is no visible definition of the tip location, and the presence of extracellular cobalt is just discernible as a pale yellow haze. Optimal times of 1 h injection and 2 h diffusion give similar results. After optimal injection–diffusion, neurons are resolved contrasted against a paler background, using the intensification methods in Chapter 20.

The Gestalten of cobalt-silver-impregnated neurons is identical to the Gestalten of Golgi-impregnated neurons. Unlike Golgi methods, though, the present technique resolves entire populations of one or more morphologic classes of nerve cells but never resolves glia.

Injection–diffusion periods can be adjusted to control uptake by few or many neuron populations. These uniquely express the cytoarchitecture of the neuropil in terms of nerve cell assemblages that show up as variously colored elements after, for example, 2 h injection and 4 h diffusion, using method D, Chapter 20 for *Calliphora* or method E for locusts and intact *Drosophila*.   Color differences are caused by various densities of CoS differentially catalyzing silver reduction, so that weakly filled elements are resolved as pale brown, green, or gray profiles (Strausfeld and Hausen, 1977).   This feature allows the investigator to trace the course of many elements.   In Golgi preparations the majority of neurons are uniformly toned, and usually the passage of a single nerve cell among many others can only be followed with uncertainty.

Patterns of Uptake by Neurons

When cobalt is introduced into cut thoracic ganglia of *Calliphora,* it passes from dendrites of neurons that are derived from the brain into relay neurons that invade the lobula and lobula plate.   In *Calliphora* and Odonata the constellation of cells so resolved is consistent from one animal to the next.   Likewise, when cobalt is introduced via a micropipette into an antenna, its passage is over specific sets of interneurons to the ventral nerve cord.   The cephalic dendrites of some of these donate cobalt ions to one set of lobula small-field interneurons (Figs. 18-3 and 18-4).   With extended diffusion periods, cobalt can even be seen in some motor neurons of the thoracic ganglia.   It seems, therefore, that the route taken by Co^{2+} ions is a most specific one and that although many populations of interneurons are postsynaptic to antennal sensory fibers, only certain of these will incorporate cobalt ions.

Injections into the brain behave in a similar fashion, revealing some populations of interneurons more often than others under conditions where the diameter of the tip is constant and its location is approximately within the same region.   These observations suggest that some elements are less susceptible to Co^{2+} uptake than others, although in the case of *Calliphora* it cannot be claimed that one or another species of interneuron is wholly refractive to the technique.   A hundred successful injections have, for example, resolved the population of several morphologic classes of interneurons in the lobula which had previously been recognized by Golgi techniques (Fig. 18-1,A,B; 18-5).

In the locust, however, injections into the optic peduncle have resolved identical populations of intrinsic medulla neurons within twenty animals but have failed to resolve most classes of columnar elements (Fig. 18-6).   Thus the conditions of x hours injection and y hours diffusion from a tip diameter of about 2 μm are optimal only for specific classes of elements; in order to reveal others it is necessary to change one or more of the injection-diffusion parameters.   Experiments on Diptera have shown that too large a cobalt reservoir may inhibit Co^{2+} uptake by, or CoS intensification of, neurons.   Too small a reservoir resolves only a few single nerve

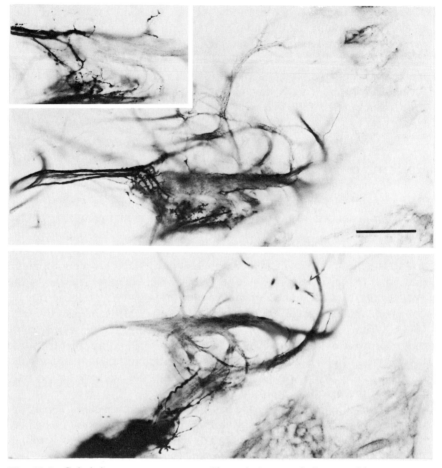

Fig. 18-3. Cobalt ions pass across specific and characteristic sets of interneurons (Strausfeld and Obermayer, 1976).   This figure illustrates the results of three separate fillings into antennal mechanosensory fibers (small-diameter, dark profiles) that have donated cobalt ions into identical sets of large descending neurons in the posterior deuterocerebrum.   A total of 50 identical fills have been prepared: Some of them, in which the diffusion period was extended from 1 to 2 h, reveal cobalt in lobula columnar neurons, derived from the dendrites of these descending interneurons (see Fig. 18-4).   Likewise, back-filling these species of descending interneurons from the thoracic ganglia gives rise to the same pattern of trans-synaptic cobalt in the lobula.

cells but not their population.   Thus, it can be taken as a general rule that optimal tip diameters and injection times should be held constant.   Diffusion times, temperature, and Co^{2+} concentration are varied according to the rate of uptake by neurons.   However, substitution of the anion is perhaps the most useful experimental manipulation.   In Diptera, cobalt

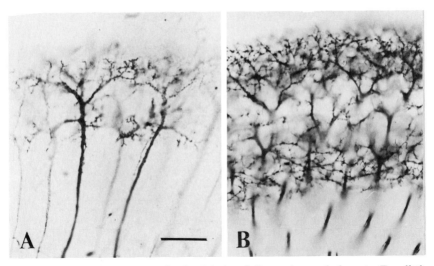

Fig. 18-4. Columnar neurons in the lobula of *C. erythrocephala*. **A.** Too little cobalt will not resolve all neurons of an assembly with equal intensity. Intensification enhances the contrast of some at the expense of others. In **B,** however, neurons of an assembly are equally filled and intensified to the same contrast. Both **A** and **B** represent a small fraction of a single cell population.

bromide pools are smaller than those of chloride, other factors being equal, and for some yet inexplicable reason uptake of Co^{2+} from the bromide pool is achieved by some morphologic classes of interneurons that seem to be less susceptible to $CoCl_2$. Uptake from a cobalt acetate pool favors yet other classes of neurons; in particular, small-field medulla elements that have, as yet, rarely been resolved after cobalt chloride injection.

The range of applications of the present technique has only been partly explored, and a routine use of the method has been somewhat restricted to dipterous insects and locusts. However, the technique has been attempted on several other species of insects where it functions well, e.g., *Apis, Aeschna,* and *Notonecta glauca.* Injections into the optic lobe neuropil of crayfish have also met with success, resolving parts of neuropil assemblages in the lamina and medulla neuropils, and the passage of neurons from the internal medulla to optic foci of the protocerebrum (Nässel, in preparation). Its use on larger brains, such as those of vertebrates, has been seldom employed. An earlier account by Willmore *et al.* (1975) illustrates that Co^{2+} is incorporated into pyramidal cells of the rat cerebral cortex. In this instance cobalt ions were applied to the pial surface from 300 μm diameter electrodes by iontophoresis. Recent experiments on mouse brains, using micropipettes and block intensification, have revealed regular arrangements of pyramidal cells in the cerebral cortex and assemblages of Purkinje cells in the cerebellum. The method has also proven a powerful one for tracing axon pathways and terminal

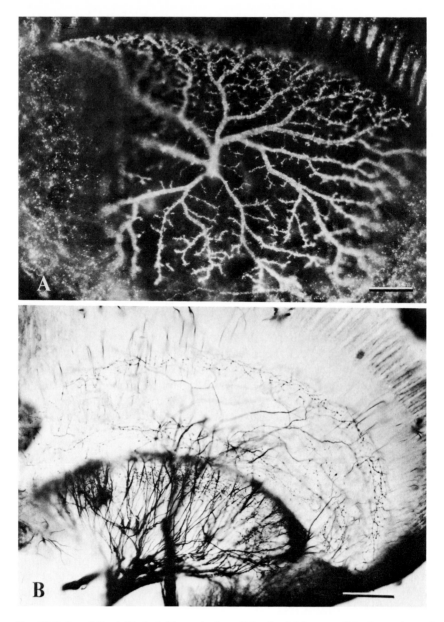

Fig. 18-5, A and B. **A** Dark-field resolution of the dendritic tree of the heterolateral motion-sensitive neuron of the lobula plate (Hausen, 1976a) revealed after 1 h injection, 1 h diffusion into the contralateral optic peduncle. **B** Medulla and lobula neuronal assemblages revealed after 1 h injection, 2 h diffusion into the posterior ventral deuterocerebrum, in the region of the cuccati bundle (see Strausfeld, 1976). Scale = 100 μm.

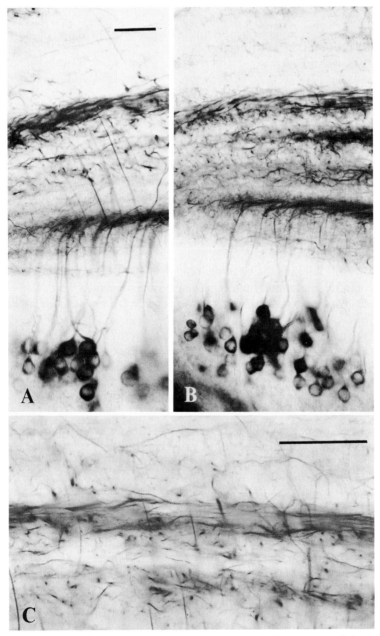

Fig. 18-6, A–C. A,B Identical uptake of Co²⁺ by intrinsic neurons of the medulla of *Locusta* within its deepest stratum. Some variations between the two fillings can be discerned at more peripheral levels. 20% CoCl₂, 2 μm diameter electrode tip: ½ h injection, ½ h diffusion. **C** Amacrine neuron fibers within the medulla of *Locusta* (scale, 10 μm). In all three examples cobalt ions were introduced from a 2 μm diameter electrode implanted in the lobula neuropil.

radiations, such as mossy fiber endings (Obermayer, 1978 and in preparation; see Fig. 18-1,C).

A Simple Comparison between the Golgi Method and the Cobalt Injection Technique

Usually, Golgi methods are modified in order to bias the impregnation of neurons toward one, or another, general class of neural elements. Neurons with large axons may, for example, be resolved at the expense of smaller cells by specific modifications of the fixative, whereas diffuse anaxonal neurons are particularly susceptible to impregnation by mercuric nitrate, which replaces the equivalent silver salt. The rationale for such modifications are summarized in Chapter 9 and demonstrate that for single-cell morphology the Golgi technique remains the most powerful analytic method. An abundance of nerve cells may be resolved by recycling the chromation and metal impregnation routines, and the method is even applicable for screening specific brain regions by the constrained influx of silver ions.

Improvements of the Golgi method, albeit legion, do not alter the basic expression of its mechanism, which approaches that of a random sampling device. Improvements merely alter the constraints under which the mechanism operates so that even if some methods show up nonrandomly distributed clusters of elements, the selection of cell location in that cluster is, nevertheless, unpredictable.

Moreover, no matter which method is employed, the actual relevance of geometric relationships between two adjacent neurons can be easily misinterpreted, because innumerable profiles that have escaped the Golgi reaction reside between impregnated fibers. This feature is all too often ignored, or forgotten, with the consequence that Golgi methods offer a fata morgana of structural relationships that may have no relevance at all to functional contiguities.

Golgi methods are used to determine functional interconnections when employed in conjunction with electron microscopy (see Strausfeld and Campos-Ortega, 1977). However, statements about the absolute populations of neurons or about graded differences in morphology have, until the present, relied upon circumstantial evidence—either from the study of many single impregnated cells or from reduced silver preparations.

The inability of Golgi methods to select all elements of a single morphologic class of neurons is unlikely to be overcome. However, synchronous uptake of cobalt by an entire neuronal population fills this gap in structural analysis and reveals novel features about the structure of a cell assembly, such as the presence of gradients that may have a direct bearing on the physiology of behavior (Fig. 18-7).

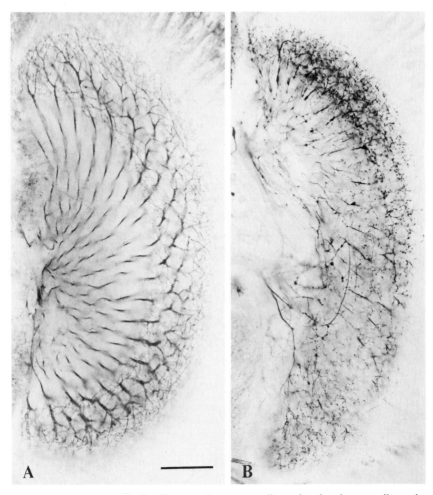

Fig. 18-7, A and B. A,B Gradients and contragradients in visual neuropil can be resolved by injection–diffusion into the optic peduncle or lateral deuterocerebrum. In **A** columnar lobular neurons of *Sarcophaga* reveal a gradient of dendritic density that increases from the upper to lower areas of the retinotopic mosaic, showing density maxima beneath the equator of the eye. In **B**, however, centrifugal columnar terminals exhibit a reverse gradient where the minimal terminal densities overlap maximal dendritic densities, and maximal terminal densities reside within the dorsal part of the retinotopic mosaic. These sets of elements are not discernible by other techniques and reveal structural assemblages that, in this case, meet requirements for some types of positional detector systems that could theoretically subserve fixation behavior (Strausfeld, in preparation). Scale **A,B** = 5 μm.

The unique feature of the cobalt injection technique is its capacity to reveal, in a reproducible fashion, an assemblage of single neurons, all of which demonstrably belong to the same morphologic class, and which can be described by sound geometric criteria. When two cell types are resolved, they are unambiguously revealed as two distinct assemblages. This degree of resolution allows descriptions of specific modes of assembly provided by any one species of neuron and thereby gives us another view of the brain that cannot be derived from either Golgi or reduced silver methods.

Acknowledgments

The techniques for backfilling contiguous neurons and cobalt injection into brains were originally devised in conjunction with M. Obermayer (E.M.B.L., Heidelberg) and Dr. K. Hausen (Max-Planck-Institut für biologische Kybernetik, Tübingen).

Chapter 19

Filling Selected Neurons with Cobalt through Cut Axons

J. S. Altman

The University of Manchester
Manchester, England

N. M. Tyrer

The University of Manchester
Institute of Science and Technology
Manchester, England

The introduction of cobalt chloride as an intracellular marker (Pitman *et al.*, 1972) and the demonstration that it could be introduced into neurons through the cut ends of their axons (Pitman *et al.*, 1973; Sandeman and Okajima, 1973) have established cobalt staining as an extremely powerful tool for the analysis of connectivity in insect neuropils. Both central and peripheral ends of cut axons can be filled and, by careful choice of nerve branches, single functionally identified neurons can be demonstrated (Tyrer and Altman, 1974; Altman and Tyrer, 1977a). In the short time since its introduction, the use of cobalt chloride has enormously advanced our understanding of the organization of insect neuropils, especially tangled ones and, together with improvements in microelectrode recording techniques, is providing the basis for new concepts of functional organization in the insect central nervous system.

The basic axonal filling technique, introduced by Iles and Mulloney (1971) for Procion yellow, involves placing the cut ends of selected axons in a pool of dilute cobalt chloride solution, which is encapsulated to isolate it totally from the rest of the nervous system and any surrounding tissues. After a sufficient time has elapsed for the migration of cobalt ions into the neurons, the cobalt in the tissue is precipitated as cobalt sulfide, which is brownish-black and insoluble, and cells that have taken up cobalt are displayed as blackish silhouettes on a clear background (Figs. 19-1, 19-3, 19-6, 19-7).

With Procion yellow, a negative charge applied to the cut end of the axon was required for the dye to migrate into neurons (Iles and Mulloney, 1971). The axonal method consequently became known as "axonal iontophoresis." Cobalt ions, however, migrate whether current is applied or not (see, e.g., Tyrer and Altman, 1974), so that the term iontophoresis is misleading. Similarly, "back-filling," derived from early applications where motor neurons were filled antidromically, is also incorrect, because cobalt ions move with equal facility in both directions along axons. In the absence of definitive information about the transport mechanism, we consider that the neutral title of *axonal filling* should be adopted.

Cobalt filling has many advantages over other selective staining methods. Uptake and migration in insects appear to be better than for Procion dyes, so that a more complete picture of the cell is obtained (Altman and Tyrer, 1974). Cobalt ions will penetrate into fine branches and terminals of a neuron so that the branching of filled cells shows up in great detail (Figs. 19-1, 19-3, 19-6, 19-7, 19-8). Although the axonal technique cannot supplant intracellular filling with microelectrodes for definitive functional identification of a cell, it requires little apparatus, and large numbers of specimens of the same neuron can be produced relatively easily for comparative work. Some neurons that are difficult to inject intracellularly, such as primary sensory afferents, are accessible to axonal filling. Histologic processing is quick and simple, compared to autoradiography of labeled amino acid tracers (Cowan *et al.*, 1972) or to horseradish peroxidase (LaVail *et al.*, 1973). Preparations can be viewed as whole mounts, or sectioned for light or electron microscopy, allowing correlation of the overall morphology of a neuron with details of its connectivity. The recently introduced block intensification methods (Strausfeld and Obermayer, 1976; see also Obermayer and Strausfeld, Chap. 20 of this volume; Bacon and Altman, 1977) give excellent resolution and great sensitivity for displaying details of neurons in both whole mounts and subsequent sections; or material can be intensified on the section (Tyrer and Bell, 1974; see also Chap. 21 of this volume).

Axonal filling can be used for two levels of analysis: first, for delineating tracts and locating cell groups, providing a basic anatomical framework for study of a ganglion (Gregory, 1974a; Iles, 1976a) and for positioning of electrodes in physiologic experiments (O'Shea *et al.*, 1974); second, for a detailed study of the morphology and connections of single, identified neurons. In insects, such *connectivity studies* already made with cobalt include locust flight neurons (Tyrer and Altman, 1974; Altman and Tyrer, 1977a; Tyrer *et al.*, 1979; Bacon and Tyrer, 1978); ocellar interneurons (C. Goodman, 1974, 1976a; L. Goodman, *et al.*, 1975; C. Goodman and Williams, 1976); tritocerebral neurons (Aubele and

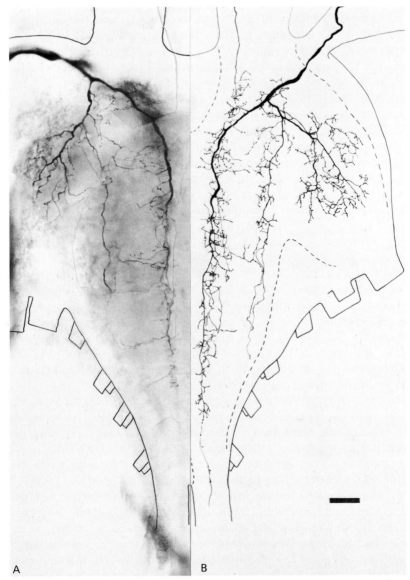

A   B

Fig. 19-1, A and B. The metathoracic wing stretch receptor (III SR) in the locust demonstrated by axonal filling with cobalt. **A** A photomicrograph montage at three focusing levels of the metathoracic ganglion, which has been intensified in whole mount (cf. Chap. 21; Bacon and Altman, 1977). **B** An unintensified preparation drawn with a drawing tube attached to the microscope. Many details that are missed by the montaging technique are much clearer when drawn, although some fine branches that could not be seen before are apparent after intensification. Scale = 100 μm.

Klemm, 1977); locust subesophageal motor, sensory, and salivary neurons (Altman and Kien, 1979); cricket cercal giant fibers (Mendenhall and Murphey, 1974); and sensory fibers in the cricket brain (Honegger and Schürmann, 1975; Honegger, 1977). The *development of identified neurons* has been studied in locust (Altman and Tyrer, 1974), cricket (Bentley, 1973), and a moth, *Manduca sexta* (Taylor and Truman, 1974; Truman and Reiss, 1976), and the *response of neurons to experimental manipulation* investigated in cockroach (Tweedle *et al.*, 1973) and cricket (Murphey *et al.*, 1975; Bentley, 1975; Clark, 1976a; McClean and Edwards, 1976). *Constancy and variability in neuron morphology* have been examined in locust motor neurons (Altman and Tyrer, 1974; Tyrer and Altman, 1974), wing stretch receptor (SR) neurons (Altman and Tyrer, 1977b), and ocellar interneurons (C. Goodman, 1974, 1976b); some "mistakes" in neuron branching have been described (C. Goodman, 1974; Altman and Tyrer, 1977b). *Neurosecretory pathways* have also been described in locust by cobalt filling (Mason, 1973; Aubele and Klemm, 1977; Rademakers, 1977).

Although first described for insects, the cobalt filling method has been adapted for a range of other groups, including crustacea (Sandeman and Okajima, 1973; Mittenthal and Wine, 1973; Wine *et al.*, 1974), *Octopus* sp. (Budelmann and Wolff, 1976), gastropod mollusks (S.-Rózsa and Salanki, 1975; Rock *et al.*, 1977), lower vertebrates (Prior and Fuller, 1973; Fuller and Prior, 1975; Székély, 1976), and mammals (Mason, 1975; Iles, 1976b; Pearl and Anderson, 1975; Mason and Lincoln, 1976; Martin and Mason, 1977; Mason *et al.*, 1977).

In this chapter, we do not intend to discuss the possible mechanisms of uptake and transport of dyes and metal ions by neurons. Although speculations are rife, there have so far been no experimental investigations specifically designed to test mechanisms such as passive diffusion or the involvement of microtubules. Information on mechanisms would be of great value in determining parameters for obtaining good and reliable fillings, but in its absence we will be reporting only on empirically determined methods, which have been developed experimentally and proved successful with much practice. In our experience, the technique can be fickle and often requires much perseverance to achieve initial success. The basic technique has therefore been modified to suit different species and purposes as well as the tastes of individual workers. We suspect that many of the "tricks" reported in the literature have become incorporated as part of the mystique of success rather than for sound physiologic reasons. In this review we will try to outline the main variations of the method, give a general guide to filling parameters, and indicate which tricks we consider to be of dubious value. For specific applications, readers should refer to the original descriptions.

Finally, considerable care must be taken with the interpretation of axonally filled cobalt preparations. Cobalt ions can be taken up by neurons other than those with cut ends in the cobalt chloride solution (Tyrer and Altman, 1974), and a recent report (Strausfeld and Obermayer, 1976) suggests that under certain conditions trans-synaptic staining can be obtained. The use of the very sensitive Timm's method for intensification may well reveal such "unexpected" neurons, especially if their cobalt chloride content is too low to be visible in unintensified material. Certain restrictions must therefore be understood and appropriate precautions taken to avoid the incorrect identification and description of neurons.

Methods for Introducing Cobalt Chloride into Cut Axons

The basic requirement for the axonal filling technique is that the pool of cobalt chloride, into which the cut nerves are placed, be completely isolated from all other tissues and bathing fluids. In the original method (Iles and Mulloney, 1971; Sandeman and Okajima, 1973), the ganglia of interest are removed from the animal and maintained in an appropriate saline solution (in vitro method), but in vivo methods have also been developed, using the animal itself as a preparation chamber (Mason, 1973). Choice of method depends on the purpose of the preparation and on the topography of the system to be filled. An account of these two methods is given first, followed by recommendations for filling particular sets of neurons.

In Vitro Preparations

The ganglia required are dissected out of the animal, with the nerve containing the axons of interest left as long as possible. Care must be taken to keep the preparation moist and not to kink the nerve or pull on its branch points. Metal instruments are best avoided for handling this nerve, and we find fine glass hooks very convenient. The tracheae of the ganglia should be left intact and if possible opened with a fine needle. A typical preparation dish is illustrated in Fig. 19-2. The ganglia are placed in one depression, which contains an appropriate bathing medium, and the cut end of the selected nerve dips into the cobalt chloride solution in the second depression. A gutter containing mineral oil (liquid paraffin) separates the two depressions, which are joined by a shallower groove in which the nerve lies. Ganglia should be arranged with the tracheae open at the surface of the bathing medium and with little tension on the nerve

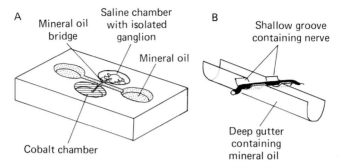

Fig. 19-2. A The preparation dish used for in vitro filling of neurons through the peripheral nerves. This is made from a Perspex block in which four depressions have been drilled. The ganglion is dissected from the animal and is placed in saline solution in one depression. The selected nerve is carried over a bridge consisting of a narrow gutter filled with mineral oil (liquid paraffin), into a second depression filled with cobalt chloride solution. The level of the oil is adjusted from reservoirs at either end of the gutter. **B** Detail of the mineral oil bridge. A shallow groove crossing the deeper gutter carries the nerve from the saline solution to the cobalt. This achieves a reliable mineral oil seal around the nerve and prevents desiccation.

root. Vaseline may be used instead of mineral oil but we find it is less effective as a seal. The preparation dish is placed in a moist chamber.

Some workers recommend placing the cut end of the nerve in distilled water for a minute before placing it in cobalt chloride solution, to open the sealed ends of axons (Mason, 1973). It is doubtful whether this is really necessary, and we prefer to avoid the extra handling of the nerve. Dimethyl sulfoxide has also been recommended, mainly for Procion filling, but again it is of doubtful value.

Bathing medium for the ganglia is usually the appropriate physiologic saline solution. Better results have been obtained with Chen and Levi-Montalcini's tissue culture medium (1969; Tyrer and Altman, 1974). This is tedious to make and expensive to buy, and its effectiveness is probably because it is isotonic and contains an energy source. A normal insect saline solution with the correct amount of trehalose or glucose added is very satisfactory (Eibl, 1976).

Current can be applied between the two depressions, with the cobalt chamber positive. In general this appears to be unnecessary and can produce anomalies such as filling contiguous neurons and glial cells (Fig. 19-3; Tyrer and Altman, 1974). Without current, preparations generally show less leakage and better contrast. Small neurons with fine axons, however, seem to fill better with a low concentration of cobalt chloride (1%–1.5%) and a current of less than 1 μA (J. Kein, personal communication).

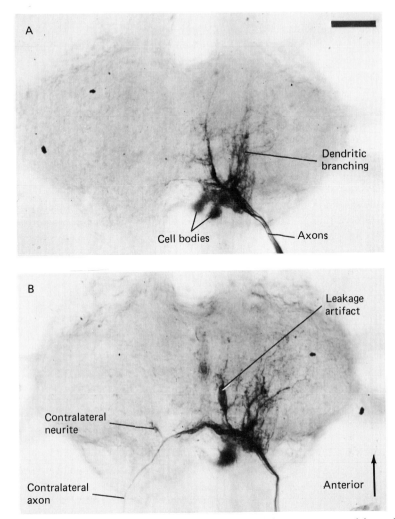

Fig. 19-3, A and B. Artifacts that can result from applying a current to drive cobalt into the peripheral nerve. **A** The mesothoracic ganglion of the locust, *Chortoicetes,* in which the four mesothoracic motor neurons innervating the metathoracic dorsal longitudinal muscles (112) have been filled from the periphery without imposed current. **B** A different preparation in which filling was assisted by a current of 0.2–0.4 μA. Two artifacts are seen: a flare surrounding some branches, apparently where cobalt has leaked from the neurons, perhaps into glial elements; and an axon in the contralateral nerve, together with parts of neurons on the contralateral side of the ganglion, has been filled. There is no physiologic evidence for direct contralateral connections for these motor neurons, and fillings without current never show these features. Scale 100 μm. [Reprinted with permission from Tyrer and Altman, 1974.]

In Vivo Preparations

Dissected Preparations

The insect is opened up and the gut removed or deflected. Cut ends of the gut must be sealed with Vaseline to prevent leakage of its contents into the body cavity. The required nerve is placed in a container of cobalt chloride solution, which is sealed off from the surrounding body tissues. This can be done by building a cup of Vaseline around the nerve using a hypodermic syringe (Fig. 19-4) (Mason, 1973), flooding the body cavity with mineral oil into which a drop of cobalt chloride solution is injected (Iles, 1976b), making a small cup of fine polythene tubing closed with Parafilm and coated with Vaseline (C. Goodman, 1974), or by blowing a bubble at the end of a fine glass tube. A hole for the nerve is made in the side of the bubble, which contains the cobalt chloride solution. The bubble is coated with Vaseline and positioned with a micromanipulator (Steinecker, in Mulloney, 1973; J. Nelson, personal communication). The cut end of the nerve may also be drawn into a cobalt chloride-filled suction electrode (Mason and Nishioka, 1974). The exposed tissues must be sealed off to prevent decomposition, which is easily done with a layer of Vaseline. Only highly purified white Vaseline should be used, as standard Vaselines tend to contain toxic impurities. If mineral oil is used, it must not enter the tracheal system, because this causes rapid death of the preparation. Current is not normally applied with in vivo preparations,

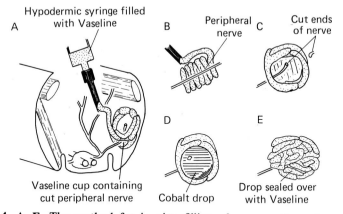

Fig. 19-4, A–E. The method for in vivo filling of neurons through peripheral nerves. **A** A small quantity of Vaseline, dispensed from a hypodermic syringe with shortened needle, is used to build a cup around the selected nerve within the animal. The sequence of the construction of the cup is shown in **B–E**. **B** Vaseline is injected under the intact nerve. **C** The nerve is cut, and a wall of Vaseline is built up to surround the cut end to be filled. **D** A drop of cobalt is pipetted into the cup. **E** The drop is sealed over with more Vaseline. Exposed tissues in the animal are also covered with Vaseline to prevent their drying out.

which should be maintained in a moist atmosphere. Hoyle (1978) describes an implantation method using a sandwich of Parafilm sealed with Vaseline to hold the cut nerve in chronic preparations which remain mobile for several days during filling.

Undissected Preparations

Sensory projections can be filled by placing a drop of cobalt chloride in a cap of insect wax over a group of sensory hairs, cut short to allow access for cobalt ions (Tyrer *et al.*, 1979) (Fig. 19-5) or by introduction of cobalt chloride into the the sense organ itself, e.g., locust ocelli (C. Goodman, 1976a). Motor neurons may also be filled by injecting cobalt chloride into the appropriate muscle.

Choice of Parameters for Filling

Parameters depend on the method of preparation, on the type of neuron to be filled, on the length and diameter of the axon and size of the arborization, and lastly on the subsequent method of examination (i.e., whole mount, with or without intensification; sections for light microscopy, or for electron microscopy). The main variables are length of filling time, temperature, and concentration of cobalt chloride solution.

General rules for unintensified whole mounts are:

1. *Concentration.* The larger the diameter of the axon to be filled, the higher the concentration that can be used, with an upper limit of about

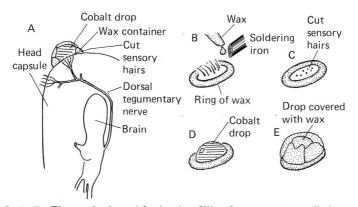

Fig. 19-5, A–E. The method used for in vivo filling from receptor cells innervating cuticular hairs. The construction sequence: **A** diagram of a completed preparation for filling a single field of wind-sensitive hairs on the locust head; **B** a ring of insect wax (3:2 beeswax:rosin) is made around the hairs to be examined, using a miniature soldering iron; **C** the hairs are removed close to their insertion in the cuticle; **D** a drop of cobalt chloride solution is placed within the wax ring; **E** a layer of wax is built up to cover the cobalt drop, to prevent evaporation.

15% cobalt chloride (weight:volume, equivalent to 630 mM). Fine sensory fibers fill better at lower concentrations (1%–5%). Strausfeld and Obermayer (1976) found that the addition of bovine serum albumen (0.013 g/100 ml in 6% $CoCl_2$) gave darker and faster fills of descending visual interneurons in the fly. Ocellar neurons in the bee appear to fill better when a drop of stannous chloride is added to the cobalt chloride (Pan and Goodman, 1977). These are isolated reports and may not be generally applicable.

2. *Time*. Large-diameter axons fill faster than small-diameter ones, but time also depends on the concentration of cobalt chloride used, the length of the nerve, and the size of the neuron's arborization. With higher concentrations, large axons fill faster but small axons (less than 2 μm) may fail to fill at all. A complete fill of a neuron with a large arborization, e.g., intersegmental interneurons or locust wing stretch receptor neuron (see Figs. 19-1 and 19-8) requires much longer than a compact neuron such as a motor neuron (Fig. 19-6). It generally takes longer to fill the same neuron in vivo than in vitro.

3. *Temperature*. For filling times up to 8 h, room temperature (approximately 20°C) is recommended, provided that the preparation remains viable. At lower temperatures, equivalent preparations take longer; e.g., locust wing SR neuron in vivo requires 2–3 times as long at 7°C as at 20°C. Some neurons seem to fill better at lower temperatures (e.g., Bentley, 1973), but this is not a general rule and any new preparation should be tested at a range of temperatures. In our experience, in vitro preparations work better at room temperature for short periods (up to 8 h), whereas in vivo-dissected preparations are better at low temperatures for longer periods (up to 48 h). Recently, we have obtained good fillings in 3–4 h by running in vivo locust preparations at normal environmental temperatures for locusts, i.e., 27°C. If preparations are left too long at room temperature or above, the filled neurons may have large (2–3 μm diameter) swellings on fine branches and the background stains heavily on intensification. We assume that these swellings, which are much larger than the typical "blebs" seen on the same neurons, are artifacts, probably related to loss of cobalt ions from the neuron.

4. *Preparation for intensification as whole mounts or on section*. Using high concentrations of cobalt chloride for long periods can cause artifactual fillings and uptake into glia that are not always visible in unintensified whole mounts (q.v.) but are revealed by intensification (Chap. 21). To obtain a clean, unambiguous picture of the selected neuron after intensification, lower concentrations of cobalt chloride and shorter times should be used (Tyrer and Altman, 1974; Altman and Tyrer, 1977a; Tyrer *et al.*, 1979). Typically we use between 1% and 5% cobalt chloride in distilled water, which gives excellent pictures of sensory neurons after intensification. Motor neurons are often less well filled when low concentrations

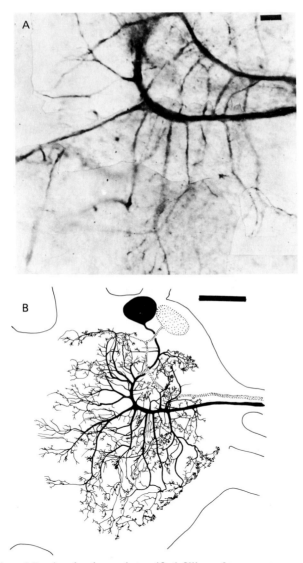

Fig. 19-6, A and B. A selective unintensified filling of two motor neurons to the locust mesothoracic wing elevator muscles 83/84. **A** A montage of high-power photomicrographs at three different focusing levels of a part of the preparation shown in **B,** which has been drawn using a drawing tube attached to the microscope. Only one neuron has been drawn in detail, and the cell body, neurite, and axon of the other have been indicated by stippling.   Scales: **A** 10 μm; **B** 100 μm.

are used, possibly because the cell body acts as a sink for cobalt ions, and we recommend concentrations of 5% or above.

When the required neuron is only faintly visible in the ganglion of interest before intensification, a complete picture is normally obtained after intensification. For example, the locust SR neuron can be filled in vitro in about 7 h with 15% cobalt chloride at room temperature. After intensification, sensory and motor neurons with axons in other branches of the same nerve are clearly visible. Filling for 2–3 h with 5% produces a faintly visible SR axon in the nerve root but intensification reveals the whole neuron without contamination.

Filling Selected Neurons

The identity of a neuron or group of neurons can be established and the morphology and connections described by filling them selectively. This makes use of the animal's anatomy by choosing a peripheral nerve branch containing axons of the neurons of interest, or where the other axons have completely separate and identifiable projections (e.g., Altman and Tyrer, 1977a; Bacon and Tyrer, 1978). Alternatively, neurons can be filled from their dendrites in sense organs (Tyrer et al., 1979) or at their terminations on a muscle (Tyrer and Altman, 1974). Neurons that have no axons in peripheral nerves are more difficult to fill selectively and functional identification always has to be made by physiologic recording. Small groups of intersegmental interneurons may be filled by splitting connectives and selecting small bundles of axons (Rehbein, 1976). Intrinsic interneurons have been stained by passive diffusion from cobalt chloride-filled glass micropipettes implanted into the neuropil (Strausfeld and Hausen, 1977; see also Bacon and Strausfeld, Chapter 18 of this volume). All selective preparations must be interpreted with care, because cobalt ions can be taken up by other than the target neurons.

Motor Neurons (Fig. 19-6)

To fill the motor neurons of a particular muscle, it is necessary to cut the nerve branch as close to the muscle surface as possible. Either in vitro or in vivo methods can be used, but in vitro may be better for more inaccessible muscles. In vitro preparations are easier to set up if a small piece of muscle is left attached to the nerve ending and placed in the cobalt bath. For unintensified whole mounts of motor neurons with a large-diameter axon (5–10 μm) and limited arborization, 10%–15% cobalt chloride for 5–8 h at room temperature is a good starting point for in vitro fillings, overnight at 5–7°C for in vivo preparations. For whole-mount intensification, 4–5 h in vivo at 27°C with a 5% solution should be sufficient.

Sensory Neurons (Fig. 19-7)

The central projections of sense organs can be demonstrated either by filling the central end of the cut sensory nerve or by bringing cobalt chloride into contact with the sensory structure.

Sensory nerves usually contain a large number of axons, and it is here in particular that artifactual fillings are caused by using high cobalt chloride concentrations and long filling times (q.v.). The problem is greater

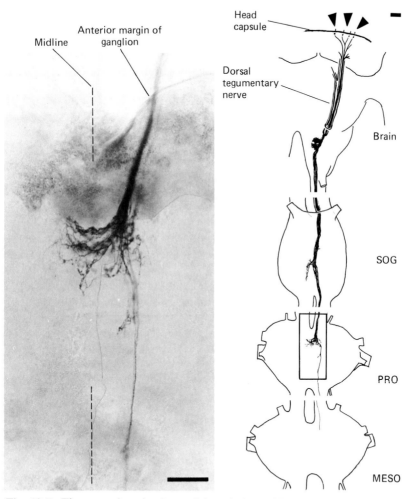

Fig. 19-7. The central projections of the wind-sensitive hairs on the locust head filled by the *in vivo* method shown in Fig. 19-5. The whole projection is drawn on the *right* and the photomontage on the *left* shows details of the terminals in the prothoracic ganglion, in an intensified whole-mount preparation. *SOG* subesophageal ganglion; *PRO* prothoracic ganglion; *MESO* mesothoracic ganglion. Scales: *Left,* 50 µm; *right,* 100 µm. [Preparation courtesy of J. P. Bacon.]

where the nerve contains a large number of small-diameter axons, as in the locust dorsal tegumentary nerve (Tyrer *et al.*, 1979) and the locust wing nerves (Altman *et al.*, 1978). Sensory axons with diameters of 5 μm or more, such as those from the locust wing tegula, stretch receptors, and chordotonal organs (Altman *et al.*, 1978) are best filled with 15% cobalt chloride for 5–8 h at room temperature in vitro, for unintensified preparations. Small-diameter fibers fill more reliably with concentrations below 5%, for 18 h at 5°C. Intensification greatly increases the scope of fine axon projections, which can be demonstrated using very low concentrations and long periods (Tyrer *et al.*, 1979).

Cut Sensory Nerves (Fig. 19-1)

In vitro and in vivo methods are both suitable, and choice of method depends mainly on the geometry of the nerve branches. Selective fillings of the locust SR neuron appear to be obtained more cleanly and reliably in vitro, but the in vivo method is simpler and therefore recommended for determining the central pathways of large sensory bundles.

Sense Organs (Fig. 19-7)

Sensory projections can be filled in undissected preparations as described previously. As well as the ease of preparation, this method is of advantage (1) for demonstrating part of a field where the sensory nerve contains axons from the whole field or from a number of sensory structures, for example, the projections of individual wind hair patches on the locust head, the axons of which all run in the dorsal tegumentary nerve (Albrecht, 1953; Tyrer *et al.*, 1979), and (2) where the sensory nerve is too fragile or inaccessible to be dissected out.

In these preparations, care has to be taken with localization and interpretation because cobalt ions can migrate under the cuticle ("cobalt creep") and be picked up by other neurons, both motor and sensory. Cobalt chloride concentration should therefore be kept low, and the use of crystals implanted into sense organs is most inadvisable.

Intersegmental Interneurons (Fig. 19-8)

Intersegmental interneurons are easily filled through the cut end of connectives. Selection of particular neurons is difficult because of the numbers of fibers in the connectives, and the method is most useful for extending observations on neurons identified by microelectrode staining (Fig. 19-8). Aubele and Klemm (1977) observed that only a small proportion of the axons in the connectives, in general those with larger diameters, are filled successfully. Many axons in the connectives only run between adjacent or subadjacent segments, so filling from a distant segment selects out those neurons that travel a long way. The best pictures

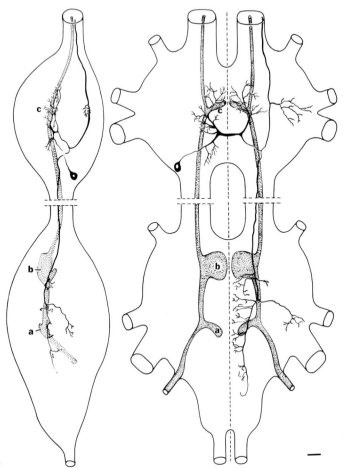

Fig. 19-8. A T-shaped auditory interneuron (G neuron) in *Locusta migratoria* identified by extracellular microelectrode recording and marking. The detailed morphology has been determined by filling a bundle dissected from the interganglionic connective. The *stippled tract* represents the projection of the primary tympanal nerve fibers and the three acoustic neuropils are labeled *a*, *b*, and *c*. The neuron is seen from dorsal (*right*) and lateral (*left*). Scale 100 μm. [Reprinted with permission from Rehbein, 1976.]

of individual neurons are, however, obtained by partial crushing of the connective (Mendenhall and Murphey, 1974) or by splitting connectives and filling small bundles (Rehbein, 1976). These preparations can be done in vivo or in vitro, but it is easier to handle a long chain of ganglia in vivo. To determine the origin or termination of a neuron in a particular ganglion by connective filling, the connective must be cut as far from the ganglion as possible, because neurons terminating in the ganglion may become stained if the ganglion becomes flooded with extraneuronal cobalt ions (q.v.). D. Bentley (personal communication) reports that naked axons can be filled without cutting, using en passant suction electrodes, but this is a difficult technique.   Potentially it could be used to demonstrate the morphology of functionally identified neurons in conjunction with stimulating and recording through the same electrode.

If the positions of particular neurons in a ganglion are known, they can be filled by placing a drop of cobalt chloride in a Vaseline cup over a hole in the ganglionic sheath (O'Shea *et al.*, 1974; Strausfeld and Obermayer, 1976).   The required neuron has to be superficial, and truly selective fills can only be obtained where the neuron's geometry is particularly favorable.   In the locust brain, the descending contralateral movement detector neuron is distinguished from other descending neurons because its cell body and neurite lie superficially in the dorsal neuropil on the side contralateral to its arborization and descending axon.   Its projection to the thoracic ganglia has been determined by filling through a small hole over the cell body region (O'Shea *et al.*, 1974).

Intrinsic Interneurons

Intrinsic interneurons have no access through extraganglionic processes. They may be filled through a hole in the ganglion sheath, as described above, or by introducing a pool of cobalt chloride into the neuropil with a micropipette (see Strausfeld and Hausen, 1977 and Chapter 18 of this volume).   Selected neurons cannot be filled in this way, but discrete placing of the cobalt pool should allow repeated sampling of particular regions, giving more control of the neurons filled than with Golgi methods (see Chap. 9, this volume).

Neurosecretory System

Neurosecretory pathways between the brain, subesophagal ganglion, and retrocerebral complex in the locust have been demonstrated by filling the nerves linking the various bodies (Mason, 1973; Aubele and Klemm, 1977; Rademakers, 1977).   The preparations are made in vivo.   Because the nerves are fine and must be dissected without saline solution, the cut end is placed in a drop of distilled water to open the axons before applying cobalt chloride solution.   Best preparations have been obtained at low temperatures for 1–3 days, without current.

Peripheral Nervous System (Fig. 19-9)

By filling the distal ends of cut peripheral nerves, both the origin of sensory neurons (Fig. 19-9A–C) and the terminations of motor neurons (Fig.

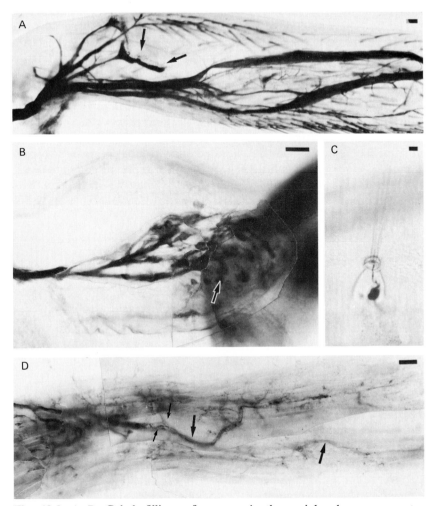

Fig. 19-9, A–D. Cobalt fillings of neurons in the peripheral nervous system.
A The prothoracic tibia of the cricket *Gryllus bimaculatus*. The course of the
nerves in the segment and the sensory cells of the tympanal organ (*arrowed*) are
clearly shown. [Photograph courtesy of E. Eibl.] **B** The sensory cells of the metathoracic tegula in the locust *S. gregaria*. The wing base is seen as a dark shadow on
the right and part of the knob-like tegula is seen through it. The tegula bears
about 40 short hairs, each innervated by a single sensory neuron. **C** The sensory
neuron innervating one of the wind-sensitive hairs on the head of *S. gregaria*.
Intensified preparation. **D** The innervation of two neck muscles (50 and 51) in the
locust, which share two fast motor neurons. The two neurons can be seen running in parallel (*arrows*) and neuromuscular endings are visible as small speckles.
[Preparation by J. Kien and J. Nelson.]   Scales: **A,B,D** 100 μm; **C** 10 μm.

19-9D) can be displayed in excellent detail (Eibl, 1976; Altman and Kien, 1979). This method is also useful for tracing peripheral nerve branches. Preparations may be done in vitro when the muscle, appendage, or body segment is placed in the saline bath. A current bias may produce faster results but better preparations are obtained without current, by filling for up to 80 h at 4–7°C (Eibl, 1976; J. S. Altman, unpublished). The in vivo method can also be used, by placing the distal end of the nerve in the cobalt cup. This is more suitable for large structures such as wings (Altman *et al.,* 1978). Simultaneous central and peripheral fills can be made in such preparations if the proximal end of the nerve is removed from the cup after the first day. Cobalt chloride concentrations of up to 15% may be used, because there is little chance of misinterpretation where single branches are filled. Peripheral preparations may be intensified with Timm's method for whole mounts (Bacon and Altman, 1977) (Fig. 19-9C; Chap. 20).

Processing after Filling

Unfortunately, no single method of processing is satisfactory for all subsequent methods of examination of cobalt-filled neurons. Light microscopy will be described first, followed by special modifications for electron microscopy.

Light Microscopy

Precipitation of Cobalt Sulfide
In vitro preparations are removed from the tissue bath and transferred to a dish of clean saline solution. In vivo preparations may be placed in the deep freeze for a few minutes to harden the Vaseline, which may then be removed in a lump together with the cobalt chloride. The animal is rinsed with saline solution, and some dissection may be necessary to allow sulfide ions access to filled structures but this should be kept to a minimum and touching the nerve cord should be avoided. Preparations are then transferred to saline solution containing approximately one drop of concentrated (44%) ammonium sulfide per milliliter, for 5–10 min. Wash well in several changes of saline solution before fixation. Ammonium sulfide deteriorates with exposure to air, because NH_4 is lost as ammonia. In the solution, first polysulfides are formed and later sulfur is precipitated. Since polysulfides can combine with cobalt to produce a soluble compound, it is essential to use fresh ammonium sulfide. Satisfactory substitutes for ammonium sulfide are a 1%–5% solution of sodium sulfide in saline or saline saturated with hydrogen sulfide (bubble H_2S through

100 ml of saline for 15 min). The latter can cause very heavy precipitates in preparations with many neurons filled.

Fixation

Choice of fixative depends on the subsequent method of examination and is summarized in Table 19-1.

Dehydration and Clearing

Fixed preparations are dehydrated through a standard alcohol or acetone series, starting at 30% after aqueous fixatives and 70% following alcoholic fixatives. Ten minutes in each step is sufficient. Recommended clearing agents are methyl benzoate, methyl salicylate, and cedarwood oil. The last is particularly good for peripheral sensory preparations. Creosote and styrene (Pitman *et al.*, 1972; Rehbein *et al.*, 1974) have also been employed but both appear to cause rapid bleaching of the cobalt deposit. Methyl benzoate occasionally causes bleaching, perhaps because of contaminants, but this is minimized by transferring to an embedding medium as soon as the tissue becomes transparent. Whole mounts may be intensified straight after fixation, after storage in 70% ethanol, or after dehydration and rehydration (Strausfeld and Obermayer, 1976; Bacon and Altman, 1977; see also Chap. 20).

Whole Mounts and Thick Sections

Genuine neutral Canada balsam appears to make more stable permanent preparations than artificial mounting media. Preparations may be mounted on cavity slides, but Eibl (1976) has introduced two alternatives that allow better all-around observation: thin metal slides with a central hole covered top and bottom with coverslips; and glass capillary tubes, into which the preparation in Canada balsam is sucked. This tube can be mounted in a bath of glycerine ($r.i.n._D = 1.4740$) and turned about its long axis with a simple screw micrometer.

To give a better view of the distribution of branches in depth, ganglia may be sectioned for light or electron microscopy after they have been examined as whole mounts in either clearing agent or Canada balsam. For wax sections and electron microscopy a compromise has to be made over fixation if good tissue preservation is to be obtained (see Table 19-1). For light microscopy, whole-mount intensification (see Chap. 20) overcomes this problem since material fixed in alcoholic Bouin is good for both whole mounts and sections. Preparations may be sliced by hand (C. Goodman, 1974) or sectioned with a steel knife after embedding in soft Araldite (Strausfeld and Obermayer, 1976). This provides a rapid method for assessing the relationship of a filled neuron to tracts and other landmarks in the neuropil (Bacon and Tyrer, 1978; Altman and Kien, 1979). Plastic sections can be counterstained with toluidine blue if background detail cannot be obtained by stopping down the microscope's condensor iris diaphragm.

Table 19-1. Recommended Fixation Procedures for Cobalt Preparations

Fixative	Time (h)	Recommended for	Advantages	Disadvantages
1. Baker's formaldehyde-calcium	6–12	Unintensified whole mounts	Very clear and least distorted ganglia; O.K. for subsequent sectioning	Fixes fat in ganglionic sheath, so that whole mounts may not be completely transparent; can't intensify
2. 70% alcohol	2–overnight	Unintensified whole mounts	Removes fat from sheath; ganglia clear well	Shrinkage; poor tissue preservation for subsequent sectioning
3. Alcoholic Bouin[a]	1–2	Intensified whole mounts; sections for intensification	Best tissue preservation; little distortion	Unintensified whole mounts do not clear well
4. Carnoy[a] (make fresh before use)	1	Whole mounts (intensified or unintensified)	Rapid; very clear ganglia	Poor tissue preservation for subsequent sectioning
5. Buffered glutaraldehyde	2	Electron microscopy	Preservation of fine structure	Whole mounts do not clear well; whole-mount intensification difficult; LM wax sections poor

[a] See Pantin (1946).

Embedding for Wax Sectioning

After clearing briefly in methyl benzoate or methyl salicylate, transfer to a warmed mixture of benzene[1] and wax for 15 min, infiltrate in two changes of clean wax (30 min each), and block out as usual.

Alternatively, a brief dip in xylene may be used as an intermediate step between clearing agent and wax, but xylene is not recommended as a primary clearing agent because it makes insect neural tissue very brittle and difficult to section well.

Electron Microscopy

Details of electron-microscopic (EM) methods for cobalt-filled neurons, particularly the problems of identification of cobalt deposits and methods of intensification, are dealt with in detail in Chapter 21. Here we give only the special modifications of the precipitation and fixation methods necessary to achieve good tissue preservation for electron microscopy.

Cobalt Sulfide Precipitation

Preparations should not be treated with saline solution before fixation. Best results have been obtained by developing in situ with 1%–2% sodium sulfide in buffer for 5–15 min at pH 7.4. Ganglia are then rinsed well with clean fixative and left to fix for 1–2 h. Alternative methods are to fix briefly with buffered glutaraldehyde, followed by treatment for 5–10 min with ammonium sulfide either in the fixative (Tyrer and Bell, 1974) or in buffer (Mobbs, 1976). For frog spinal cord, Székély and Kosaras (1976) recommend a total of 2 min immersion in a di-sodium hydrogen phosphate solution saturated with hydrogen sulfide.

Fixation

Routine EM fixation in buffered glutaraldehyde is used (see Chap. 10). Osmium postfixation may be omitted so that the neuron can be visualized in the plastic block for orientation during sectioning but membranes and synaptic ultrastructure are not well preserved. We use 1% osmium tetroxide in buffer for 1–2 h but this sometimes gives rise to problems if a silver intensification method is subsequently employed (see Chap. 21). Mason and Nishioka (1974) recommend sketching the neuron after glutaraldehyde fixation but before osmification.

Examination and Analysis

In good, unintensified whole-mount preparations, the filled neurons appear dark against a clear background (see Figs. 19-3, 19-6). Sometimes

---

[1] Health and Safety regulations in the U.K. recommend that benzene be withdrawn from use because of potential carcinogenic properties.

the ganglion has a brownish coloration, assumed to be caused by cobalt ions leaking from dead neurons. Such preparations should be eliminated from critical analyses, because the leakage may give rise to artifactual filling (see p. 397). As an additional safeguard, the reflexes of in vivo preparations may be tested at the end of the filling period to check that the preparation is still alive. In intensified whole mounts the neuron is very black against a golden-brown background (see Figs. 19-1, 19-7).

Whole-mount preparations can be viewed with a stereomicroscope and a combination of transmitted and reflected light, or with a compound microscope and transmitted light. Stereomicroscopy and stereo-pair photographs are valuable for forming an impression of the three-dimensional shape of a neuron or the depth at which tracts and branches lie, but we find the compound microscope essential for resolving very fine detail. The limited depth of focus is in many ways an advantage because it reduces confusion and allows single branches to be traced.

Photography. Depth of focus and resolution limit the usefulness of photographs of neurons in whole mount. Resolution in stereo-pair photographs is poor, while photomicrographs taken with a compound microscope give good resolution of the detail of single branches, but the extent of a neuron is difficult to portray, even by montaging (see Figs. 19-1, 19-6, 19-7, 19-9B,D). Kodak Plus-X with a green filter to improve contrast gives best results with unintensified whole mounts. For intensified whole-mount preparations, Kodak Pan-X is employed and a green filter is required only at high magnification. For intensified wax sections Ilford Pan-F is recommended, while thick plastic sections of intensified whole mounts are better taken on Kodak Pan-X with a green filter.

Drawing. The only satisfactory method of displaying all the details of a neuron's branching is to draw it using a compound microscope with camera lucida or drawing tube attachment (Tyrer and Altman, 1974). This, however, collapses the arborization into a two-dimensional picture and loses all information about the layering of branches in the neuropil.

Analysis of the three-dimensional distribution of a neuron and its relationship to other components of the neuropil are essential for determining the connectivity of identified neurons. Some information can be obtained from whole mounts using interference contrast microscopy to cut optical sections through the tissue. The relationship between the filled neuron and unstained features of the neuropil can then be visualized, but resolution of fine branches is impaired. Related x, y, and z plots can also be made from whole mounts using a mechanical device attached to the stage controls and fine focus of the microscope (e.g., McKenzie and Vogt, 1976).

Both these methods only give a first approximation of branch distribution. Detailed information about the tracts in which filled neurons run and the contacts they make with other neurons can only be obtained from serial sections.   Plastic sections (25–50 μm) of whole-mount intensified ganglia provide a rapid method, but more detail can be obtained from thinner wax sections intensified on the section (Tyrer and Bell, 1974).   For resolution of background details in sections, a light counterstain can be used (see Chap. 21) or the condensor iris diaphragm on the microscope may be stopped down.   With phase contrast the resolution of fine cobalt-filled profiles is very poor but interference contrast microscopy can be useful (Altman and Tyrer, 1977a).

Reconstruction in one or two planes in addition to the whole-mount view is often necessary to display all the features of a neuron (Tyrer and Altman, 1974; Altman and Tyrer, 1977a).   The most promising solution to this tedious procedure is to use a computer to make three-dimensional reconstructions from digitized x, y, z coordinates, which are obtained by tracing branches in whole mounts or serial sections (Glaser *et al.*, 1977).

Errors and Artifacts: Cautions for Interpretation

To identify a neuron or tract by selective axonal cobalt filling requires that the cobalt ions migrate only up the axons to which they are applied.   It is now clear that these desirable conditions are rarely fulfilled and the cobalt ions are both highly mobile in the nervous system and easily taken up by neurons other than those intended.   This pitfall is a serious trap for the unwary when determining the origin and identity of filled cells.   The problem is amplified when using the Timm's intensification procedures, which are very senstive and can pick out very small quantities of cobalt, invisible in unintensified material.   As artifacts are more likely to occur in material filled with high concentrations of cobalt for long times, it is essential that filling parameters are clearly stated in the methods sections of all papers employing the cobalt method.

Parallel Filling

Parallel filling in peripheral nerves provides the best demonstration of this difficulty.   Where selective fillings of one nerve branch are attempted, axons running in other branches of the same nerve root may also become filled (Fig. 19-10).   For example, the locust metathoracic nerve 1 has three major branches, $1C_1$ and $1D_2$, which are entirely sensory, and $1D_1$, containing five large motor axons and a few smaller ones (Campbell, 1961; Altman and Tyrer, 1974).   Filling either of the sensory branches often produces quite dense fills of the motor neurons in nerve $1D_1$, although the cut end of the nerve filled may be several millimeters from the junction of the branches.   Here the artifactual filling is easy to identify because

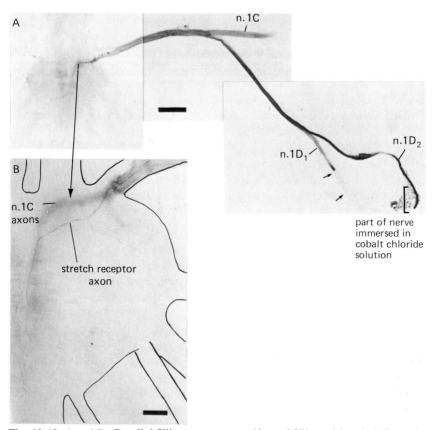

Fig. 19-10, A and B. Parallel filling: common artifact of filling with cobalt from the periphery. **A** The metathoracic ganglion of a preparation in which the cut end of nerve $1D_2$ has been immersed in cobalt chloride (indicated by bracket) to fill the stretch receptor neuron. Cobalt has spread into axons of other branches of nerve 1. In the nerve $1D_1$ centrifugal spread has been blocked by crushing (unfilled portion indicated by *arrows*). Scale 1 mm. **B** Detail of the central projections of the same preparation. Although only nerve $1D_2$ was selected in order to fill the stretch receptor neuron, axons in nerve 1C from the wing sense organs (e.g., tegula, see Fig. 19-9B) are also filled. This artifact can be avoided by reducing the concentration of cobalt chloride and by shortening the filling time. Scale 100 μm.

the components of the various branches are well known, but it has not yet been possible to separate satisfactorily the projections of the corresponding nerves $1C_1$ and $1D_2$ in the mesothorax (Tyrer and Altman, 1974). Similar problems have arisen in the dorsal tegumentary nerve, where using 3 M cobalt nitrate to fill small groups of head-hairs resulted in the staining of all the axons in the nerve (Tyrer, Bacon, and Davies, 1979). Cross sections of cobalt-filled nerves reveal considerable deposits of cobalt in the glial layers and extracellular spaces, and cobalt ions must be taken up from this pool through the membranes of the axons. The extracellular microelectrode staining technique (Rehbein *et al.*, 1974; Chap. 18) demonstrates that physically intact axons can easily take up and transport cobalt ions, although the mechanism is unknown. The problem may be worse where the large numbers of small-diameter axons found in many sensory nerves provide a high resistance pathway for the cobalt ions, which therefore concentrate in the glia. In the cockroach, filling the tarsal nerve or the dendrites of the campaniform sensilla on the trochanter results in staining the motor neurons of both proximal leg and thoracic muscles (Krämer, personal communication). A simple safeguard is always to check whether cobalt has migrated back along the other branches of the filled nerve (Fig. 19-10).

Overfilling in the CNS (Fig. 19-11)

The observation of uptake into other axons in peripheral nerves immediately raises the question of whether neurons within the CNS can pick up cobalt ions from an extraneuronal pool. In certain experiments, such as filling the optic stalk between medulla and lobula in the cricket (Honegger and Schürmann, 1975) or filling connectives, the adjacent neuropil may become flooded with cobalt ions. Neurons leaving this pool may well be intrinsic to the flooded area without an axon in the tract or connective filled. Filling a peripheral nerve with a high concentration of cobalt chloride for too long has a similar effect because it produces a pool of cobalt ions in the ganglion in the area around the nerve root (Fig. 19-11A). This can easily be controlled for by running preparations under milder conditions. These situations are easily identified and reservations can be built into the interpretation of such material.

Overfilling of peripheral nerves also gives rise to curious distortions of neuron morphology (Fig. 19-11B), which may be caused by the filled neurons becoming leaky as they are poisoned by cobalt. Although overfilling may be necessary to demonstrate the details of distal parts of long projections, e.g., interganglionic sensory projections (Tyrer and Altman, 1974), features in the more proximal parts of the ganglion should be ignored unless they appear routinely in fills made with lower concentrations and shorter filling times.

Trans-synaptic Staining

The recent report that cobalt ions can be transported into presynaptic neurons (Strausfeld and Obermayer, 1976) deals with neurons in the fly brain that were already known to be contiguous. It is essential to be aware of conditions that could give rise to a spurious appearance of trans-

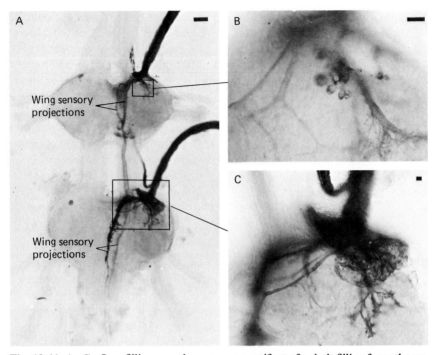

Fig. 19-11, A–C. Overfilling: another common artifact of cobalt filling from the periphery, caused by high cobalt concentration or by too long a filling time. **A** Meso- and metathoracic ganglion from a locust with nerve 1 of each ganglion filled. This unintensified preparation was deliberately overfilled in order to demonstrate the distal parts of the interganglionic sensory projections from the wings. There is an accumulation of extraneuronal cobalt sulfide at the roots of the nerves. Scale 100 μm. **B** Detail from **A** to show balloon-like swellings associated with filled neurons in the mesothoracic ganglion. Scale 10 μm. **C** Detail from **A** of the cobalt sulfide accumulation at the root of metathoracic nerve 1. This is probably due to cobalt in the glia, and if overfilling is prolonged the whole ganglion may go black. Scale 10 μm.

synaptic staining. First, intact neurons may take up cobalt ions from an extraneuronal pool within the neuropil. Even in ganglia where the neuropil appears uncontaminated in whole mounts, electron microscopy of Timm's intensified sections often shows a low concentration of cobalt ions in glial cells (J. S. Altman, C. A. Davies, and N. M. Tyrer, unpublished), a problem that increases with the filling times and cobalt ion concentration. Just as in the peripheral nerves, there is a danger that neurons may acquire cobalt ions from glial cells in common with the filled neuron. In the ganglion this could be misinterpreted as trans-synaptic staining. C. Goodman (1976a) found that filling the ocellar cup in locust with 1 *M* cobalt chloride invariably results in staining the same set of second-order neurons, which are never demonstrated when 250 m*M* cobalt chloride is used. Although this may be true trans-synaptic staining, it may just as easily be due to the second-order neurons picking up cobalt ions from a local extraneuronal pool. Filling higher order cells cannot by itself imply functional contiguity.

The second source of error comes from the filling of sensory neurons from the body surface. As mentioned previously cobalt ions may migrate under the cuticle and be taken up by other sensory and motor neurons, not necessarily in the same peripheral nerve as the target structure. Unexpected fills of neurons with branches in the same neuropil as the target neuron could be interpreted as trans-synaptic fills, especially if direct sensory-motor contacts are predicted from physiologic evidence. If there is an alternative route through which the second set of neurons could have been filled, one cannot prove that trans-synaptic staining has occurred.

Tactics for Identifying the Origin of a Neuron or Tract

To obtain a complete, detailed fill of a particular neuron or tract may require a long filling time and higher concentration of cobalt chloride. These are exactly the conditions most likely to produce spurious fillings and artifacts. To avoid misinterpretation, identification of a neuron and determination of the details of its morphology must be two separate steps. Short fills with low concentrations should be used to identify the major features of the neuron, sufficient for it to be recognized subsequently in heavier, mixed fills. The details of the branching pattern can then be studied in mixed fills, provided that there is no overlap of branches. A large neuron such as the locust wing SR (Altman and Tyrer, 1977a) can rarely be displayed in its entirety in a single preparation, with no contamination from other neurons; thus, details have to be pieced together from several preparations made under different conditions. Skillful manipulation of the advantages provided by the animal's own anatomy can solve many of these problems.

Conclusions

Judiciously used, axonal filling with cobalt can be a major tool for untangling the structure of the insect nervous system. Like all inventions the limitations of which are not well understood, there is a serious risk of misuse. Provided that certain precautions are adopted, both in application and interpretation, cobalt staining can provide information that has not previously been obtainable.

Perhaps the most important insight it has given us so far concerns the regularity in the organization of the tangled thoracic neuropils. Structures in these neuropils can now be identified functionally, and the comparison of a number of preparations of the same neuron reveals consistency and, hence, order in areas that lack a recognizable pattern. Furthermore, the interganglionic neurons that have so far been described fall into a few distinct morphologic types, suggesting that we will find a repetition of connections according to function. Interneurons with similar output functions but dissimilar inputs appear to have similar patterns of terminals in the thoracic ganglia (Kalmring *et al.*, 1974). As more neurons are identified both structurally and functionally, their classification into types should begin to reveal rules about the overall organization of the insect nervous system.

The cobalt method is significant, not only for linking the structure of neurons to their functions, but also for providing a bridge between neuroanatomy and neurophysiology. Physiologists now not only know where to look for their neurons when recording but also have a more direct appreciation of the possibilities the structure of the nervous system affords the interpretation of their results (Burrows, 1975). For anatomists, the assignment of functions to neurons enlarges their ability to interpret organization and enables them to predict how systems might work in neuronal terms (Altman and Tyrer, 1977a). Such theoretical approaches are important for demonstrating the possible complexities that the limited sampling techniques currently available to physiologists are insufficient to reveal.

However, the startling beauty of some cobalt preparations is not enough. By revealing the complexity of some insect neurons, the cobalt method has dispelled the idea that the organization of the insect nervous system is simple. Recent physiologic and anatomical evidence suggests that many of these neurons are multifunctional or multimodal. Now we need to know not only which cells are linked together but the exact spatial distribution of the contacts between two neurons. This is a vast challenge for which great perseverance at both the light and electron-microscopic levels will be necessary. Cobalt staining is providing a spearhead for this attack but it will require imaginative manipulation and combination with other staining methods to produce satisfactory answers. The first flush of excitement is over but there are far larger rewards for those with perseverance.

Addendum

Further experience with whole mount intensification has demonstrated that the cleanest and most selective fills are obtained with 1.5% cobalt chloride. Addition of *l*-lysine to a 0.37 *M* (approximately 9%) cobalt chloride solution is recommended by Lázár (1978) for filling fine axons in frogs. A complex cobalt-containing ion is formed and the effect may merely be to reduce the free cobalt ion concentration.

Other heavy metals may give results comparable to cobalt when used in low concentrations. In particular, we have had acceptable results with nickel and cuprous chlorides. Using rubeanic acid (dithio-oxamide) (Quicke and Brace, 1978), a different colored complex is formed with each heavy metal so that it is theoretically possible to study two inter-digitated neurons. Unfortunately, cobalt forms a yellow complex and branches are hardly visible, but nickel appears blue and copper green, giving reasonable resolution. These preparations can subsequently be intensified but then the distinction between metals is lost. Rubeanic acid has the advantage that it does not smell, but the complexes it forms are acid soluble, so tissue has to be treated more carefully and cannot be stored in 70% alcohol.

Pitman (1979) gives a recipe for decoloring overintensified ganglia, using potassium ferricyanide and sodium thiosulfate to remove silver deposited on the ganglionic sheath. The same effect can be achieved with 1% iodine dissolved either in potassium iodide or in absolute alcohol. These methods are inadvisable because it is easy to erase parts of the filled neurons. In Pitman's case, silvering of the sheath occurs because formalin is used as a fixative, which acts as a reducing agent and accelerates inten-sification (Bacon and Altman, 1977). If fixatives such as Bouin or Carnoy are used and the pH of the developer solution is correct (see Chapter 21), a silver mirror should not form. The best remedy, if ganglia are too dark after intensification to use as whole mounts, is to embed in soft plastic and cut thick sections (50–100 μm).

Recently, trouble has been reported with whole mount intensified preparations going black in Canada balsam. This only appears to happen if the Canada balsam is old and may be due to acidity.

Filling motor neurons by injecting cobalt chloride into muscles (p. 381) is highly unreliable (J. Kien and J. S. Altman, unpublished). Although good fills of motor neurons can be obtained this way, they are not always those of the muscle injected and so can be very misleading. Cobalt ions appear to leak into the hemolymph from the muscle so that the ganglia often darken excessively when intensified and tracheae may stain very heavily.

Acknowledgments

Much of the experimental work on which this chapter and Chapter 21 are based was done at the Universität Konstanz, Federal Republic of Germany, supported by DFG grant Ku 240/3 and 240/5 to Dr. Wolfram Kutsch. We thank Dr. Kutsch for making this stay possible. Many people have contributed their experiences and we are particularly grateful to Jon Bacon, Elizabeth Bell, David Bentley, Jenny Kien, Joy Nelson, Malu Obermayer, Vicky Stirling, Nick Strausfeld and Les Williams for discussions which may have been included in these chapters without due acknowledgments.

Chapter 20

Silver-Staining Cobalt Sulfide Deposits within Neurons of Intact Ganglia

M. Obermayer
N. J. Strausfeld

European Molecular Biology Laboratory
Heidelberg, F. R. G.

Among procedures most commonly used to fill selected neurons, cobalt chloride diffusion or injection is routinely employed for showing up projections that emanate from or project to nerve roots and fasciculi in the insect and vertebrate central nervous system (Altman and Tyrer, 1974; Altman, 1976; Gallyas, 1971; Goodman, 1976; Székély and Gallyas, 1975). By comparison, the organic dye Procion yellow has been consistantly favored for demonstrating the shapes of single neurons after electrophysiologic recordings (see, e.g., Kater, and Nicholson 1973, for collected · articles; Hausen, 1976a; Dvorak *et al.,* 1975a). Horseradish peroxidase has only recently been applied to insects and is used for tracing the origin of single motor neurons from identified muscles (Coggeshall, 1978) or the destination of primary efferents from receptors (Ghyssen and Deak, 1978).

Comparisons between the forms of neurons resolved by Procion dyes and the shapes of neurons resolved by Golgi impregnations reveal that in invertebrates the organic dye passes to all the nerve cell's ramifications. Even very small profiles can be resolved by light microscopy, and their resolution is aided by the fluorescent property of the dye emitting rather than absorbing light when irradiated in ultraviolet light (cf. Hausen, 1976a; Strausfeld, 1976).

403

On the other hand, comparisons between Golgi-impregnated nerve cells and neurons that have been filled with cobalt sulfide after initial diffusion with cobalt chloride or cobalt nitrate leave much to be desired. In most cases cobalt sulfide shows up cell bodies, cell body fibers, large-diameter dendrites, axons, and telodendria, often with great clarity. However, even if this degree of resolution may allow statements about consistancies of neuronal form, resolution of processes that contain trace amounts of cobalt sulfide can be made only with uncertainty. Such details include the specializations from smaller profiles such as spines, blebs, varicosities, and tuberescences, and many small-diameter arborizations cannot be resolved at all. These disadvantages become critical if the unmodified technique is to be used for screening small nervous systems, such as those of *Drosophila* and its mutants. In the case of large nerve cells, such as motor neurons or orthopterian ventral ganglia, the available resolution after sulfide treatment may show up sufficient detail for determining some characteristic distributions of branches or dendritic spines. However, by comparison with Golgi-impregnated neurons or with cells stained by the intra vitam methylene blue method, these elements appear sparsely branched and have smaller domains than would be expected.

The resolubility of a neuron depends on the density of the cobalt sulfide that it contains. Although higher concentrations are liable to improve contrast between the cell and the background, there is a limit to the initial cobalt intake by the nerve cell. The salt is toxic, and concentrations of more than 25% usually poison the tissue and cause rapid degeneration of the nerve cell and leakage of $CoCl_2$ into the surrounding extracellular spaces. The compromise between a low concentration in an apparently "healthy" neuron is unavoidable even though it allows only poor resolution.

Tyrer and Bell (1974) employed Timm's (1958a) intensification method for enhancing cobalt sulfide-containing profiles by the deposition of metallic silver. Their illustrations implied that possibly as little as 50% of a neuronal arborization could be unambiguously resolved before intensification. Tyrer and Bell derived their results from back-fills of relatively large neurons and intensification of paraffin-embedded and sectioned material.

The Tyrer and Bell procedure is a powerful one, but relies on prior embedding and sectioning. This treatment carries with it certain disadvantages, in addition to the shrinkage incurred by paraffin treatment. The major drawbacks of intensification on sections are, first, that background coloration can be extensive. This will obscure intensification of profiles that contain only trace amounts of cobalt sulfide. Second, there is evidence that cobalt sulfide can shift its location during paraffin infiltration. This is no hindrance when investigating large-diameter processes. However, there is a need for an alternative procedure when examining very small nerve cells or nerve cells that have taken up trace quantities of cobalt, such as occurs during the so-called trans-synaptic diffusions (Strausfeld and Obermayer, 1976; Goodman and Williams, 1976). Another constraint of thinly sectioned material is its natural limitation for resolution in the third dimension. An intensification procedure that can be performed prior to thick sectioning or whole-mount resolution is therefore of particular advantage.

This chapter describes some variants of intensification procedures that were designed for intact central nervous systems and which permit resolution of large and small neurons either by direct or by dark-field illumination. The procedures also allow material to be embedded in resins and

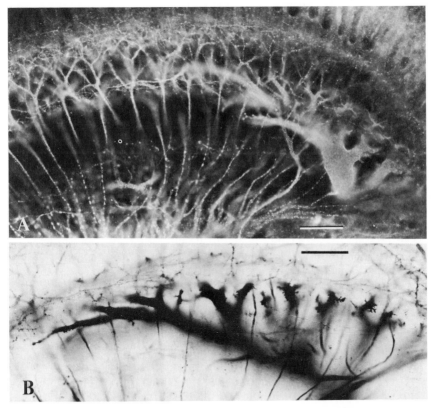

Fig. 20-1, A and B. Two ways of visualizing identical structures filled with cobalt sulfide. **A** Dorsal unique neurons are resolved by dark-field illumination against columnar cells of the lobula. **B** The same species of giant neuron (*Calliphora* male) is resolved by transmitted illumination. Both preparations were intensified according to method D but the first developed an early silver mirror which then inhibited proper intensification of deeper tissue. Unless otherwise stated the scale for this and all other figures is 20 μm.

sectioned at between 40 and 100 μm. This is an essential requirement for subsequent electron microscopy of a previously identified element. The present methods also resolve arrays of contiguously filled neurons whereby nerve cells containing trace amounts of cobalt sulfide can be differentiated from other elements by their coloration or by dark-field illumination (Fig. 20-1). However, before describing the principal schedules for the methods, it is convenient to outline the basic concepts of the silver intensification procedure.

Silver Intensification Procedures

Liesegang Physical Development for Histology

In 1911 Liesegang described a photographic rescue operation, and its application to neurohistology, in which he demonstrated that the so-called physical development procedure could retrieve a latent image on an exposed, fixed, but undeveloped photographic plate. This method, for long a standard in textbooks of photographic techniques, relies on the catalytic action of atoms of metallic silver that are initially derived from photodissociation of silver bromide. If the photographic plate is washed after fixation (which dissolves away all undissociated silver halides) and is then treated simultaneously with a reducing agent, such as formalin or hydroquinone, in the presence of silver nitrate, the silver specks catalyze the reduction of silver nitrate from the solution. The original silver nuclei become covered with metallic silver. This reaction is autocatalytic and ultimately builds up the latent image. Liesegang (1928) showed that the speed and graininess of the image formation could, to some extent, be controlled by the use of a protective colloid that was either gum arabic or gelatin. This decelerates the interaction between silver nitrate and the reducer within the aqueous solution.

Liesegang (1928) used the intensification procedure in neuroanatomy. As he pointed out, both the block Cajal and Bielschowsky reduced silver procedures are often unsatisfactory in that deposits of metallic silver at argyrophilic sites are patchy. It is commonly observed that little or no staining has taken place deep within the tissue. In the block silver procedures the silver is initially introduced in the form of nitrate or ammoniacal nitrate. The capricious nature of the method is, to a large extent, due to the diffusion of the reducer into the tissue being counteracted by the diffusion of silver cations out of it. Thus, near the tissue surface there is usually dense staining at the expense of deeper structures.

Liesegang overcame this difficulty by initially treating tissue for block impregnation; but prior to reduction the tissue was fixed, sectioned, and then treated by the physical development method. He used as a reducer formalin or hydroquinone and obtained an even coloration of nerve fibers at the surface and at some depth into the section. Likewise, sections of reduced silver-stained tissue can be treated by Liesegang's method in order to enhance the contrast of silver-containing profiles.

James Metal Sulfide Phenomenon

Metallic silver decreases the induction period of reactions between hydroquinone and silver nitrate and thence catalyzes the reduction of silver ions to further deposits of metallic silver. Other catalysts are known—for example, colloidal gold (Querido, 1948) as well as sulfides. The action of the latter was first recognized by Sheppard and Mees (1907; see also Mees and James, 1966; Sheppard, 1944) during investigations on the sulfide contents of commercial photographic emulsions. Later, James (1939c) demonstrated that there was a marked increase of the reduction rate of silver nitrate by a developer, such as hydroquinone, in the presence of silver sulfide or metallic silver with another species of, or an equivalent, sulfide (see also Spinelli, 1975). In addition to sulfides, sulfhydral oxidation products or the presence of quinone decrease the induction period of the hydroquinone-silver nitrate reaction (James, 1939a,b).

Voigt and Timm's Intensification Procedure

Physical development and its application to tissue containing metallic sulfides were exploited by Voigt (1959) and Timm (1958a) for investigating the sites of heavy metal deposits in pathologic tissue and has since become a standard procedure for detecting the natural occurrence of heavy metals in the vertebrate brain (Timm, 1958b; Haug, 1973). The Timm procedure demands pretreatment of tissue with a weak solution of ammonium sulfide followed by treatment with colloid-carried silver nitrate and hydroquinone. The rate of reaction depends upon the temperature, the concentration of hydroquinone, the pH, the colloid concentration, and the silver ion concentration.

In a set of model experiments James (1939a–c) showed that the reaction rate (1) is dependent upon $\frac{2}{3}$ the power of the Ag^+ ion concentration, (2) is inversely proportional to the concentration of gum arabic, and (3) is proportional to the hydroquinone concentration. The reaction rate is also dependent on the pH of the solution, being accelerated in the alkaline. Thus, judicious addition of a weak acid will decrease the rate of reduction. Tyrer and Bell's modification of Timm's procedure for paraffin sections used a low concentration of silver nitrate and hydroquinone within a pH range of 3.8–4.0, which was adjusted by the addition of citric acid. However, using their concentrations and schedule on whole ganglia, there is a tendency for an accelerated reaction with the consequence that reaction rates are fastest near the tissue surface. This usually results in what is here termed an "early silver mirror," which has an auto-catalytic action during further development and thus inhibits physical development deep within the tissue.

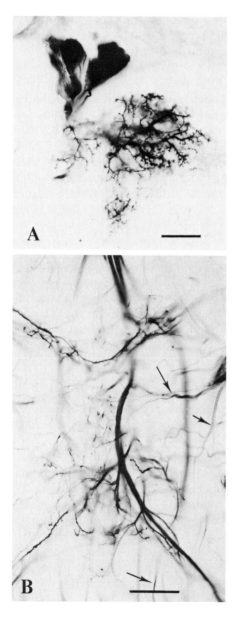

Fig. 20-2. A A group of descending neuron dendrites that invade the anterior optic tubercle, revealed after intensification with glycine-formalin as the developer (*C. erythocephala,* male). **B** Terminal arborizations of descending neurons in the pro- and mesathoracic ganglia of *Calliphora.* Tracheae (*arrowed*) are also resolved if the tissue has been treated by a clearing agent such as methyl salicylate prior to intensification (method D).

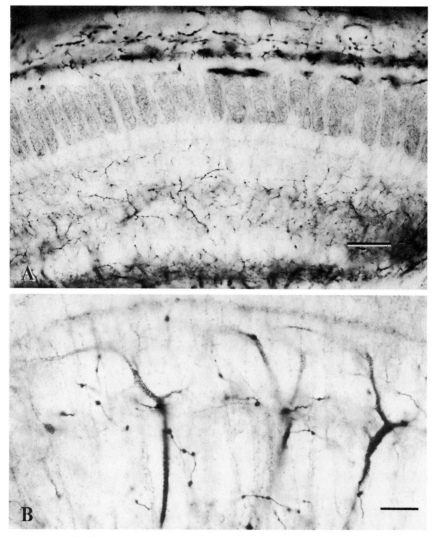

Fig. 20-3. A Sagittal section through the lobula and lobula plate of *Calliphora*, intensified by Querido's silver-sulfite procedure, revealing features of background fibroarchitecture with intensified neurons and fibers of the second optic chiasma. **B** Supraperiodic columnar neurons of male *Sarcophaga*, with associated climbing centrifugal fibers in the lobula, intensified by method F, showing background structures of small medulla inputs of the retinotopic array.   Scale = 5 μm.

Several basic schedules have been devised to overcome problems encountered during whole-tissue intensification, such as (1) spontaneous reduction of silver at the tissue surface, (2) the short duration of the induction period before the reaction between hydroquinone and silver nitrate, (3) compensation for silver extrusion from the tissue, and (4) introduction of silver and/or hydroquinone to potential autocatalytic sites (cobalt sulfide) prior to the induction period. Multiple step intensifications, using at first very low concentrations of the reactants, are used to lay down metallic silver with sulfide sites.

The methods are listed in the order of their first application. There is a general tendency by investigators to simplify the procedure. The original method A employed lactic acid as one means of prolonging the induction period. Other slow developers, such as glycine-formalin admixtures, have been used instead of hydroquinone in method B (steps 2.1 and 2.2) during the preincubation stage in order to deposit trace amounts of silver at the sites of cobalt sulfide and thus to give rise to two catalysts, CoS + metallic silver. After glycine-formalin reduction the tissue is washed in distilled water, incubated as in steps 2.3 and 2.4 (Fig. 20-2A).

Although they are difficult to execute, complex multistep intensifications have been found useful. In some instances, they may be employed to show up background features of neuropil that can be usefully compared to reduced silver preparations. Buffered hydroquinone-sodium sulfite developers with colloid and silver nitrate (Querido, 1948) may also be employed to resolve background fibroarchitecture. A silver complex compound substitutes silver nitrate and is prepared by adding 1 vol of saturated sodium sulfite to 1 vol of 10% silver nitrate. The precipitate is redissolved by vigorous shaking. Querido recommended the addition of 4 vol of 15% gum arabic and 2 vol of 2% hydroquinone to the silver-sulfite complex. However, for block intensification the strength of the silver solution should be reduced by one-half and citric acid shoud be added to the silver-developer to obtain pH of 3.8 to 5. The end point of the reaction is indicated by a gray-green suspension, and tissue is placed in a fresh silver-developer for as often as is prudent to reach full development of sulfide-containing profiles (Fig. 20-3A).

Methodology

Uptake of Cobalt Chloride and Its Conversion to Cobalt Sulfide

Neurons within the thoracic ganglia and brains have been filled successfully with cobalt chloride, cobalt nitrate, cobalt acetate and cobalt bromide through cut peripheral axons, interganglionic connectives, or by electrode injection (see Chap. 18). The techniques have been used on the following species: *Locusta migratoria, Schistocera gregaria, Gryllus campestris, Apis mellifica, Drosophila melanogaster, Eristalis tenax, Calliphora erythrocephala, C. vomitoria, Musca domestica,* and *Coenagrion puella.* Cobalt ions are subsequently converted to cobalt sulfide by immersing tissue in 0.2% ammonium sulfide in distilled water or in Ringer's solution. Solutions are adjusted to an appropriate molarity by the addition of sucrose. A 2% solution of saturated sodium sulfide can also be usefully employed. Sulfide solutions are most effective within a pH range of 8.0 to 10.5 adjusted by the addition of acetic acid, or in PIPES buffer (see Chap. 9, Appendix II).

Cobalt chloride is introduced into the nervous system of large insects, such as Orthoptera, via their sensory nerves or split ventral nerve cord connectives that are isolated from other tissue by a Vaseline cup (nontoxic grade Vaseline: Vaselinum Album. DAB 7. E. Scheurich Pharmwerk GmbH, Appenweier, G. F. R.). The usual cobalt chloride solution is between 1% and 6% in an insect Ringer's solution, KCl, or in distilled water. Uptake periods of 3 to 8 h, at room temperature, through cut nerve bundles from a cobalt pool were sufficient to fill locust neurons 3 mm distant from the source of $CoCl_2$. However, in species of *Aeschna,* Co^{2+} migration is faster (5 mm in 1 h, at the same initial concentration of 6% in distilled water) and in the cricket, *G. campestris,* Co^{2+} migration is slower (5–12 mm/h). Odonatous neurons are usually filled after diffusion times of between 8 and 24 h at 4°C and between 2 and 6 h at 20°C (distances in excess of 2 cm).

Parts of interneurons from or to the brains of dipterous insects can be filled retrogradely from thoracic ganglia severed from their peripheral connections. A small lesion is made in the ventral or dorsal surface and the fused ganglia, connected to the brain by the ventral nerve cord, are supported on a sliver of glass, parafilm, or polythene. The ganglia are isolated from other tissue by a Vaseline cup. Four percent $CoCl_2$, in Case Ringer's solution (Case, 1954) or in distilled water, is allowed to diffuse into the ganglion for 3 to 12 h at 20°C or 4°C, respectively (Strausfeld and Obermayer, 1976). Peripheral nerves of *Drosophila* and *Calliphora* are filled by immersion in a cobalt pool contained in a glass pipette (see Chap. 19). Cobalt is introduced to the site of uptake by a microsyringe. Sensillae, such as hair plates and bristles, can be filled by coating the cuticle with nail varnish, shaving off the protruding hairs, and isolating their stumps in a cobalt pool surrounded by Vaseline.

Only results from living animals have been used to evaluate the techniques. The effects of cell death during diffusion on the intensification procedure are described later.

After diffusion, the tissue is immersed in an alkaline Ringer's solution or in 2% sucrose containing between 0.2% and 2% ammonium or sodium sulfide, for 2 to 10 min. Tissue is then rinsed in water or Ringer's solution and subsequently fixed either in Carnoy, alcoholic Bouin, or AAF (Lillie, 1965) or paraformaldehyde fixative (phosphate-buffered, pH 7.0). After 2 h fixation, the specimen is rinsed in 70% alcohol or in water and then glued to a petri dish with rapid-polymerizing two-component dental glue [Scutan (R): Epimine Plastic, ESPE GmbH, Seefeld, G. F. R.]. Brains and ganglia are dissected free of connective tissue and fat bodies before they are intensified. *Drosophila,* early instar Orthoptera, and other small insects can be treated, intact, with one of the intensification procedures (methods C, D, and E are recommended).

Intensification Procedures: Original Schedules

Brains or ganglia are brought to water through descending alcohols and are then treated by one of the following procedures.

Schedule A: For Low pH, Weak Organic Acid Intensification
(pH 2.8–3.4) (Fig. 20-5B)

   1. Immerse tissue overnight in a stock colloid solution consisting of 20% gum arabic, 10 g sucrose, and 4–10 ml lactic acid (96% Merck, analytic grade).
   2. Immerse for at least 1 h, and not more than 3 h, at 20°C, in stock colloid containing 0.2 g hydroquinone per 100 ml.
   3. Decant reducer and replace with 9 vol fresh colloid/reducer to which has been added 1 vol 1% $AgNO_3$. Brains are incubated at 40°C to 60°C until the tissue takes on a pale green or brown appearance. Place in a fresh solution and incubate until the tissue is tobacco brown or dark green. This should be intermittently and briefly controlled under a binocular equipped with dark-field illumination.
   4. After development (60–120 min), tissue is placed in 10 ml stock colloid containing 2 drops of concentrated acetic acid, for 5 min (stop bath).
   5. Fix tissue in 5% sodium thiosulfate for 10 to 20 min.
   6. Wash in distilled water, dehydrate, and embed in Araldite.

   All intensification procedures are carried out in red light (Kodak light red safety filter, series 1A), on a hot plate, or in an oven excluded from light.

Schedule B: For Low pH, Strong Organic Acid (pH 2.7–3.0)
(Figure 20-4B)

   1. Preincubation: Tissue is immersed overnight at 20°C in stock colloid solution consisting of 20 g gum arabic and 10 g sucrose in 100 ml distilled water.

2. Development
  2.1. Preincubate (3 h at 40°C) in stock colloid containing 0.15 g hydroquinone per 100 ml and between 0.5 and 0.8 g citric acid to bring to the required acidity (pH 3.0–3.2).
  2.2. Next, place tissue in stock acidic colloid-hydroquinone (9 vol) to which is added 1 vol 0.1% $AgNO_3$ (40 min at 50°C).
  2.3. As above, increasing $AgNO_3$ concentration to 0.2% (40 min).
  2.4. As above, increasing $AgNO_3$ concentration to 1%. Place in a fresh solution each half hour until the tissue is tobacco brown.
3. Wash in 1% citric acid for 10 min.
4. Dehydrate as in schedule A.

Fixation in thiosulfate is optional. After dehydration, tissue is rinsed in propylene oxide (twice for 10 min) and then left overnight in a 1:1 mixture of propylene oxide and "soft" Araldite. After most of the propylene oxide has evaporated, brains are blotted and immersed in fresh Araldite. They are oriented in appropriate casts (e.g., inverted BEEM capsules) and are polymerized at 60°C to 90°C. Serial Araldite sections can be conveniently cut on a Reichardt sliding microtome with normal steel "C" knives. Sections may be wet mounted in araldite under coverslips, sealed with nail varnish, or mounted in Permount.

All reagents except the gum arabic were analytic grade. Pure Merck gum arabic has been found to be inferior to commercial gum arabic, possibly because the former contains trace contaminents from the milling procedure. These may interfere with the physical development. Gum arabic is usually hydrated for two days prior to the experiment. A crystal of thymol prevents fungal growth.

The preceding schedules give minimum incubation and development times for dipterous material but are optimal for Orthoptera and Odonata. The use of lactic acid on tissue fixed in Carnoy results in less background coloration, possibly because the intensification reactions proceed at a slower rate and are localized only at the initial sites of induction. This method has proved advantageous for resolving descending neurons in dipterous protocerebra as well as some contiguous axons that have taken up the cobalt from descending neuron dendrites. However, low concentrations of cobalt in trans-synaptically filled small neurons will be resolved by direct illumination only after extended incubations in sechedule A, step 4, thus risking the development of a silver mirror at later stages of the procedure (see section "General Considerations"). Trace amounts of silver-plated cobalt sulfide in secondarily filled neurons can be visualized by dark-field illumination (see Fig. 20-1A). Schedule A may be combined with B. Stage 1 (A) is omitted: Preincubation at 60°C rather than at 50°C at stage 2.2 A precedes stage 2.4 (B). Tissue is removed when the solution becomes a dirty olive green.

Method B is good for dipterous and orthopterous material fixed either in

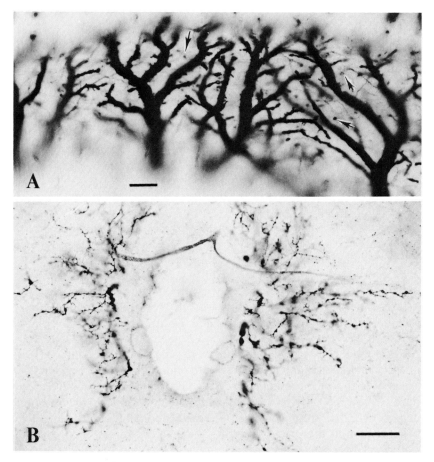

Fig. 20-4. A Dendrites of horizontal motion-sensitive neurons (Hausen, 1976) of the lobula plate of *Calliphora,* intensified by method D.   Prior to stage 2 of this technique, tissue was preincubated in $^1/_{10}$ strength hydroquinone and $^1/_{10}$ strength silver nitrate in the acidic colloid for 1 h.   This sensitizes elements with very low sulfide content and has here revealed slender mimetic processes with horizontal cell dendrites, as well as the columnar mappings of retinotopic pathways (*arrowed*).   Scale 5 μm.   **B** A pair of tritocerebral dendritic trees revealed after injection of cobalt bromide into the neuropil and intensified by method B.

monoaldehydes, Carnoy, or Bouin and reveals neurons that have primarily incorporated cobalt as well as contiguous arrays of trans-synaptically filled elements that have taken up cobalt from other neurons.   The latter appear paler in color compared with initially filled elements.

   Background staining is minimal in Bouin-fixed material using schedules A or B.   Aldehyde-fixed tissue is usually darker and elements containing a low concentration of Co^{2+} are difficult to discern against the background.   It is essential that tissue is washed for several hours to remove

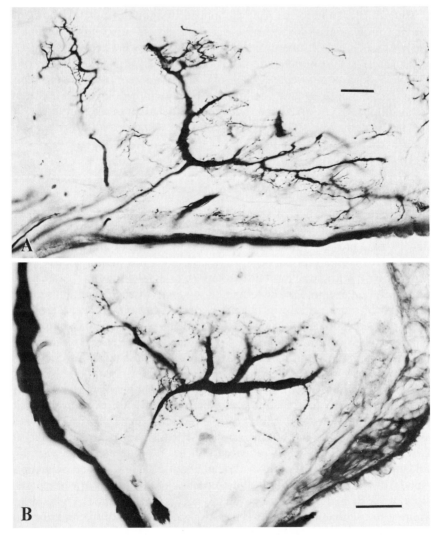

Fig. 20-5. **A** Descending neurons of the lateral deuterocerebrum of *Schistocerca migratoria* (method C). **B** Giant descending neuron of the subesophageal ganglion of *Musca domestica*, back-filled by injection from a glass micropipette embedded in the metathoracic ganglion (method A). Note that in both preparations there is a late silver mirror at the tissue surface that would totally obscure whole-mount viewing.

excess aldehyde after fixation since these groups normally act as reducing agents and can contribute to a background silver reduction. Dialdehydes should not be used as fixatives before intensification.

Method B is superior to A when applied to odonatous material. Such findings indicate that different species show various susceptibilities to the basic procedures. If one or another procedure is found to give poor results, then the first modification of it should be an adjustment of the protective colloid, either by a decrease of concentration or by its replacement with, for example, gelatin, celloidin, or polyethylene glycol (Singh and Strausfeld, 1980). Differences between the behavior of the method in different tissue may be related to their colloidal properties after fixation and their permeability to reactants.

Intensification: Simplified Techniques

Fixation and the choice of species demand manipulation of the method, including timing, adjustment of the pH, and concentration of the reducer. One problem met by the authors is that some neuropil promotes a faster silver reduction than others even though it has undergone identical pretreatment with cobalt, sulfide, fixation, and dehydration. In Orthoptera, the brain promotes slow reduction, whereas intensification of profiles takes place more rapidly in the pro-, meso-, and metathoracic ganglia, with the subesophageal ganglion the first to reach full development. However, the reverse occurs in *Drosophila,* a species whose cerebral neuropil promotes rapid silver reduction unless it is specially protected by agarose infiltration (see section, "General Considerations").

The basic schedules A and B have been much simplified. The intensification procedure of Strausfeld and Obermayer (1976) uses an alcoholic colloid that proves useful insofar as Ag^+ reduction generally proceeds at a slower rate and the method can be applied to many species, including Crustacea (Wine, 1977). A recent modification of the method by Strausfeld and Hausen (1977) employs a preincubation stage followed by incubation in 30% or 40% alcoholic colloid, containing relatively high concentrations of both hydroquinone and citric acid. This method is altogether simpler to use than either A or B because the endpoint of the reaction can be determined by the color of the solution rather than by examing the tissue itself.

Method C (Strausfeld and Obermayer, 1976) (Fig. 20-1B)

1. Tissue is hydrated to 30% alcohol and incubated for 1 h in 25% gum arabic dissolved in 30% alcohol containing 1.6 g citric acid and 1 g hydroquinone/100 ml (at 40°C).

2. Tissue is transferred to the same solution, containing 0.5 g silver ni-

trate/100 ml, and is incubated at 40–50°C until tobacco brown. If the solution darkens before the tissue, it should be immediately replaced by a fresh one.

3. The reaction is stopped in 1% acetic acid in 30% alcohol, and then washed in 30% alcohol, dehydrated, and embedded in Araldite.

Method D (Strausfeld and Hausen, 1977) (Fig. 20-2B, 20-4A)

1. Tissue is incubated in a solution consisting of 10 vol 30% or 40% gum arabic in 30% alcohol to which has been added 1 vol 1% hydroquinone and 2 vol 10% citric acid, both dissolved in 30% alcohol. Three volumes of 30% alcohol are finally added to adjust the colloid concentration to that of stage 2. Incubation is at 50°C for 30 min.

2. Tissue is transferred to a solution containing 10 vol 30% or 40% gum arabic, 1 vol 10% hydroquinone, 1–3 vol 10% citric acid, and 1 vol 1% silver nitrate. All solutions are made up in 30% alcohol. Incubation is at 40–60°C and proceeds until the solution becomes khaki green. This takes 1–3 h, and is extended for a further 0.5–2 h after this endpoint. Tissue may also be incubated for 1 h with $1/10$ silver nitrate concentration prior to stage 2.

3. Development is stopped as in method C, and the tissue is embedded in Araldite. No advantage is gained by increasing silver concentrations: Its effect is merely to enhance background coloration at the expense of silver intensification of sulfide-containing profiles. Increasing the concentration of the reducer must be counteracted by a corresponding increase of citric acid within the pH range 2.6–3.0 to avoid granular deposits of silver as well as artifactual encrustation of fibers. Increasing the concentration of the developer alone accelerates the reaction rate and shortens the induction period. Preincubation in colloid with silver, as opposed to colloid with developer, results in the intensification of trachea at the expense of cobalt sulfide.

Bleaching

Various attempts have been made to bleach out excessive background, in particular when material was fixed in buffered paraformaldehyde or glutaraldehyde. Bleaching in dilute H_2O_2 should be avoided because it is almost impossible to control and there is the risk of unevenly bleaching intensified profiles. Bleaching with copper sulfate and potassium bromide was suggested by Liesegang (1928), but in our experience it is difficult to control and causes lateral diffusion of the silver. Liesegang suggested differentiation in a 2% solution of iron chloride. This can be employed on brains that appear excessively dark after intensification and that are required for whole-mount viewing. Brains are immersed in chloride after fixation and after rinsing in water for 10–20 min at 40°C. After differentiation, tissue is rinsed in distilled water and then fixed a second time in sodium thiosulfate. At least in principle, it is possible to suppress argyrophilia of the tissue prior to intensification, thereby making it less susceptible to the formation of extraneous silver nuclei. This pretreatment imitates methods that demonstrate the presence of degenerating neurons by reduced silver. In part, successful staining of degenerated elements is reputed to rely on the suppression of argyrophillic elements within nondegenerated cells by treatment with potassium permanganate followed by oxalic acid and hydroquinone. This treatment will also suppress the background of block-intensified brains but must be used with extreme caution. The schedule is as follows:

1. Bring tissue into water.
2. Wash in 0.01% potassium permanganate for 5–30 min (20°C).
3. Wash in distilled water for 30 min.
4. Bleach in 0.5% oxalic acid, 0.5% hydroquinone (3–10 min).
5. Wash in several changes of distilled water over 30 min. Continue as for schedule A, B, C, or D.

Stages 2 and 4 are critical. Slight overexposure to permanganate will cause extensive lateral spread of cobalt sulfide out of neurons: It might be speculated that the initial site of Co^{2+} is its incorporation by assembled neurofibrillar proteins and that spreading is due to their disassembly.

Pitman (1979) recently published an effective way of removing silver deposits that obscure at the surface of whole ganglia. This method employs a modified ''Farmer's'' solution to partly destain after block intensification.

Usually, careful control of method B and stopping the reaction at the appropriate intensity will resolve intensified elements against a pale brown background that can later be viewed in whole-mount preparations. An even simpler modification of Timm's method was recently devised by Bacon and Altman, which is optimal for whole-mount Orthoptera neuropil. For *Drosophila* the colloid concentration may be increased to 5% and citric acid decreased to 0.75%. In our experience it is useful to precede stage 3 with a weak incubation in 0.5 ml 1% $AgNO_3$ (1 vol) in 10 vol developer for 30 min.

Method E (Bacon and Altman, 1977) (Fig. 20-6A)

1. Tissue is warmed to 60°C in distilled water for 5 min and then transferred to a developer stock solution containing 3 g gum arabic, 0.8 g citric acid, 0.17 g hydroquinone, 10 g sucrose, dissolved in 100 ml distilled water.

2. Tissue is soaked in the stock developer at 60°C for 1 h and is then transferred to 10 parts developer and 1 part 1% silver nitrate.

3. Development is in the dark, at 60°C. The tissue is transferred to a fresh solution every 30 min or as soon as a silver mirror begins to appear on the surface of the tissue or the solution.

4. After 1–2 h, tissue is transferred to warm distilled water, cooled, dehydrated, cleared in methylbenzoate and whole-mounted in Canada balsam. The method is particularly useful for Orthoptera and intact *Drosophila,* and for neuropil in opened, but otherwise intact, heads of large Diptera. Material is embedded in Araldite and sectioned at 10–50 μm.

Finally, and by way of contrast, a method is described that is used to resolve both CoS profiles and neuropil background in order to relate single neurons to brain regions and tracts. Initially filled trans-synaptically filled elements are resolved as black profiles against a brown or gray background, depending on the acidity of the developer.

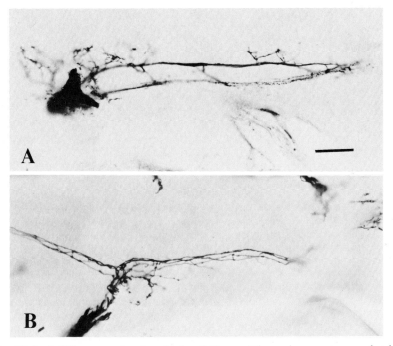

Fig. 20-6. **A** Wing receptor terminals of *Drosophila melanogaster* resolved by method E. Cobalt injection into the wing base invariably resolves this pattern in the wild-type animal. **B** By comparison, single sensory fibers stochastically resolved by the local application of silver nitrate after Golgi chromation of the intact animal.

Method F (Strausfeld and Hausen, 1977) (Fig. 20-3B)

1. Tissue is preincubated in 5% gum arabic (9 vol), 1.4% hydroquinone (1 vol), 10% citric acid (1 vol) for 1 h.

2. Transfer to the above solution containing 1 vol 0.2% silver nitrate; incubate for 30 min.

3. Transfer to a 10% gum arabic solution mixed to the same proportions of developer, acid, and silver nitrate (30 min).

4. Transfer to a 20% gum arabic solution (10 vol) with 2.0% hydroquinone (1 vol), 16% citric acid (1 vol), and 0.6% silver nitrate (1 vol). After

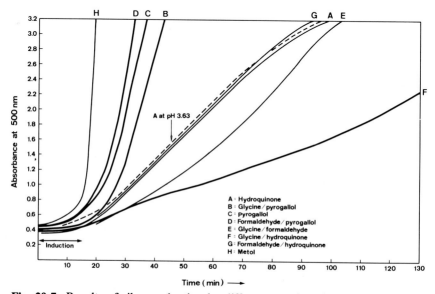

Fig. 20-7. Results of silver reduction by different species of developers listed in Table 20-1. Glycine-hydroquinone developers are the slowest species for silver intensification and are useful to lay down small quantities of metallic silver with the sulfide catalyst, and can be used as the first stage of a method that should show up neuropil background.   Formalin-hydroquinone or hydroquinone alone is, however, only slightly faster than the combined gylcine-formalin developer. Admixtures of formaldehyde with another developer accelerate the reaction, admixtures of glycine with another developer decelerate the reaction compared with the reaction rate of the second developer alone (with the exception of pyrogallol-glycine). Note also that a slight shift in pH toward the alkaline markedly reduces the induction period (developer A: hydroquinone at pH 3.59 and 3.63). Commercial developers such as Metol, as well as pyrogallol, are wholly unsuitable for intensification procedures in intact tissue but can be used for sectioned material (Table 20-2). Incubation of developer-silver admixtures was in red light at 60°C. The extent of a silver reduction in the metallic-colloid suspension was measured every 3 min as a function of absorbance, at 500 nm, on a Beckman double beam-type M VI spectrophotometer.

30 min transfer to a fresh solution containing 30% gum arabic (10 vol), 5% hydroquinone (1 vol), 10% citric acid (1 vol), and 1% silver nitrate (1 vol).

5. After 30 min place in warm distilled water, cool in the dark, and dehydrate. Embed in Araldite and section at between 20 and 40 μm.

The timing of stages 4 and 5 is critical. Prolonged incubation results in a dark brown background. For some species it is useful to increase the silver concentration of stage 3 to 0.4% (1 vol) and to decrease the period of incubation of stages 4 and 5. Incubation is carried out at 60°C, in the dark.

The Reducer

Many developers have been tried as substitutes for hydroquinone, initially in dummy experiments and in order to discover possible ways of controlling the rate of Timm's reaction. The initial experiments were designed as follows: 9-ml aliquots of 30% gum arabic, dissolved in 30% alcohol, were made up with 2 vol 10% citric acid. Just before use, 2 vol of a developing agent were added, and the whole solution was then mixed with 1 ml of 1% $AgNO_3$. The rate of autocatalytic reduction of silver nitrate to metallic silver is measured as a function of time and opacity (silver

Table 20-1. Different Species of Developers[a]

Stock colloid	
30% Gum arabic in 30% ethanol, 9 vol	pH 3.67
10% Citric acid in 30% ethanol, 2 vol	

Developers (added to 9 vol stock colloid with 1 vol silver nitrate)	Final pH
A. 1% Hydroquinone in 30% ethanol: 1 vol 30% Ethanol: 1 vol	3.59
B. 1% Glycine in 30% ethanol: 1 vol 1% Pyrogallol in 30% ethanol: 1 vol	3.62
C. 1% Pyrogallol in 30% ethanol: 1 vol 30% Ethanol: 1 vol	3.57
D. 1% Pyrogallol in 30% ethanol: 1 vol 5% Formaldehyde in 30% ethanol: 1 vol	3.59
E. 1% Glycine in 30% ethanol: 1 vol 5% Formaldehyde in 30% ethanol: 1 vol	3.63
F. 1% Glycine in 30% ethanol: 1 vol 1% Hydroquinone in 30% ethanol: 1 vol	3.62
G. 1% Hydroquinone in 30% ethanol: 1 vol 5% Formaldehyde in 30% ethanol: 1 vol	3.58
H. 1% Metol in 30% ethanol: 1 vol 30% Ethanol: 1 vol	3.58

[a] Refer also to Fig. 20-7.

density) in a densitometer. Rapid commercial developers containing pyrogallol (or pyrogallol alone), p-methylaminophenol (Metol), or diaminophenol result in practically no induction period before spontaneous and rapid reduction to silver and are thus wholly unsuitable for block intensification. On the other hand, addition of sodium sulfite (0.5 g/10 ml) to the developer will decelerate the reduction but results in neurofibrillar background staining near the surface of the tissue. This inhibits intensification in depth. The classic slow developers such as glycine and formalin can usefully replace hydroquinone, but at concentrations that are at least twofold. Addition of pyridine or its derivatives with hydroquinone, glycine, or formalin, markedly decreases the induction period and accelerates spontaneous reduction to silver to cause an early silver mirror at the tissue surface. The reaction rates of some useful reducer solutions are shown in Fig. 20-7 (Table 20-1).

pH and Restainers

As in photography, slow developers used at high pHs tend to produce "fogging" or, in neural tissue, background coloration. Alkaline pHs also accelerate reduction to silver (James, 1939c) and should thus be avoided. Likewise, the stronger acids, such as boric acid, favored by Liesegang, markedly reduce the induction period. A 0.1 M solution of boric acid has a pH of 5.3.

Commonly, citric acid is used to adjust the pH of the developer, and in most methods the optimal pH range is 2.8–3.8. The exception is method D, in which large concentrations of hydroquinone and citric acid are used in the pH range 2.6–3.4. Below pH 2.6 intensification does not occur unless the silver concentration is raised to 0.4% (total percentage) of the developing solution. This, however, gives rise to "coarse grain" development. For all methods, the lower the pH, within the working range, the slower is the development. Rapid intensification can be stopped by addition of acetic acid. The tissue is then washed in distilled water and placed in a fresh developer solution.

Restraining agents added to the physical developer should, if possible, be avoided, although some tissues may demand deceleration of the autocatalytic reaction. For each 100 ml of the developer 0.01 g of potassium bromide may be added after addition of silver nitrate. Also, 0.005 g of 5-nitrobensimidazol or 1 H-benzotriazol/100 ml stock colloid can be added. Both are relatively insoluble at temperatures below 30°C. Both should be added to the warm stock solution before adding silver nitrate.

Colloids

Gum arabic (acacia) is usually chosen as the protective colloid whose diffusion into the tissue is supposedly slow. Other naturally occurring colloids are gelatin, egg white, and starch, which could be considered as substitutes for gum arabic. Egg white or casein is used in the proteinaceous silver Bodian procedures. These substances have been tested but were found (with the exception of gelatin) to be unsuitable for whole tissue intensification. Gelatin (6%) can be used, but the rate of intensification is significantly slower and the contrast between profiles and background is poor. Possibly silver nitrate and hydroquinone are strongly bonded to hydrated gelatin and, as a consequence, intensified cells invariably appear pale. However, this may be advantageous for future development of combined intensification and electron microscopy. Commercial colloids, such as Dextran (mol. wt. 32,000–48,000 and 60,000–90,000), are not recommended as substitutes.

Gum xanthan, gum karaya, and gum guar (Sigma chemicals) are useful substitutes for gum arabic and each promotes a characteristic background coloration of neuropil (pink, brown, or gray) (Table 20-2). Purified agarose-infiltrated tissue has been intensified in developer that is carried in low colloid concentrations and has proved useful for the intensification of profiles in small pieces of tissue such as brains of *Drosophila*. Polyethylene glycol (mol. wt. 20,000) is an excellent protective colloid (to replace gum arabic in method D) but requires incubation for up to 4 h. The solution should be changed twice. Tissue intensified in developer with polyethylene glycol characteristically shows a completely transparent background (Fig. 20-8, A–C).

Tungstosilicic Acid Intensification (Gallyas, 1971)

Székély and Gallyas (1975) recently developed a novel procedure for intensifying cobalt sulfide profiles in frozen sections of the frog spinal cord. This method has produced stunningly clear preparations in their hands and may find an application in insect material. It is particularly attractive because it is used after fixation in glutaraldehyde.

The schedule, taken from their account, is as follows.

Stock solutions:
  A. 10% tungstosilicic acid
     0.2% silver nitrate
     0.2% ammonium nitrate   in 100 ml distilled water
     0.063% formalin
  B. 5% sodium carbonate

Tissue is washed after fixation in distilled water and then incubated in 2.0% sodium hydroxide for 1 h. Tissue is next washed in three changes of 0.5% sodium acetate after which it is placed in 0.1% copper sulfate solution for 10 min. Tissue is again washed 3 to 4 times in sodium acetate (0.1%) before it is placed in the developer. This is prepared by adding equal amounts of A, drop by drop, to B. Intensification begins after an

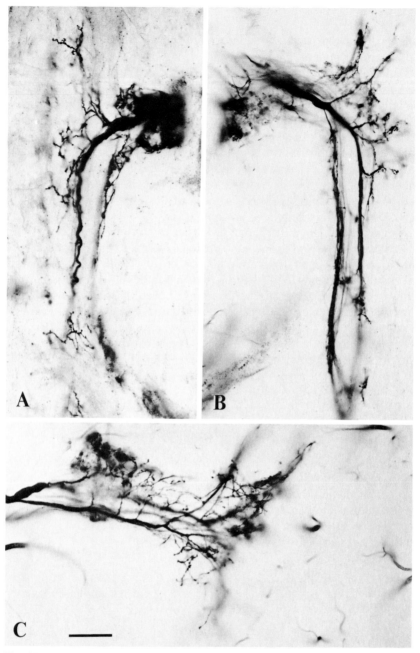

Fig. 20-8, A–C. Comparisons between different colloids. This figure illustrates wing sensory fibers to the mesothoracic and metathoracic ganglia of *Calliphora*, intensified by method D in the presence of gum karaya (**A**), gum xanthan (**B**), and polyethylene glycol (**C**)—PEG: mol. wt. 20,000 (40% in 30% alcohol). Intensification in PEG requires at least two successive incubations at stage 2, method D, each for 2–3 h.

Table 20-2. Colloids and Developers that can be Routinely Used for
Intensification Procedures

Colloids	Developers
Gum arabic	Hydroquinone
Gum Guar	Hydroquinone, formalin
Gum xanthan	Formalin-glycine,[b] hydroquinone
Gum karaya	Formalin, hydroquinone
Gelatin	Pyrogallol,[a] hydroquinone
Agarose	Pyrogallol-formalin,[a,b] hydroquinone
Polyethylene glycol	Pyrogallol-glycine,[a,b] hydroquinone

[a] All pyrogallol developers are rapid and are more suitable for sectioned material.
[b] Compound developers.

induction period of 10 min at room temperature, and on frog frozen sections, is completed after 20 min. Development is stopped by adding 1% acetic acid. The authors exert some control over the speed of development by the concentration of formalin (their developer) and by temperature.

The great advantage of the method is its use at room temperature rather than at 50–60°C. Possibly, for block intensification, the additions of restraining agents or combined glycine-formalin developers may satisfactorily prolong the reaction time.

General Considerations

Like most histologic procedures, the silver intensification methods are capricious, whether used on sections or on intact ganglia. The success or failure of block intensification is, in part, dependent on avoiding catalysis of silver reduction at the brain surface during the early part of the physical development. Success is entirely dependent on both components, silver nitrate and hydroquinone, penetrating the brain in an acid medium during an induction period before the onset of silver reduction. If reduction of silver takes place before the tissue is permeated with the physical developer, then intensification will invariably be omitted from the depth of the tissue. One advantage of the present techniques is that, by treating intact tissue, the forms of nerve cells and their dispositions can be reliably compared to results derived from other block techniques, such as the Golgi methods or block reduced-silver impregnations. Also, neural tissue need not be wholly removed from the specimen, and in small insects such as *Drosophila* intact animals may be block intensified if an appropriate opening is made in the body and head cuticle. The greater the amount of surrounding tissue, such as muscle, the lower is the concentration of colloid needed for the reaction.

An important reason why the technique sometimes fails is the presence of catalytic sites other than cobalt sulfide profiles. One that has already been mentioned is the presence of aldehydes that will enhance background intensification; two others are excess sulfides or sulfhydrals within the neuropil. The simple expedient of thoroughly washing the tissue after treatment with sulfides and after fixation largely overcomes such drawbacks. The disadvantage of extraneous catalytic sites cannot be overemphasized, since these will invariably catalyze silver reduction at the expense of filled neurons. Thus, it follows that the initial procedure of filling must be meticulously executed in order to avoid leakage of cobalt salts into the body cavity. Likewise it is essential to carry out the subsequent stages on animals that have clearly survived the initial filling operation.

Three minor, but useful, tips for achieving uncontaminated intensification are as follows: During incubation tissue is handled with thin, blunt, glass rods and is transported to a new solution on glass loops. These, like plastic petri dishes used for intensification, should be used once only. It is also recommended (Hausen, personal communication) that the tissue is supported on filter paper during intensification. This insures proper circulation of the developer and achieves even staining.

Death of the animal during the filling period will cause clumped aggregates of CoS in what are probably degenerating fibers. Death may also result in cobalt leakage into the surrounding extracellular space. Clumped CoS deposits will result in patchy intensification and patchy background coloration presumably because the discontinuous and sometimes large CoS aggregates preferentially trigger silver reduction so that autocatalysis proceeds at a faster rate at a "large grain" than elsewhere. It is therefore not surprising that the more evenly distributed the cobalt sulfide deposits, the greater the chance of success of subsequent intensification.

Comparisons between block and sectioned material of dipterous brains show that after block treatment cobalt can be detected in contiguous assemblages of at least four interneurons. However, after the same filling conditions with $CoCl_2$ (Strausfeld and Obermayer, 1976), but using intensification on paraffin sections, cobalt can, in our hands, only be intensified between primarily and secondarily filled neurons if secondarily filled neurons are of large diameter.

Observations that paraffin treatment may displace intensifiable deposits (Brunk and Sköld, 1967) agree with our own observation that cobalt sulfide shifts in densely impregnated material. However, it is doubtful that block staining holds a singular advantage over section staining when a single neuron has been filled with cobalt from an electrode (Kien, 1976). In such cases the observer can ill afford to lose the single cell, and the somewhat less sensitive intensification of paraffin sections is preferred, at least until the block procedure has been mastered for that species. Once this has been achieved, the advantages of the latter become apparent. These are (1) resolution in depth; (2) lack of tissue distortion; (3) extreme clarity of background, which permits the observer to relate parts of the neuron's domain to regions of neuropil by means of Normaski interference contrast.

Last, it should be pointed out that some experimental conditions preclude simple block intensification. These conditions are rare and up to the time of writing have occurred only in the case of *Drosophila*. It was mentioned earlier that in this species the neuropil of the brain intensifies at a faster rate than that of the thoracic ganglia. Thus in order to resolve the fine details of neurons that invade both the brain and the ventral ganglia, it is necessary to intensify on sectioned material avoiding simple paraffin embedding. One method is to embed in celloidin and intensify the partly hydrated celloidin section (using method C). However, the technique is tiresome, and thin celloidin sections (15–20 μm) are difficult to obtain. The present tactic is to preincubate intact *Drosophila* as in method E, using a 5% colloid solution and then to intensify as in method E for 30 min, using 5% colloid developer with 0.5 vol of 1% silver nitrate. The specimen is then rinsed in warm distilled water and infiltrated with 2% agarose. After orienting the specimen, the cooled agarose block is infiltrated with paraffin wax and serial sections are cut at between 10 and 20 μm. Sections are dewaxed and intensified as in method E, including the preincubation stage. After 30 min to 1 h, the sections are rinsed in warm distilled water, fixed for 10 min in a 5% sodium thiosulfate solution, washed, dehydrated, and mounted under Permount. This method allows the resolution of neurons with a clarity that is as good as any of the block procedures, and it is presently employed for the resolution of neuronal assemblies in *Drosophila,* after cobalt injection into the brain.

Acknowledgments

We thank Dr. J. Altman, University of Manchester, Dr. Carol Mason, the Bardeen Medical Laboratories, Madison, and Drs. J. Bacon and L. Williams, Max-Planck-Institut für Verhaltensphysiologie, Seewiesen, for correspondence concerning some of the vagaries of the cobalt procedures. We are especially grateful to Dr. G. Székély, Department of Anatomy, University of Debrecen, Hungary, for correspondence concerning features of intensification on vertebrate material and to Dr. N. R. Singh, Tata Institute, Bombay, for advice concerning the use of natural gums as colloids. Dr. K. Hausen, Max-Planck-Institut für biologische Kybernetik, Tübingen, is likewise thanked for the many discussions about refinements of techniques.

Chapter 21

Intensification of Cobalt-Filled Neurons in Sections (Light and Electron Microscopy)

N. M. Tyrer
M. K. Shaw

The University of Manchester
Institute of Science and Technology
Manchester, England

J. S. Altman

The University of Manchester
Manchester, England

Cobalt-filled neurons in whole mounts of cleared ganglia are dramatic and extremely useful for giving a total picture of the extent and complexity of a neuron (Chap. 19). The distribution of branches in three dimensions is, however, difficult to determine, and one can get virtually no information about the relationship of the filled neuron to other known features in the neuropil. Ganglia must be sectioned to obtain a complete picture of a neuron and details of its connections. Our first attempts to examine cobalt-filled neurons in 10-μm wax sections were very disappointing: The cobalt sulfide deposits were pale brown and very difficult to see, especially if any background stain was used. Small profiles were impossible to identify (Fig. 21-1A). The early electron micrographs of cobalt material were also unrewarding, for only profiles containing large concentrations of cobalt sulfide could be recognized with certainty (Pitman *et al.*, 1973; Mason and Nishioka, 1974). It was clear that some method of amplification was needed if cobalt sulfide were to become a really useful tool for the analysis of connections between neurons.

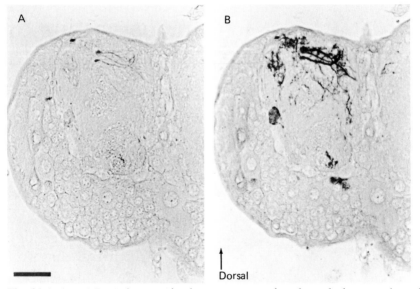

Fig. 21-1, A and B. A An unstained transverse section through the mesothoracic ganglion of the locust, *Chortoicetes terminifera,* viewed with the condenser slightly defocused to show up the background.   Branches of a group of four motor neurons that were filled with cobalt are just visible, mostly in the dorsal neuropil. **B** The same section after intensification by the Timm's sulfide-silver method.   The filled profiles have become intensely black with precipitated silver and numerous fine branches that were not previously apparent are now visible.   The background is stained a pale brown.   Scale = 100 μm.

We are indebted to Professor R. Menzel (Berlin) for the inquiries he made that resulted in a reference from Professor W. Vogel (Konstanz) to the sulfide-silver method of Timm (1958b).   This was first introduced as a light-microscopic method for the detection of indigenous heavy metals, such as zinc in pancreatic islet cells, and subsequently modified for use in electron microscopy (Pihl and Falkmer, 1967).   Insoluble heavy metal sulfides resulting from treatment of the tissue with ammonium or hydrogen sulfide, reduce silver nitrate (applied as a physical developer solution) to silver, which is deposited on the heavy metal nuclei.   The reaction is catalyzed by hydroquinone in the developer.

A colloidal developer is used to control the rate of silver deposition, which prevents the normal silvering of background tissue.   With small modifications, this method is excellent for intensifying the cobalt sulfide precipitate in sections of filled neurons (Fig. 21-1B; Tyrer and Bell, 1974).   The method is extremely sensitive and, since the cobalt concentration is much greater than that of the indigenous heavy metals, cobalt can be intensified selectively in wax and thin plastic sections (Tyrer and Bell, 1974), in thick frozen sections (Iles, 1976b; Mason and Lincoln,

1976), and in whole ganglia (Strausfeld and Obermayer, 1976; Bacon and Altman, 1977; see also Chap. 20).

Other methods for identifying cobalt in sections have been developed. Cobalt may be precipitated with 3,3′-diaminobenzidine (DAB) to form an osmiphilic polymer that can be detected in the electron microscope (Gillette and Pomeranz, 1973, 1975; Atwood and Pomeranz, 1974, 1977). Kirkham *et al.* (1975) used X-ray dispersive analysis with the electron microscope to identify cobalt sulfide deposits from the X-ray energy spectrum. Neither of these methods is well suited to light microscopy and the advantage of Timm's method is that it allows detailed examination of the same neuron with both light and electron microscopy.

Experience with intensification methods over the past few years has led to several improvements that have made the original method more reliable. We now have a much better understanding of the pitfalls and precautions to be taken, both in application and for interpretation. We use the light-microscopic techniques routinely (Tyrer and Altman, 1974; Altman and Tyrer, 1977a; Tyrer *et al.*, 1979; Altman and Kien, 1979). Methods for electron microscopy are still not entirely satisfactory, but recent innovations, such as block intensification (Chap. 20), indicate that useful ultrastructural information can be obtained.

Timm's Sulfide-Silver Intensification Method for Cobalt

The basic method for sectioned material is given in Tyrer and Bell (1974).

Stock Solution

>   30.00 g powdered gum acacia (or gum arabic)
>   0.43 g citric acid
>   0.17 g hydroquinone
>   10.00 g sucrose
>   100.00 ml distilled water

1. Heat the distilled water to 60°C, add citric acid, hydroquinone, and sucrose, and stir until dissolved.
2. Slowly add the gum arabic, stirring continually until dissolved (30–45 min).
3. Cool to room temperature and adjust pH to 3.9–4.0.

This solution should be 2–3 days old before use and may be kept indefinitely in the refrigerator, but silver precipitation is more rapid in older solutions, presumably because of changes in the properties of the colloid. We routinely make the stock solution on a Friday and use it from Monday to Friday of the following week.

Development

The stock solution is warmed to development temperature and freshly made silver nitrate solution is added immediately before use at a concentration depending on section thickness and the embedding medium (see below). Sections are incubated in this developer in the dark at 40–60°C.

Reliability

The method given by Tyrer and Bell (1974) tends to be capricious, because the rate of silver precipitation is unpredictable. Experience has shown that the following factors all affect reliability.

1. *Quality of gum acacia (arabic):* Several laboratories report that commercially powdered gum acacia causes very rapid silver precipitation, presumably because it is contaminated with metal during grinding. The crude, uncleaned product ("sorts"), ground with a pestle and mortar, gives much better results. Alternative stabilizers are polyethylene glycol (N. J. Strausfeld, personal communication) or cationic surfactants (P. Mobbs, personal communication).

2. *pH:* The stability of a developer solution containing a gum increases as the pH becomes more acid but the more stable the solution, the slower the specific deposition of silver. For whole mounts, it has been found that pHs between 2.6 and 2.4 give most controllable intensification (Bacon and Altman, 1977). The section method may therefore benefit from a slight reduction in pH. The pH should be altered by changing the citric acid content of the solution.

3. *Dirt* may provide nuclei for silver deposition, causing a generalized precipitate over the whole section. It is important that all glassware be scrupulously clean and metal equipment should not be used.

4. *Cold slides* transferred into warm developer solution produce a rapid darkening of the developer and silver precipitation. This is caused partly by the bubbles that form on the slides acting as nuclei and can be prevented if the slides are prewarmed in distilled water to the same temperature as the developer. Similarly, a slide removed for inspection must be rewarmed before it is returned to the developer.

5. *Solution preparation:* C. Goodman (1976a) reported that development is much more rapid (7–10 min at 60°C) and precipitation problems are avoided if the hydroquinone is added to the developer solution immediately before use.

With adequate precautions for cleanliness and constant temperature, we find the method can be used as a routine histologic batch process.

Times and Temperatures

Development time depends on temperature, cobalt concentration in the material, and the embedding medium. At lower temperatures, the reaction takes longer but is more easily controlled because the end point is approached more gradually, but below about 35°C the reaction may fail completely; 50–55°C is a good compromise. Martin and Mason (1977) reported that shrinkage of thick frozen sections of mammalian brain is least at 37°C.

Development may be checked under the microscope at intervals, slides being washed in warm distilled water both before and after inspection. The end point is reached when cobalt-filled profiles appear black on a pale yellow or pale brown background. Further development will produce either a granular and muddy background or a general precipitation of silver over the whole section. As silver is deposited, it in turn causes further reduction of silver nitrate, so that the whole process is accelerated as the reaction proceeds. This means that heavy or extensive cobalt fillings cause generalized silver precipitation sooner. Consequently development time must be reduced and fine detail sacrificed.

Silver may precipitate out of the developer solution before a satisfactory end point is reached. In this case, slides should be removed immediately, washed well in warm distilled water, and transferred to fresh developer at the same temperature. Provided that slides have not been fixed in sodium thiosulfate, they can be developed a second time if a reasonable end point is not obtained.

When development is complete, sections are washed well in warm water to remove the colloidal developer and to halt the reaction. An acid stop-bath may be used for critical work. Further treatment is dealt with in subsequent sections.

Light Microscopy

Wax Sections

Ganglia containing cobalt-filled neurons are developed with ammonium sulfide in saline solution, fixed in alcoholic Bouin (Pantin, 1946), dehydrated, embedded in wax, and sectioned at 10 μm (see Chap. 19). Slides are dewaxed, and rehydrated, finishing with distilled water warmed to development temperature. Developing may be done in a petri dish, coplin jar, or in bulk in a staining trough, all with periodic agitation. Single slides in a petri dish can be controlled more accurately but the end point is reached more quickly. Coplin jars may give uneven results because of poor mixing.

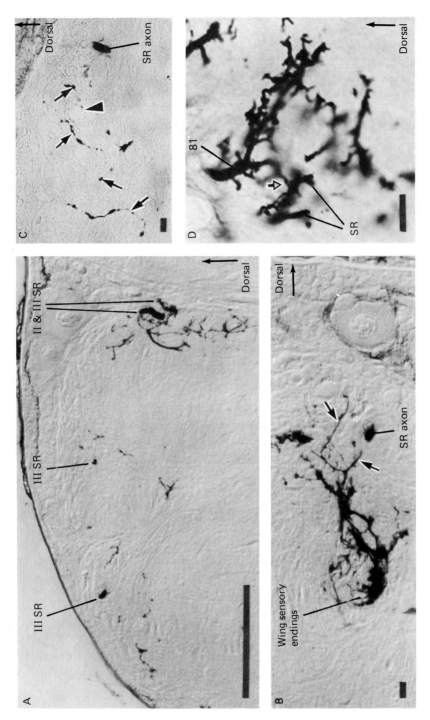

The developer contains one part of 1% silver nitrate to nine parts of stock solution. At 50°C, the end point is usually reached after 25–30 min incubation. Wash in warm water for 5 min, wash in distilled water briefly, fix in 5% sodium thiosulfate for 5 min, wash, dehydrate, and mount in Canada balsam. These preparations should not fade.

Background resolution. Usually the background tissue stains a light yellowish-brown and further treatment is unnecessary (Figs. 21-1B, 21-2C). If resolution is unsatisfactory, sections can be counterstained with eosin, methyl green, or toluidine blue. Alternatively, unstained sections can be examined with interference contrast microscopy (Fig. 21-2A,B), but the finer details of intensified neurons are then difficult to resolve.

Background staining during intensification depends on the intensity of the initial cobalt fill. With heavy fills, small concentrations of cobalt sulfide that are found in the glia (see Fig. 21-7) are presumably responsible for the brownish background. In control ganglia developed at the same time, the background remains quite pale. The background color of control ganglia is not, however, the same in all laboratories, which leads us to suspect that chemicals from different sources, with variations in quality or impurities, are responsible, but we have not investigated this systematically. H.-W. Honegger (personal communication) suggests that intensity of background staining may depend on the exact pH value of the developer solution.

Plastic Sections

One can develop 1- to 2-μm sections of material embedded in any epoxy resin in exactly the same way except that the concentration of silver nitrate in the developer is increased to one part of 5% silver nitrate to nine parts of stock solution. Satisfactory results have also been obtained with the whole mount intensification formula (Bacon and Altman, 1977) using 1 part of 1% silver nitrate to 9 parts of stock solution. Slides must be warmed before incubation, which is usually done in a petri dish. Times

---

Fig. 21-2, A–D. The value of Timm's method is demonstrated by these transverse sections of the thoracic ganglia of a locust showing details of the central arborization of the mesothoracic stretch receptor (*II SR*) and metathoracic stretch receptor (*III SR*) and their associations with other neurons. **A** Distribution of the branching of the two neurons running in parallel through the metathoracic ganglion. **B** Stretch receptor branches entering the dorsal wing sensory tract and other branches entering the dorsal neuropil of the metathoracic ganglion. **C** Stretch receptor branches in the dorsal neuropil of the metathoracic ganglion (*small arrows*) contacting through fibers (*large arrow*). **D** Contacts (*arrowed*) between the mesothoracic stretch receptor and a branch of a motor neuron to the mesothoracic dorsal longitudinal muscle (*81*). Scale **A** = 100 μm; scale **B–D** = 10 μm. [**A–D** from Altman and Tyrer, 1977a.]

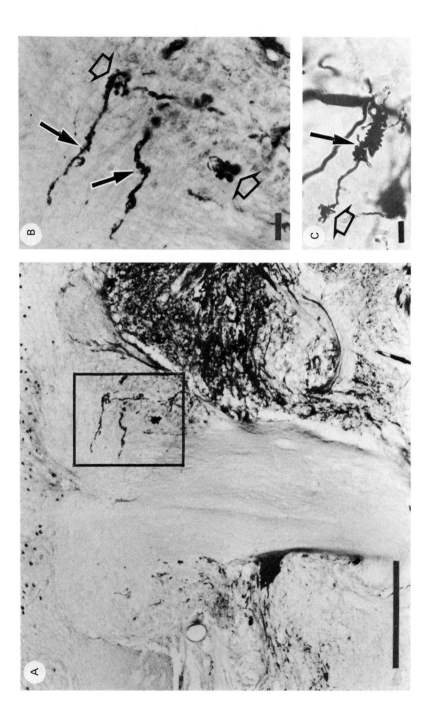

and temperatures are as for wax sections. Brief counterstaining with 1%
toluidine blue in borax may be required to display background detail.
After counterstaining, slides are dried on a hot plate and may be mounted
in Canada balsam or examined under immersion oil.

Frozen Sections

Frozen sections have been used mainly in vertebrate work (Prior and
Fuller, 1973; Iles, 1976b; Mason and Lincoln, 1976). Brains are fixed
conventionally, and frozen sections 30–90 μm thick are cut. Sections
may be intensified individually before mounting on slides with gelatin
(Mason and Lincoln, 1976) or attached and then intensified (Iles, 1976b).
The developer contains 5% silver nitrate 1:9 with stock solution (Mason
and Lincoln, 1976).

Results

Timm's intensification on sections reveals a great deal of fine detail, most
of which was invisible before treatment (Figs. 21-1A,B, 21-2, 21-3).
Small branches of single neurons can be traced to their terminations in the
neuropil (Fig. 21-2A–C) and contacts between identified neurons can be
detected (Fig. 21-2D). The morphology of individual neurons, especially
terminal branches, is displayed in detail comparable with that of Golgi
preparations (Fig. 21-3).

Electron Microscopy

The main problem encountered in electron microscopy (EM) of cobalt-
filled neurons is poor tissue preservation, arising from damage caused by
cobalt filling and precipitation. There are also handling problems involved with intensification of ultrathin sections. Although some difficul-

---

Fig. 21-3, A and B. Comparison of cobalt and Golgi-stained terminals in the mushroom body of the cricket brain. **A** Frontal section through the ipsilateral hemi-
sphere of *Gryllus campestris* showing cobalt-filled extrinsic fibers terminating in
the mushroom body. 10-μm wax section, Timm's intensified. [Courtesy of H.-
W. Honegger and F.-W. Schürmann.] **B** Detail of endings with invaginated boutons from **A**. Cylindrical boutons are indicated by *closed arrows,* knob-like ones
by *open arrows*. These can be compared with the Golgi-impregnated terminals
seen in a 100-μm section in **C**, showing the same type of endings in the mushroom
body of *Acheta domesticus*. [**B** from Honegger and Schürmann, 1975; **C** from
Schürmann, 1973.] Scale **A** = 100 μm; scale **B,C** = 10 μm.

ties still remain, useful information can be obtained with the methods described here.

Cobalt Sulfide Precipitation and Fixation

Ammonium sulfide treatment causes considerable ultrastructural damage, but this can be minimized by transferring ganglia directly to fixative before treatment with ammonium sulfide, or by using 1%–5% sodium sulfide in buffer (see Chap. 19). Routine EM fixation in phosphate-buffered glutaraldehyde or paraldehyde/glutaraldehyde (see Chap. 19) for 1–2 h follows precipitation. For material to be silver intensified, it is better to avoid cacodylate buffers, which contain arsenic. Osmium postfixation has been a subject of debate. Earlier workers omitted it, so that the neuron could be visualized in the block while sectioning (Pitman *et al.*, 1973) or to prevent reaction with the silver developer (Tyrer and Bell, 1974), but without osmium, preservation of membranes and synaptic vesicles and their counterstaining with uranyl acetate and lead citrate are often poor. It is now standard practice to postfix with 1%–2% osmium tetroxide for 1–2 h (Székély and Kosaras, 1976; Rademakers, 1977; Altman *et al.*, 1979, 1980). Provided that intensification is not carried on for too long, there is no undue precipitation of silver onto osmiphilic structures.

Timm's Development

We have used three procedures for intensifying cobalt for electron microscopy: development of thin sections; development of 1-μm plastic sections on glass slides and subsequent resectioning; and intensification of whole ganglia after glutaraldehyde fixation. Each method has advantages and disadvantages and has different applications.

   1. *Development of ultrathin sections* (Fig. 21-5). There are two problems with silver intensification of ultrathin sections. First, the developer must not come into contact with copper grids, and, second, difficulties are frequently encountered with counterstaining after treatment with Timm's developer. We have found that this is due to the persistence of gum acacia on the surface of the section, because it is very difficult to wash off entirely. Several techniques have been used to avoid developer solution reacting with copper grids. Nylon grids can be used or copper grids can be completely coated by immersion in a formvar solution before covering the upper surface with a formvar film in the conventional way. Alternatively, thin sections can be collected and stained in a drop of liquid in the center of a polyethylene ring (Fig. 21-4) (Tyrer and Bell, 1974). The rings are punched from polyethylene sheeting and flamed briefly to smooth the

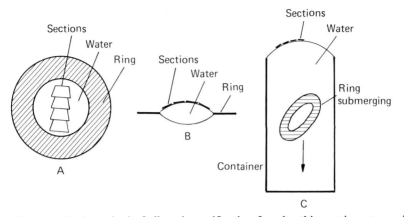

Fig. 21-4, A–C. A method of silver intensification for ultrathin sections to avoid contact between copper grids and the Timm's developer. **A** Plan view of polyethylene ring with a ribbon of sections held on a captive drop in its center. **B** Side view of the polyethylene ring, sections, and captive drop. **C** Side view of polyethylene ring and sections in the staining vessel. After intensification is finished, the relatively deep vessel allows the ring to be submerged, leaving the sections on the surface to be picked up on conventional electron microscope slot-grids for staining with uranyl acetate and lead citrate.

edges. Staining solutions are held in narrow plastic tubes, e.g., Beem capsules, filled to give a gently curving meniscus, and the polyethylene ring containing the sections is transferred from one tube to the next with forceps (Fig. 21-4). At the end of the process, the ring is forced below the surface to release the sections, which are then picked up from below on formvar-coated slot grids. Sections may be handled singly or as ribbons. Another solution is that devised by Rademakers (1977), who used a Plexiglas slide with holes to hold sections. This enabled him to monitor intensification under the light microscope.

Removal of gum acacia after intensification has been facilitated by reducing the gum acacia concentration in the developer to 3% (see Method 2). Thorough washing then removes it adequately. An alternative, after either Method 1 or 2, is to air-dry sections overnight before counterstaining, when the gum acacia forms inconspicuous crystals, but this can make the formvar very brittle.

Method 1 (Tyrer and Bell, 1974; Rademakers, 1977): Use 1 part of 1%–5% silver nitrate solution to 9 parts of stock solution, pH 3.9–4.0 (recipe, p. 431).

Method 2 (Altman *et al.*, 1979): Use the whole-mount developer stock solution recipe (p. 442), which contains only 3% gum acacia at pH 2.6; 1 part of 1% silver nitrate solution to 9 parts of stock solution.

For both methods, incubate in the dark at 50°C. Incubation time may be estimated from the time taken to intensify a 1-μm-thick plastic section

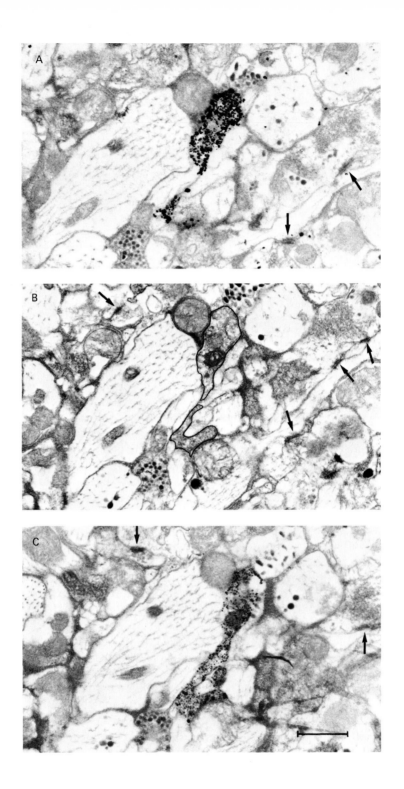

from the same area on a glass slide. For method 1, development of thin sections will take about the same time; for method 2 half as long is usually sufficient. Development time varies from 20 to 120 min according to the amount of cobalt in the profiles. It is difficult to get a consistent cutoff point for development, and as little as 10 min extra can make a large difference in the intensity of the silver precipitate (cf. Fig. 21-5A, C). After development, transfer to an acid stop-bath before washing in 3 changes of distilled water for 2–3 h or overnight. If the sections have not been mounted on grids during staining they can now be transferred to grids. For air-drying, wash thoroughly in distilled water for 1–2 h, mount sections on grids, and dry overnight.

Sections are contrasted with saturated aqueous uranyl acetate and lead citrate in the conventional way. Alcoholic solutions of uranyl acetate can cause any gum acacia remaining on the sections to form an insoluble deposit.

Although long series of sections can be dealt with by this method, the amount of handling and capriciousness of intensification makes some loss of sections inevitable. The deposit on the cobalt-filled profiles is heavy (Fig. 21-5A, C), with a large grain size that obscures ultrastructural details within the profiles, making synaptic contacts difficult to identify. Most useful information can be obtained by intensifying every second or third section and comparing with the intervening, conventionally stained sections (Fig. 21-5). In tangled neuropils, fine processes that are not running normal to the plane of section can be difficult to interpret in such series because the dramatic change in profile from one section to the next makes matching difficult (Fig. 21-5). Synaptic regions on fine processes may also only be visible on one or two sections. For these reasons the following two methods may be more useful in some material.

2. *Resectioning of intensified 1-µm plastic sections* (Fig. 21-6A). One-micrometer sections are mounted on glass slides and developed as described previously. They can then be examined with the light microscope to trace axons and select areas for examination. Sections of interest are drawn or photographed and then lifted off the slide and remounted

---

Fig. 21-5, A–C. Intensification of ultrathin sections. A series of three sections of motor neuropil in the locust mesothoracic ganglion. Motor neurons of the metathoracic dorsal longitudinal muscles have been filled with cobalt and here a terminal in the anterior end of the ganglion is shown. Sections **A** and **C** have been intensified on the section by method 2 followed by saturated aqueous uranyl acetate and lead citrate counterstaining. **A** was developed for 1 h 45 min, **C** for 1 h 35 min. Section **B** is not intensified and is contrasted with alcoholic uranyl acetate and lead citrate. The cobalt-filled profile has been outlined. This illustrates the difficulty of following fine profiles in the unstained sections that are necessary for examination of the fine structure of the stained axon. Synaptic structures are marked. Scale **A–C** = 1 µm. ×20,000.

on faced, epoxy-filled specimen capsules. This procedure is difficult after contact with immersion oil, which interferes with adhesion to the blank. Instead, a concentrated sugar solution (89.4%) or a commercially available syrup (e.g., Tate and Lyle Golden Syrup, refractive index 1.498) should be used with oil immersion lenses (Tyrer and Altman, 1976). A ribbon of approximately 13 thin sections can be cut from one such section. These are mounted on slot grids and contrasted with uranyl acetate and lead citrate in the normal way.

This method is of necessity a sampling technique. Some loss and slight changes in orientation are inevitable when a 1-μm section is recut so that long, complete series for EM cannot be obtained. It does allow, however, good correlation between light and electron microscopy, so that particular branches or terminals can be selected for ultrastructural examination. For a large neuron, or specific contacts between neurons, a sampling method of this type may allow better localization than the examination of a long series of thin sections at high power, which is in any case a mammoth task. A further advantage is that the intensification only penetrates the surface layer of the 1-μm section. The neuron can be identified in the superficial thin sections and traced into subsequent sections where ultrastructural details are not obscured by the silver precipitate. Grain size is also smaller than that produced by development on thin sections (Fig. 21-6A).

3. *Electron microscopy of ganglia intensified as whole mounts* (see Fig. 21-6B). The whole-mount modification (see Obermayer and Strausfeld, this volume, Chap. 20; Bacon and Altman, 1977) has only recently been introduced and procedures for electron microscopy are not yet fully worked out. From our preliminary observations, however, it seems that adequate preservation and ease of manipulation will establish this as the routine method for the electron microscopy of cobalt-filled neurons.

After cobalt precipitation and standard glutaraldehyde/osmium fixation (Chap. 19), ganglia are brought to 70% ethanol for at least 1 h, then returned to distilled water and warmed to 50°C. They are incubated for 1 h in a weak Timm's stock solution containing 3 g of gum acacia, 0.8 g citric acid, 0.17 g hydroquinone, and 6 g of sucrose, all dissolved in 100 ml distilled water. The pH should be between 2.4 and 2.6 (Bacon and Altman, 1977). Ganglia are then transferred to a freshly prepared warm developer solution containing one part of 1% silver nitrate in nine parts of weak stock solution and incubated at 50°C for a further 30–90 min until silver begins to precipitate on the surface of the ganglion. Wash well in warm water, dehydrate in an alcohol series, and embed in Spurr's resin or other plastic medium.

In 1-μm sections of block intensified material counterstained with toluidine blue, cobalt-filled neurons appear as golden or gray profiles. Neurons can be traced through 1-μm series and appropriate areas selected for

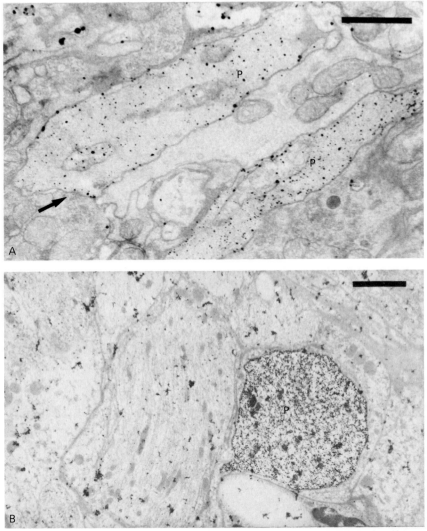

Fig. 21-6, A and B. Electron micrographs to illustrate the results of different methods of intensification. **A** An intensified 1-μm section that has been resectioned for electron microscopy. A fine deposit of silver that does not obscure the fine structure is clearly visible in two profiles (*P*). A possible synaptic site is *arrowed*. ×20,000. **B** A section of a cobalt-filled profile, *P*, which has been intensified in block, showing fine deposits of silver throughout the neuron. Scale **A,B** = 1 μm. ×15,000.

EM, or long series of thin sections can be prepared. Sections are counterstained with uranyl acetate and lead citrate as usual.

The silver precipitate is usually fine grained (Fig. 21-6B) and small profiles have to be examined at high magnification to locate the marker. One disadvantage is that the filled neurons cannot be visualized in whole mount if an osmium postfix is used. This may not, therefore, be ideal for critical studies, but it cuts out much of the tedious preparatory work and uncertainties involved in the other two methods, thus allowing results to be obtained with reasonable speed.

Sensitivity of Timm's Method

Timm's method was designed to detect metal ions at very low concentration and consequently will intensify cobalt deposits that are too faint to see in untreated material. This can lead to erroneous interpretations when neurons other than the selected ones take up a low concentration of cobalt through parallel diffusion or overfilling (see Chap. 19); or neurons in the same nerve branch, which fill too weakly to be detected in unintensified whole mounts, suddenly appear in intensified material. In addition, heavy cobalt fills can cause sufficient background contamination to make the interpretation of intensified sections ambiguous in the light microscope, and electron micrographs of intensified material demonstrate that weak deposits of cobalt sulfide appear in glial profiles around neurons containing large amounts of cobalt sulfide (Fig. 21-7).

Both these problems can be avoided if only a small quantity of cobalt is introduced into the selected neurons, by reducing both the cobalt chloride concentration used and the filling time (Altman and Tyrer, 1977a; see Chap. 19). As a rule of thumb, neurons that are faintly visible as they enter the ganglion of interest will be shown up in detail in intensified material and most artifactual staining will be avoided. A further advantage of short filling times and low cobalt concentrations is markedly improved tissue preservation.

The Need for Sections

Cobalt filling of neurons has two main functions: first, as an aid to physiologic experiments and their interpretation (e.g., Burrows, 1973; Pearson and Fourtner, 1975; Rehbein *et al.,* 1974) and, second, as a tool in neuroanatomical research (Tyrer and Altman, 1974; Altman and Tyrer, 1977a,b; C. Goodman 1976). For physiology the principal requirement is to visualize the extent and location of the branches of cells from which

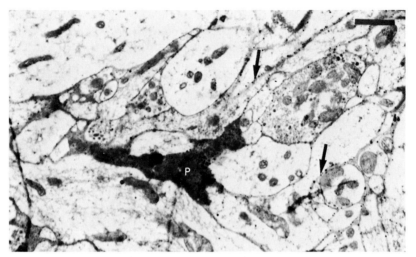

Fig. 21-7. In heavily filled specimens, intensification shows a fine deposit of cobalt in the glia (*arrowed*) around the filled profile (*P*). This section of a locust thoracic motor neuron was intensified on the section by method 1 and air-dried before counterstaining.   Scale = 1 μm.   ×15,000.

records are made.   This can be achieved using whole mounts, either with or without intensification.   More detailed neuroanatomical information about precise positions of branches within neuropils and contacts between terminals of the stained neuron and other elements in the ganglia cannot be obtained from whole mounts alone.   Whole mount methods are excellent for determining the overall morphology of a neuron, but serial sections are necessary to study connectivity.   The combined analysis of cobalt-filled neurons in whole mounts and serial light microscope sections with ultrastructural studies should prove very powerful for describing synaptic relationships of identified neurons.

Addendum

Electron microscopy of cobalt-filled neurons is now being used routinely by several groups of workers.   Intensification of ultrathin sections is the most common method.   For serial sections, we intensify one section in ten to identify the filled profiles, which can then be traced through the unintensified, normally counterstained intervening sections to provide details of synaptic ultrastructure (Altman *et al.,* 1979, 1980).   Good results have also been obtained with the resectioning method (C. Phillips, personal communication).   Here the deposit is so fine and sparse that ultrastructural details are not obscured.

Problems with fixation for electron microscopy have been alleviated by using a low concentration of cobalt chloride to fill the neuron and by extremely careful handling of the tissue (Altman *et al.*, 1979).

Acknowledgments

We thank Christine Davies and Robert Fellows for their patient help in developing the techniques for electron microscopy.

References

Adams, C. W. M.: Disorders of neurones and neuroglia. In: *Neurohistochemistry* (C. W. M. Adams, ed.), pp. 403–436. Elsevier, Amsterdam, 1965

Adams, C. W. M., Sloper, J. C.: The hypothalamic elaboration of posterior pituitary principles in man, the rat and dog. Histochemical evidence derived from a performic acid-Alcian blue reaction for cystine. J. Endocrinol. *13*:221–228 (1956)

Adams, J. C.: Technical considerations on the use of horseradish peroxidase as a neuronal marker. Neuroscience *2*:141–145 (1977)

Aghajanian, G. K., Bloom, F. E.: The formation of synaptic junctions in developing rat brain: A quantitative electron microscopic study. Brain Res. *6*:716–727 (1967)

Akert, K., Moor, H., Pfenninger, K., Sandri, C.: Contribution of new impregnation methods and freeze etching to the problem of synaptic fine structure. Prog. Brain Res. *31*:223–240 (1969)

Akert, K., Sandri, C.: An electron microscopic study of zinc iodide -osmium impregnations of neurons. I. Staining of synaptic vesicles at cholinergic junctions. Brain Res. *7*:286–295 (1968)

Albrecht, F. O.: *The Anatomy of the Migratory Locust.* Athlone Press, London, 1953

Altman, J.: Experimental reorganization of the cerebellar cortex. III. regeneration of the external germinal layer and granule cell ectopia. J. Comp. Neurol. *149*:153–180 (1973)

Altman, J.: VI. Effects of late X-irradiation schedules that interfere with cell acquisition after stellate cells are formed. J. Comp. Neurol. *165*:65–76 (1976)

Altman, J., Anderson, W. J.: Experimental reorganization of the cerebellar cortex. I. Morphological effects of elimination of all microneurons with prolonged X-irradiation started at birth. J. Comp. Neurol. *146*:355–406 (1972)

Altman, J., Anderson, W. J.: II. Effects of elimination of most microneurons with prolonged X-irradiation started at four days. J. Comp. Neurol. *149*:123–152 (1973)

Altman, J. S.: The functional organization of the thoracic ganglia in insects. In: *Insect Neuroacoustics* (W. B. Broughton, ed.), pp. 4–10. Polytechnic Publication, London, 1976

Altman, J. S., Bell, E. M.: A rapid method for the demonstration of nerve cell bodies in invertebrate central nervous systems. Brain Res. *63*:487–489 (1973)

Altman, J. S., Kien, J.: Suboesophageal neurons involved in head movements and feeding in locusts. Proc. Roy. Soc. (Lond.) B *205*:209–227 (1979)

Altman, J. S., Tyrer, N. M.: Insect flight as a system for the study of the development of neuronal connections. In: *The Experimental Analysis of Insect Behaviour* (L. Barton Browne, ed.), pp. 159–179. Springer-Verlag, Berlin, New York, 1974

Altman, J. S., Tyrer, N. M.: The locust wing hinge stretch receptors. I. Primary sensory neurones with enormous central arborizations. J. Comp. Neurol. *172*:409–430 (1977a)

Altman, J. S., Tyrer, N. M.: The locust wing hinge stretch receptors. II. Variation, alternative pathways and "mistakes" in the central arborizations. J. Comp. Neurol. *172*:431–440 (1977b)

Altman, J. S., Anselment, E., Kutsch, W.: Post-embryonic development of an insect sensory system: Ingrowth of axons from wing sense organs in *Locusta migratoria*. Proc. R. Soc. London Ser. B *202*:497–516 (1978)

Altman, J. S., Shaw, M., Tyrer, N. M.: Visualization of synapses of physiologically identified cobalt-filled neurons. J. Physiol. Lond. *296*:2–3 (1979)

Andèn, N.-E., Carlsson, A., Häggendal, J.: Adrenergic mechanisms. Annu. Rev. Pharmacol. *9*:119–134 (1969)

Andersen, S. O.: Cuticular sclerotization on larvae and adult locust, *Schistocerca gregaria*. J. Insect Physiol. *20*:1537–1552 (1974)

Ascher, P., Glowinski, J., Tauc, L., Taxi, J.: Discussion of stimulation-induced release of serotonin. Adv. Pharmacol. *6a*:365–368 (1968)

Atwood, H. L., Morin, W. A.: Neuromuscular and axoaxonal synapses of the crayfish opener muscle. J. Ultrastruct. Res. *32*:351–369 (1970)

Atwood, H. L., Pomeranz, B.: Crustacean motor neuron connections traced by back filling for electron microscopy. J. Cell Biol. *63*:329–334 (1974)

Atwood, H. L., Pomeranz, B.: Dendrite bottlenecks of crustacean motor neurons. J. Neurocytol. *8*:251–268 (1977)

Aubele, E., Klemm, N.: Origin, destination and mapping of tritocerebral neurons of locust. Cell Tissue Res. *178*:199–219 (1977)

Autrum, H., Zettler, F., Järvilehto, M.: Postsynaptic potentials from a single monopolar neuron of the ganglion opticum of the blowfly *Calliphora*. Z. Vergl. Physiol. *70*:414–424 (1970)

Awasthi, V. B.: Studies on the neurosecretory cells of the brain of normal and starved red cotton bugs, *Dysdercus koenigii* (Fabr.) (Heteroptera, Pyrrhocoridae). Aust. Zool. *17*:24–25 (1972)

Axelrod, J.: Relationship between catecholamines and other hormones. Recent Prog. Horm. Res. *31*:1–35 (1975)

Bach, A., Chodat, R.: Untersuchungen über die Rolle der Peroxyde in der Chemie der lebenden Zelle IV. Über Peroxydase. Ber. Dtsch. Chem. Ges. *36*:600–605 (1903)

Bacon, J. P., Altman, J. S.: A silver intensification method for cobalt-filled neurones in wholemount preparations. Brain Res. *138*:359–363 (1977)

Bacon, J. P., Tyrer, N. M.: The tritocerebral commissure giant (TCG): A bimodal interneurone in the locust. J. Comp. Physiol. *126*:317–325 (1978)

Bader, C., Bertrand, D., Bertrand, G., Perrelet, A.: A pressure device for intracellular injection. Experientia *30*:1366–1367 (1974)

Baker, J. R.: *Cytological Technique*. Methuen, London, 1950

Bangle, R.: Gomori's paraldehyde-fuchsin stain. I. Physico-chemical and staining properties of the dye. J. Histochem. Cytochem. *2*:291–299 (1954)

Bank, E. W., Davenport, H. A.: Staining paraffin sections with protargol. 5. Chloral hydrate mixtures, with and without formamide, for fixing peripheral nerves. Stain Technol. *15*:9–14 (1940)

Bargmann, W.: Über die neurosekretorische Verknüpfung von Hypothalamus und Neurohypophyse. Z. Zellforsch. Mikrosk. Anat. *34*:610–634 (1949)

Barlow, J., Martin, R.: Structural identification and distribution of synaptic profiles in the octopus brain using the zinc iodide -osmium method. Brain Res. *25*:241–253 (1971)

Bartelmez, G. W., Hoerr, N. L.: Vestibular club endings in *Ameiurus:* Further evidence on morphology of the synapse. J. Comp. Neurol. *57*:401–428 (1933)

Baudry, N., Baehr, J.-C.: Etude histochimique des cellules neurosécrétrices de l'ensemble du système central de *Rhodnius prolixus* Stål. (Hétéroptère, Reduviidae). C. R. Hebd. Seances Acad. Sci., Ser. D *270*:174–177 (1970)

Bauer, P. S., Stacey, T. R.: The use of PIPES buffer in the fixation of mammalian and marine tissues for electron microscopy. J. Microsc. (Oxford) *109*:315–327 (1977)

Baumgarten, H. G., Lachenmeyer, L., Schlossberger, H. G.: Evidence for a degeneration of indoleamine containing nerve terminals in rat brain, induced by 5,6-dihydroxytryptamine. Z. Zellforsch. Mikrosk. Anat. *125*:553–569 (1972)

Beattie, T. M.: Vital staining of neurosecretory material with Acridine Orange in the insect, *Periplaneta americana*. Experientia *27*:110–111 (1971)

Bentley, D. R.: A topological map of the locust flight system motor neurons. J. Insect. Physiol. *16*:905–918 (1970)

Bentley, D.: Postembryonic development of insect motor systems. In: *Developmental Neurobiology of Arthropods* (D. Young, ed.), pp. 147–177. Cambridge Univ. Press, London, New York, 1973

Bentley, D.: Single gene cricket mutations: Effects on behavior, sensilla, sensory neurons, and identified interneurons. Science *187*:760–764 (1975)

Berek, M.: Zur Theorie der Abbildung im Mikroskop. Optik *6*:1–219 (1950)

Bern, H.: The properties of neurosecretory cells. Gen. Comp. Endocrinol., Suppl. *1:*117–132 (1962)

Bertram, E. G., Ihrig, H. K.: Zinc chromate solutions for impregnation of nervous tissue. Stain Technol. *33:*187 (1958)

Bertram, E. G., Ihrig, H. K.: Staining formalin-fixed nerve tissue with mercuric nitrate. Stain Technol. *34:*99–108 (1959)

Bertram, E. G., Ihrig, H. K.: Improvements of the Golgi method by pH control. Stain Technol. *32:*87–94 (1957)

Birks, R., Katz, B., Miledi, R.: Physiological and structural changes at the amphibian myoneural junction, in the course of nerve degeneration. J. Physiol. *150:*145–166 (1960)

Bishop, A., Bishop, L. G.: Staining of fly interneurons with Lucifer yellow. (in preparation)

Bishop, L. G., Keehn, D. G.: Neural correlates of the optomotor response in the fly. Kybernetik *3:*288–295 (1967)

Bishop, L. G., Keehn, D. G., McCann, G. D.: Studies of motion detection by interneurons of the optic lobes and brain of flies. J. Neurophysiol. *31:*509–525 (1968)

Björklund, A., Falck, B.: Cytofluorometry of biogenic monoamines in the Falck-Hillarp method. Structural identification by spectral analysis. In: *Structural Identification by Spectral Analysis. Fluorescence Techniques in Cell Biology* (A. A. Thear, M. Sernetz, eds.), pp. 171–181. Springer-Verlag, Berlin, New York, 1973

Björklund, A., Stenevi, U.: Acid catalysis of the formaldehyde condensation reaction for sensitive histochemical demonstration of tryptamines and 3-methoxylated phenylethylamines. I. Model experiments. J. Histochem. Cytochem. *18:*794–802 (1970)

Björklund, A., Ehinger, B., Falck, B.: A method for differentiating dopamine from noradrenaline in tissue sections by microspectrocluorometry. J. Histochem. Cytochem. *16:*263–270 (1968)

Björklund, A., Falck, B., Klemm, N.: Microspectrofluorometric and chemical investigation of catecholamine-containing structures in the thoracic ganglia of *Trichoptera*. J. Insect Physiol. *16:*1147–1154 (1970)

Björklund, A., Falck, B., Lindvall, O.: Microspectrofluorometric analysis of cellular monoamines after formaldehyde or glyoxylic acid condensation. In: *Methods in Brain Research* (P. B. Bradley, ed.), pp. 249–294. Wiley, New York, 1975

Björklund, A., Falck, B., Owman, C.: Fluorescence microscopic and microspectrofluorometric techniques for the cellular localization and characterization of biogenic amines. In: *The Thyroid and Biogenic Amines* (J. E. Rall, I. J. Kopin, eds.), pp. 318–368. North-Holland Publ., Amsterdam, 1972

Blackstadt, T. W.: Mapping of experimental axon degeneration by electron microscopy of Golgi preparations. Z. Zellforsch. Mikrosk. Anat. *67:*819–834 (1965)

Blackstadt, T. W.: Electron microscopy of Golgi preparations for the study of neuronal relations. In: *Contemporary Research Methods in Neuroanatomy* (W. J. H. Nauta, S. O. F. Ebbesson, eds.), pp. 186–216. Springer-Verlag, Berlin, New York, 1970

Blackstadt, T. W., Fregerslev, S., Laurberg, S., Rokkedal, K.: Golgi impregnation with potassium dichromate and mercurous or mercuric nitrate: Identification of the precipitate by X-ray and electron diffraction methods. Histochemie *36:*247–268 (1973)

Bland, K. P., House, C. R., Ginsborg, B. L., Laszlo, I.: Catecholamine transmitter for salivary secretion in the cockroach. Nature (London), New Biol. *244:*26–27 (1973)

Blaschko, H.: The natural history of amine oxidases. Ergeb. Physiol., Biol. Chem. Exp. Pharmakol. *70:*83–148 (1974)

Blest, A. D.: Some modifications of Holmes' silver method for insect central nervous systems. Q. J. Microsc. Sci. *102:*413–417 (1961)

Blest, A. D.: A new method for the reduced silver impregnation of arthropod central nervous systems. Proc. R. Soc. London, Ser. B *193:*191–197 (1976)

Blest, A. D., Davie, P. S.: A new fixative solution to precede the reduced silver impregnation of arthropod central nervous systems. Stain Technol. *52,*273–275 (1977)

Blest, A. D., Land, M. F.: The physiological optics of *Dinopis subrufus* L. Koch: A fish lens in a spider. Proc. R. Soc. London, Ser. B *196:*197–222 (1977)

Bloom, F. E., Aghajanian, G. K.: Cytochemistry of synapses: A selective staining method for electron microscopy. Science *154:*1575–1577 (1966)

Bloom, F. E., Aghajanian, G. K.: Fine structural and cytochemical analysis of the staining of synaptic junctions with phosphotungstic acid. J. Ultrastruct. Res. *22:*361–375 (1968)

Bodenstein, D.: Studies on nerve regeneration in *Periplaneta americana*. J. Exp. Zool. *136:*89–115 (1957)

Bodian, D.: A new method for staining nerve fibers and nerve endings in mounted paraffin sections. Anat. Rec. *65:*89–97 (1936)

Bodian, D.: The staining of paraffin sections of nervous tissues with activated protargol. The role of fixatives. Anat. Rec. *69:*153–162 (1937)

Bodnaryk, R. P., Brunet, P. C. J., Koeppe, J. K.: On the metabolism of N-acetyldopamine in *Periplaneta americana*. J. Insect Physiol. *20:*911–923 (1974)

Boeckh, J.: Die Reaktionen von Neuronen im Deutocerebrum der Wanderheuschrecke bei Duftreizung der Antennen. Verh. Dtsch. Zool. Ges. *66:*189–193 (1973)

Boeckh, J., Sandri, C., Akert, K.: Sensorische Eingänge und synaptische Verbindungen im Zentralnervensystem von Insekten. Z. Zellforsch. Mikrosk. Anat. *103:*420–446 (1970)

Boschek, C. B.: On the fine structure of the peripheral retina and lamina ganglionaris of the fly *Musca domestica*. Z. Zellforsch. Mikrosk. Anat. *118:*369–409 (1971)

Boulton, P. S.: Degeneration and regeneration in the insect central nervous system. I. Z. Zellforsch. Mikrosk. Anat. *101:*98–118 (1969)

Bounhiol, J.-J.: Observation de cellules neurosécrétrices chez le ver à soie vivant. Actes Congr. Assoc. Fr. Av. Sci. *74:*1–4 (1955)

Bowes, J. H., Cater, C. W.: The reaction of glutaraldehyde with proteins and other biological materials. J. R. Microsc. Soc. *98:*193–200 (1966)

Boycott, B. B., Dowling, J. E.: Organization of the primate retina: Light Microscopy. Phil. Trans. R. Soc. Lond. Ser. B *255:*109–176 (1969)

Boycott, B. B., Dowling, J. E., Fisher, S. K., Kolb, H., Laties, A. M.: Interplexiform cells of the mammalian retina and their comparison with catecholamine-containing retinal cells. Proc. R. Soc. London, Ser. B *191*:353–368 (1975)

Braitenberg, V.: Patterns of projection in the visual system of the fly. 1. Retina-lamina projection. Exp. Brain Res. *3*:271–298 (1967)

Brightman, M. W., Reese, T. S., Feder, N.: Assessment with the electron microscope of the permeability to peroxidase of cerebral endothelium in mice and sharks. In: *Capillary Permeability* (E. H. Thaysen, ed.). Munksgard, Copenhagen, 1970

Brownstein, M. J., Palkovits, M., Saavedra, J. M., Kizer, J. S.: Distribution of hypothalamic hormones and neurotransmitters within the diencephalon. Front. Neuroendocrinol. *4*:1–23 (1976)

Brunk, U., Sköld, G.: The oxidation problem in the sulfide-silver method for histochemical demonstration of metals. Acta Histochem. *27*:199–206 (1967)

Budelmann, B. U., Wolff, H. G.: Mapping of neurons in the gravity receptor system of the octopus statocyst by iontophoretic cobalt staining. Cell Tissue Res. *171*:403–406 (1976)

Bunt, A. H., Hendrickson, A. E., Lund, J. S., Lund, R. D., Fuchs, A. F.: Monkey retinal ganglion cells: Morphometric analysis and tracing of axonal projections, with a consideration of the peroxidase technique. J. Comp. Neurol. *164*:265–286 (1975)

Burckhardt, W., Braitenberg, V.: Some peculiar synaptic complexes in the first visual ganglion of the fly, *Musca domestica*. Cell Tissue Res. *173*:287–308 (1976)

Burrows, M.: Physiological and morphological properties of the metathoracic common inhibitory neuron of the locust. J. Comp. Physiol. *82*:59–78 (1973)

Burrows, M.: Monosynaptic connexions between wing stretch receptors and flight motoneurones of the locust. J. Exp. Biol. *62*:189–219 (1975)

Burrows, M., Hoyle, G.: Neural mechanisms underlying behavior in the locust *Schistocerca gregaria*. III. Topography of limb motor neurons in the metathoracic ganglion. J. Neurobiol. *4*:167–186 (1973)

Burrows, M., Rowell, C. H. F.: Connection between descending visual interneurons and metathoracic motoneurons in the locust. J. Comp. Physiol. *85*:221–234 (1973)

Cajal, S. R., Sánchez y Sánchez, D.: Contribucion al conocimiento de los centros nerviosos de los insectos. Parte I. Retina y centros opticos. Trab. Lab. Invest. Biol. Univ. Madrid *13*:1–168 (1915)

Cajal, S. R., de Castro, F.: *Elementos de Técnica micrografica del sistemo nervioso*. Tipografican Artistica Alameda, Madrid, 1933

Cameron, M. L., Steele, J. E.: Simplified aldehyde-fuchsin staining of neurosecretory cells. Stain Technol. *34*:265–266 (1959)

Campbell, J. I.: The anatomy of the nervous system of the mesothorax of *Locusta migratorioides* R and F. Proc. Zool. Soc. London *137*:403–432 (1961)

Campos-Ortega, J. A., Strausfeld, N. J.: The columnar organization of the second synaptic region of the visual system of *Musca domestica* L. I. Receptor terminals in the medulla. Z. Zellforsch. Mikrosk. Anat. *124*:561–585 (1972)

Campos-Ortega, J. A., Strausfeld, N. J.: Synaptic connections of intrinsic cells and basket arborizations in the external plexiform layer of the fly's eye. Brain Res. *59*:119–136 (1973)

Carbonetto, S., Muller, K. J.: A regenerating neurone in the leech can form an electrical synapse on its severed axon segment. Nature (London) *267*:450–452 (1976)

Carlsson, A.: Some aspects of dopamine in the central nervous system. Adv. Neurol. *5*:59–68 (1974)

Case, R.: Differentiation of the effect of pH and CO_2 on the spiracular function of insects. J. Cell. Comp. Physiol. *49*:103–113 (1954)

Caspersson, T. G., Hillarp, N.-Å., Ritzèn, M.: Fluorescence microspectrophotometry of cellular catecholamines and 5-hydroxytryptamine. Exp. Cell Res. *42*:415–428 (1966)

Castel, M., Spira, M. E.,, Parnas, I., Yarom, Y.: Ultrastructure of region of a low safety factor in inhomogeneous giant axon of the cockroach. J. Neurophysiol. *39*:900–908 (1976)

Champy, C.: Granules et substances réduisant l'iodure d'osmium. J. Anat. Physiol. Norm. Pathol. Homme Anim. *49*:323–343 (1913)

Chanussot, M. B., Pentreath, V. W.: Etude des monoamines du système nerveux stomatogastrique extracephalique de *Periplaneta americana*. C. R. Seances Soc. Biol. Ses Fil. *167*:1405–1408 (1973)

Chanussot, M. B., Dando, J., Moulins, M., Laverack, M. S.: Mise en évidence d'une amine biogène dans le système nerveux stomatogastrique des Insectes: Etude histochemique et ultrastructurale. C. R. Hebd. Seances Acad. Sci., Ser. D *268*:2101–2104 (1969)

Chen, J. S., Chen, M. G. M.: Modifications of the Bodian technique applied to insect nerves. Stain Technol. *44*:50–51 (1969)

Chen, J. S., Levi-Montalcini, R.: Axonal outgrowth and cell migration in vitro from nervous system of cockroach embryos. Science *166*:631–632 (1969)

Chowdhury, T. K.: Techniques of intracellular micro-injection. In: *Glass Microelectrodes* (M. Lavallée, O. F. Schanne, N. C. Hébert, eds.), pp. 404–423. Wiley, New York, 1969

Christensen, B. N.: Procion brown: An intracellular dye for light and electron microscopy. Science *182*:1255–1256 (1973)

Clark, R. D.: Structural and functional changes in an identified cricket neuron after separation from the soma. I. Structural changes. J. Comp. Neurol. *170*:253–266 (1976a)

Clark, R. D.: Structural and functional changes in an identified cricket neuron after separation from the soma. II. Functional changes. J. Comp. Neurol. *170*:267–277 (1976b)

Coalson, R. E.: Pseudoisocyanin staining of insulin and specificity of empirical islet cell stain. Stain Technol. *41*:121–130 (1966)

Coggshall, J. C.: Neurons associated with the dorsal longitudinal flight muscles of *Drosophila melanogaster*. J. Comp. Neurol. *177*:707–720 (1978)

Cohen, E. B., Pappas, G. D.: Dark profiles in the apparently-normal central nervous system: A problem in the electron microscopic identification of early anterograde axonal degeneration. J. Comp. Neurol. *136*:375–396 (1968)

Cohen, M. J., Jacklet, J. W.: The functional organisation of motor neurons in an insect ganglion. Philos. Trans. R. Soc. London, Ser. B *252*:561–572 (1967)

Colman, D. R., Scalia, F., Cabrales, E.: Light and electron microscopic observations on the anterograde transport of horseradish peroxidase in the optic pathway in the mouse and rat. Brain Res. *102*:156–163 (1976)

Colonnier, M.: The tangential organisation of the visual cortex. J. Anat. *98*:327–344 (1964)

Cooke, I. M., Goldstone, M. W.: Fluorescence localization of monoamines in crab neurosecretory structures. J. Exp. Biol. *53*:651–668 (1970)

Corrodi, H., Jonsson, G.: Fluorescence methods for the histochemical demonstration of monoamines: 4. Histochemical differentiation between dopamine and noradrenaline in models. J. Histochem. Cytochem. *13*:484–487 (1965)

Corrodi, H., Jonsson, G.: Fluoreszenzmethoden zur histochemischen Sichtbarmachung von Monoaminen. 6. Identifizierung der fluoreszierenden Produkte aus m-Hydroxyphenylathylaminen und Formaldehyd. Helv. Chim. Acta *49*:798–806 (1966)

Corrodi, H., Hillarp, N.-Å., Jonsson, G.: Fluorescence methods for the histochemical demonstration of monoamines. 3. Sodium borohydride reduction of the fluorescent compounds as a specificity test. J. Histochem. Cytochem. *12*:582–586 (1964)

Cotman, C., Mathews, D. W.: Synaptic plasma membranes from rat brain synaptosomes: Isolation and partial characterization. Biochim. Biophys. Acta *249*:380–394 (1971)

Cotman, C. W., Banker, G., Churchill, L., Taylor, D.: Isolation of postsynaptic densities from rat brain. J. Cell Biol. *63*:441–455 (1974)

Couteaux, R.: Principaux critères morhologiques et cytochimiques utilisables aujourd'hui pour définir les divers types de synapses. Actual. Neuro-Physiol. *3*:145–173 (1961)

Cowan, W. M.: Anterograde and retrograde transneuronal degeneration in the central and peripheral nervous system. In: *Contemporary Research Methods in Neuroanatomy* (W. J. H. Nauta, S. O. E. Ebbesson, eds.), pp. 217–251. Springer-Verlag, Berlin, New York, 1970

Cowan, W. M., Gottlieb, D. I., Hendrikson, A. E., Price, J. L., Woolsey, T. A.: The autoradiographic demonstration of axonal connections in the central nervous system. Brain Res. *37*:21–51 (1972)

Cox, W.: Imprägnation des centralen Nervensystems mit Quecksilbersalzen. Arch. Mikrosk. Anat. *37*:16–21 (1891)

Credland, P. F., Scales, M. D. C.: The neurosecretory cells of the brain and suboesophageal ganglion of *Chironomus riparius*. J. Insect Physiol. *22*:633–642 (1976)

Criegee, R., Marchand, B., Wannowius, A.: Zur Kenntnis der organishen Osmiumverbindungen. Justus Liebigs Ann. Chem. *550*:99 (1942)

Crossman, A. R., Kerkut, G. A., Walker, R. J.: Electrophysiological studies on the axon pathways of specified nerve cells in the central ganglia of two insect species *Periplaneta americana* and *Schistocerca gregaria*. Comp. Biochem. Physiol. A *43*:393–445 (1972)

Crossman, A. R., Kerkut, G. A., Pitman, R. M., Walker, R. J.: Electrically excitable nerve cell bodies in the central ganglia of two insect species *Periplaneta americana* and *Schistocerca gregaria*. Investigations of cell geometry and morphology by intracellular dye inspection. Comp. Biochem. Physiol. A *40*:579–594 (1971)

Cullheim, S.; Kellerth, J.-O.: A morphological study in the axons and recurrent collaterals of cat sciatic α-motoneurons after intracellular staining with horseradish peroxidase. J. Comp. Neurol. *178*:537–557 (1978)

Cullheim, S., Kellerth, J.-O., Conradi, S.: Evidence for direct synaptic interconnections between cat spinal motoneurones via the recurrent axon collateral: A
morphological study using intracellular injection of horseradish peroxidase.
Brain Res. *132*:1–10 (1977)

Dagan, D., Sarne, Y.: Evidence for the cholinergic nature of cockroach giant
fibres: Use of specific degeneration. J. Comp. Physiol. *126*:157–160 (1978)

Davenport, H. A., Kline, C. L.: Staining paraffin sections with protargol. 1.
Experiments with Bodian's method. 2. Use of *n*-propyl and *n*-butyl alcohol
in Hofker's fixative. Stain Technol. *13*:147–160 (1938)

Davenport, H. A., McArthur, J., Bruesch, S. R.: Staining paraffin sections with
protargol. 3. The optimum pH for reduction. 4. A two-hour staining
method. Stain Technol. *14*:21–26 (1939)

Davenport, H. A., Porter, R. W., Thomas, R. W.: The stainability of nerve fibers
by protargol with various fixatives and staining technics. Stain Technol.
22:41–50 (1947)

Day, M. F.: Neurosecretory cells in the ganglia of Lepidoptera. Nature (London)
145:264 (1940)

Del Rio Hortega, P.: Tercera aportacion al conocimiento morfologico e inter-
pretacion functional de la oligodendroglia. Mem. R. Soc. Esp. Hist. Nat.
14:5 (1928)

Dennison, M. E.: Electron stereoscopy as a mean of classifying synaptic vesicles. J. Cell Sci. *8*:525–539 (1971)

De Robertis, E. D. P.: Submicroscopic organization of some synaptic regions.
Acta Neurol. Latinoam. *1*:3–15 (1955)

De Robertis, E. D. P., Benett, H. S.: Some features of the submicroscopic morphology of synapses in frog and earthworm. J. Biophys. Biochem. Cytol.
1:47–58 (1955)

DeVoe, R. D., Ockleford, E. M.: Intracellular responses from cells of the medulla
of the fly *Calliphora erythrocephala*. Biol. Cybernetics. *23*:13–24 (1976)

Dewhurst, S. A., Croker, S. G., Ikeda, K., McCaman, R. E.: Metabolism of biogenic amines in *Drosophila* nervous tissue. Comp. Biochem. Physiol. B
43:975–981 (1972)

Dogiel, A. S.: Техника окраски нервной системы метиленовой синью.
С.-петербург. Издатель Риккер К.Л. 48 страниц. (1902)

Dogra, G. S.: Neurosecretory system of Heteroptera (Hemiptera) and role of the
aorta as a neurohaemal organ. Nature (London) *215*:199–201. (1967a)

Dogra, G. S.: Studies on the neurosecretory system of the female mole cricket
Gryllotalpa africana (Orthoptera: Gryllotalpidae). J. Zool. *152*:163–178
(1967b)

Dogra, G. S., Tandan, B. K.: Adaptation of certain histological techniques for in
situ demonstration of the neuro-endocrine system of insects and other animals. Q. J. Microsc. Sci. *105*:455–466 (1964)

Domonkos, J.: Lipid metabolism in Wallerian degeneration. In: *Handbook of
Neurochemistry* (A. Lajtha, ed.), Vol. 7, pp. 93–106. Plenum, New York,
1972

Dowling, J. E., Chappell, R. L.: Neural organization of the median ocellus of the
dragonfly. II. Synaptic structure. J. Gen. Physiol. *60*:148–165 (1972)

Drawert, J.: Histochemische und zytophotometrische Untersuchungen an neuro-
sekretorischen Zellen der Staateule Agrotis segetum Schiff. unter besonderer

Berücksichtigung der DDD-Reaktion. Acta Histochem. *24*:345–354 (1966)

Drescher, W.: Regenerationsversuche am Gehirn von Periplaneta americana unter Berücksichtigung von Verhaltensänderung und Neurosekretion. Z. Morphol. Oekol. Tiere *48*:576–649 (1960)

Duboscq, O.: Recherches sur les chilopodes. Arch. Zool. Exp. Gen. [3] *6*:481–650 (1898)

Dürig, A.: Das Formalin als Fixierungsmittel anstatt der Osmiumsäure bei der Methode Ramón y Cajal's. Anat. Anz. *10*:659–670 (1895)

Dvorak, D. R., Bishop, L. G., Eckert, H. E.: On the identification of movement detectors in the fly optic lobe. J. Comp. Physiol. *100*:5–23 (1975a)

Dvorak, D., Bishop, L. G., Eckert, H. E.: Intracellular identification of a directionally selective motion detecting neuron in the fly optic lobe. Vision Res. *15*:451–453 (1975b)

Dyck, L. E., Robertson, H. A.: Micro-estimation of monoamine oxidase activity in invertebrates. Proc. Can. Fed. Biol. Soc. *19*:182p (1976)

Ebbesson, S. O. E.: The selective silver-impregnation of degenerating axons and their synaptic endings in nonmammalian species. In: *Contemporary Research Methods in Neuroanatomy* (W. J. H. Nauta, S. O. E. Ebbesson, eds.), pp. 132–161. Springer-Verlag, Berlin, New York, 1970

Eckert, H. E., Bishop, L. G.: Anatomical and phyiological properties of the vertical cells in the third optic ganglion of *Phaenicia sericata* (Diptera, Calliphoridae). J. Comp. Physiol. *126*:57–86 (1978)

Edwards, J. S., Palka, J.: The cerci and abdominal giant fibres of the house cricket, *Acheta domesticus*. I. Anatomy and physiology of normal adults. Proc. R. Soc. London, Ser. B *185*:83–103 (1974)

Edwards, J. S., Palka, J.: Neural generation and regeneration in insects. In: *Simpler Networks and Behavior* (J. C. Fentress, ed.), pp. 167–185. Sinauer Assoc., Sunderland, Massachusetts, 1976

Ehinger, B., Falck, B.: Adrenergic retinal neurons of some New World monkeys. Z. Zellforsch. Mikrosk. Anat. *100*:364–375 (1969)

Ehrlich, P.: Über die Methylenblaureaktion der lebenden Nervensubstanz. Dtsch. Med. Wochenschr. *4*:49–52 (1886)

Eibl, E.: Morphologische und neuroanatomische Untersuchungen zur Sinnesorgananstattung der proximalen Tibienabschnitte und ihrer zentralen Projektionen bei Grillen. Doctoral Dissertation, University of Köln (1976)

Elftman, H.: Aldehyde-fuchsin for pituitary cytochemistry. J. Histochem. Cytochem. *7*:96 (1956)

Elofsson, R., Klemm, N.: Monoamine-containing neurons in the optic ganglia of crustaceans and insects. Z. Zellforsch. Mikrosk. Anat. *133*:475–499 (1972)

Emson, P. C., Burrows, M., Fonnum, F.: Levels of glutamate decarboxylase, choline acetyltransferase and acetylcholinesterase in identified motoneurons of the locust. J. Neurobiol. *5*:33–42 (1974)

Eränkö, O.: Light and electronmicroscopic histochemical evidence of granular and non-granular storage of catecholamines in the sympathetic ganglion of the rat. Histochem. J. *4*:213–224 (1972)

Eränkö, O.: Histochemical demonstration of catecholamines in sympathetic ganglia. Ann. Histochem. *21*:83–100 (1976)

Ernst, K.-D., Boeckh, J., Boeckh, V.: A neuroanatomical study on the organization of the central antennal pathways in insects. II. Deutocerebral connections in *Locusta migratoria*. Cell Tissue Res. *176:*285–308 (1977)

Evans, P. H., Fox, P. M.: Comparison of various biogenic amines as substrates for acetyltransferase from *Apis mellifera* (L.) CNS. Comp. Biochem. Physiol. C *51:*139–141 (1975)

Evans, P., O'Shea, M.: The identification of an octopaminergic neuron which modulates neuromuscular transmission in the locust. Nature (London) *270:*257–259 (1977)

Ewen, A. B.: An improved aldehyde fuchsin staining technique for neurosecretory products in insects. Trans. Am. Microsc. Soc. *81:*94–96 (1962)

Falck, B.: Observation on the possibilities for the cellular localization of monoamines with a fluorescence method. Acta Physiol. Scand. *56,* Suppl. 197: 1–25 (1962)

Farley, R. D., Milburn, N. S.: Structure and function of the giant fibre system in the cockroach, *Periplaneta americana*. J. Insect Physiol. *15:*457–476 (1969)

Fedosova, I. A.: Neuropile architectonics of the last abdominal ganglion in the crayfish *Astacus astacus*. Z. Evol. Biokhim. fiziol. *10:*503–507 (1974)

FitzGerald, M. J. T.: A general-purpose silver technique for peripheral nerve fibers in frozen sections. Stain Technol. *38:*321–327 (1963)

FitzGerald, M. J. T.: The double-impregnation silver technique for nerve fibres in paraffin sections. Q. J. Microsc. Sci. *105:*359–361 (1964)

Foelix, R. F.: Occurrence of synapses in peripheral sensory nerves in arachnids. Nature (London) *254:*146–147 (1975)

Foley, J. O.: A protargol method for staining nerve fibers in frozen or celloidin sections. Stain Technol. *18:*27–53 (1943)

Fominych, M. Ya.: Чувствителъные элементы туловишного моэга некоторых вицов аннелид: Сб. Туловищный мозг членистоногих и червей. *The Truncal Brain of Arthropods and Worms.* Works of the Institute of Biology and Soil DVNC of the Academy of Sciences of the USSR, "Nauka" (Leningrad). 136–167 (1977)

Fourtner, C. R., Pearson, K. G.: Morphological and physiological properties of motor neurons innervating insect leg muscles. In: *Identified Neurons and Behavior of Arthropods* (G. Hoyle, ed.), Chapter 5, pp. 87–100. Plenum, New York, 1977

Fox, C. A., Ubeda-Purkiss, M., Ihrig, H. K., Biogoli, D.: Zinc chromate modification of the Golgi technique. Stain Technol. *26:*109–14 (1951)

Franke, W. W., Krien, S., Brown, R. M.: Simultaneous glutaraldehyde-osmium tetroxide fixation with postosmication. An improved fixation procedure for electron microscopy of plant and animal cells. Histochemie *19:*162–164 (1969)

Fregerslev, S., Blackstadt, T. W., Fredens, K., Holm, M. J.: Golgi potassium-dichromate silver-nitrate impregnation. Histochemie *25:*63–71 (1971a)

Fregerslev, S., Blackstadt, T. W., Fredens, K., Holm, M. J., Ramon-Moliner, E.: Golgi impregnation with mercuric chloride: Studies on the precipitate by X-ray powder diffraction and selected area electron diffraction. Histochemie *26:*289–304 (1971b)

Frontali, N.: Histochemical localization of catecholamines in the brain of normal and drug-treated cockroaches. J. Insect Physiol. *14:*881–886 (1968)

Frontera, J. G.: Improved Golgi-type impregnation of nerve cells. Anat. Rec. *148:*371–372 1964. See discussion in: *Contemporary Research Methods in*

Neuroanatomy (J. H. Nauta and S. O. E. Ebbesson, eds.), pp. 50–52. Springer-Verlag, Berlin, New York, 1970

Fry, J. P., House, C. R., Sharman, D. F.: Analysis of the catecholamine content of the salivary gland of the cockroach. Br. J. Pharmacol. *51:*166–177 (1974)

Fuller, P. M., Prior, D. J.: Cobalt iontophoresis techniques for tracing afferent and efferent connections in the vertebrate CNS. Brain Res. *88:*211–220 (1975)

Furtado, A. F.: Etude histophysiologique des cellules neurosécrétrices de la pars intercerebralis de larves femelles de *Panstrongylus megistus* (Heteroptera: Reduviidae). C. R. Hebd. Seances Acad. Sci., Ser. D *283:*163–166 (1976)

Furtado, A. F.: Etude des cellules neurosécrétrices, de leur site de décharge et des corpora cardiaca chez une Punaise vivipare, *Stilbocoris natalensis* (Hétéroptères, Lygeidés). C. R. Hebd. Seances Acad. Sci., Ser D *272:*2364–2367 (1971)

Furshpan, E. J., Furukawa, T.: Intracellular and extracellular responses of the neural region of the Mauthner cell of the goldfish. J. Neurophysiol. *25:*732–771 (1962)

Fuxe, K., Jonsson, G.: A modification of the histochemical fluorescence method for the improved localization of 5-hydroxytryptamine. Histochemie *11:*161–166 (1967)

Gabe, M.: Sur quelques applications de la coloration par la fuchsine-paraldéhyde. Bull. Microsc. Appl. [2] *3:*153–162 (1953)

Gabe, M.: Données histologiques sur la neurosécrétion chez les Arachnides. Arch. Anat. Microsc. Morphol. Exp. *44:*351–383 (1955)

Gabe, M.: *Neurosecretion.* Pergamon, Oxford, 1966

Gabe, M.: Données histochimiques sur l'évolution du produit de neurosécrétion protocéphalique des Insectes Ptérygotes au cours de son cheminement axonal. Acta Histochem. *43:*168–183 (1972)

Gallyas, F.: A principle for silver staining of tissue elements by physical development. Acta Morphol. Acad. Sci. Hung. *19:*57–91 (1971)

Ganagarajah, M., Saleuddin, A. S. M.: A simple manoeuver to prevent loss of sections for performic acid-Victoria blue technique and a comparison of various neurosecretory stains. Can. J. Zool. *48:*1457–1458 (1970)

Gatenby, J. B., Beams, H. W.: *The Microtomist's Vade-Mecum.* Churchill, London, 1950

Geisert, E. E., Jr.: The use of tritiated horseradish peroxidase for defining neuronal pathways: A new application. Brain Res. *117:*130–135 (1976)

Gerolt, P.: Mode of entry of contact insecticides. J. Insect Physiol. *15:*563–580 (1969)

Gerolt, P.: Mode of entry of oxime carbamates into insects. Pestic. Sci. *3:*43–55 (1972)

Gersch, M., Hentschel, E., Ude, J.: Aminerge Substanzen in lateralen Herynerven und im somatogastrischen Nervensystem der Schabe *Blaberus craniifer* Burm. Zool. Jahrb., Abt. Allg. Zool. Physiol. *78:*1–15 (1974)

Gerschenfeld, H. M.: Chemical transmission in invertebrate central nervous systems and neuromuscular junctions. Physiol. Rev. *53:*1–119 (1973)

Ghysen, A., Deak, I. I.: Experimental analysis of sensory nerve pathways in *Drosophila.* Wilhelm Roux's Arch. Develop. Biol. *184:*273–283 (1978)

Gillette, R., Pomeranz, B.: Neuron geometry and circuitry via the electron microscope: Intracellular staining with osmiophilic polymer. Science *182*:1256–1258 (1973)

Gillette, R., Pomeranz, B.: Ultrastructural correlates of interneuronal function in the abdominal ganglion of *Aplysia californica*. J. Neurobiol. *6*:463–474 (1975)

Girardie, A., Girardie, J.: Etude histologique, histochimique et ultrastructurale de la pars intercerebralis chez *Locusta migratoria* L. (Orthoptère). Z. Zellforsch. Mikrosk. Anat. *78*:54–75 (1967)

Glaser, S., Miller, J., Xuong, N. G., Selverston, A.: Computer reconstruction of invertebrate nerve cells. In: *Computer Analysis of Neuronal Structure* (R. D. Lindsay, ed.), pp. 21–58. Plenum, New York, 1977

Glassner, H. F., Breslau, A. M., Agress, C. M.: Silver staining of nerve fibers in cardiac tissue. Stain Technol. *29*:189–195 (1954)

Glauert, A. M.: *Practical Methods in Electron Microscopy*, Vol. 3. Am. Elsevier, New York, 1975

Glauert, A. M., Glauert, R. H.: Araldite as an embedding medium for electron microscopy. J. Biophys. Biochem. Cytol. *4*:191–194 (1958)

Goetz, K. G.: Optomotorische Untersuchungen des visuellen Systems einiger Augenmutanten der Fruchtfliege *Drosophila*. Kybernetik *2*:77–92 (1964)

Goetz, K. G.: Flight control in *Drosophila* by visual perception of motion. Kybernetik. *4*:199–208 (1968)

Goetz, K. G.: Visual control of orientation patterns. 1. Processing of cues from the moving environment in the *Drosophila* navigation system. In: *Information Processing in the Visual Systems of Arthropods* (R. Wehner, ed.), pp. 255–263. Springer-Verlag, Berlin, New York, 1972

Goetz, K. G., Wenking, H.: Visual control of locomotion in the walking fruitfly *Drosophila*. J. Comp. Physiol. *85*:235–266 (1973)

Golgi, C.: Sulla struttura della sostanza grigia dell cervello. Gazz. Med. Lombarda *33*:244–246 (1873)

Golgi, C.: Un nuovo processo di technica microscopia. R. C. Inst. Lombardo Sci. Ser. 2 *12*:5 (1878)

Golgi, C.: Di una nuova reazione apparentemente nera della cellule nervose cerebrali ottenuta col bichlorura di mercurio. Arch. Sci. Med. *3*, fasc.1:1–7 (1879)

Golgi, C.: Modificazione del metodo di colorazione deli elementi nervosi col bicloruro di mercurio. Riv. Med. Nap. *7*:193–194 (1891)

Goll, W.: Strukturentersuchungen am Gehirn von Formica. Z. Morphol. Oekol. Tiere *59*:143–210 (1967)

Goodman, C.: Anatomy of locust ocelar interneurons: Constancy and variability. J. Comp. Physiol. *95*:185–201 (1974)

Goodman, C.: Anatomy of the ocellar interneurons of acridid grasshoppers. I. The large interneurons. Cell Tissue Res. *175*:183–202 (1976a)

Goodman, C.: Constancy and uniqueness in a large population of small interneurons. Science *193*:502–504 (1976b)

Goodman, C., Williams, J. L. D.: Anatomy of the ocellar interneurons of acridid grasshoppers. II. The small interneurons. Cell Tissue Res. *175*:203–225 (1976)

Goodman, L. J., Patterson, J. A., Mobbs, P. G.: The projection of ocellar neurons within the brain of the locust *Schistocerca gregaria*. Cell Tissue Res. *157:*467–492 (1975)

Graham, R. C., Karnovsky, M. J.: The early stages of absorption of injected horseradish peroxidase in the proximal tubules of mouse kidney: Ultrastructural cytochemistry by a new technique. J. Histochem. Cytochem. *14:*291–302 (1966)

Gray, E. G., Guillery, R. W.: Synaptic morphology in the normal and degenerating nervous system. Int. Rev. Cytol. *19:*111–182 (1966)

Gray, E. G.: Electron microscopy of presynaptic organelles of the spinal cord. J. Anat. *97:*101–106 (1973)

Gray, P.: *The Microtomist's Formulary and Guide*. Constable, London, 1954

Gregory, G. E.: Silver staining of insect central nervous systems by the Bodian protargol method. Acta Zool. (Stockholm) *51:*169–178 (1970)

Gregory, G. E.: Simple fluorescence staining of insect central nerve fibres with Procion yellow. Stain Technol. *48:*85–87 (1973)

Gregory, G. E.: Neuroanatomy of the mesothoracic ganglion of the cockroach *Periplaneta americana* (L.). I. The roots of the peripheral nerves. Phil. Trans. R. Soc. London, Ser. B *267:*421–465 (1974a)

Gregory, G. E.: Effect of sodium sulfite purity on color of Bodian silver staining of insect central nervous systems. Stain Technol. *49:*401–405 (1974b)

Gregory, G. E.: Alcoholic Bouin fixation of insect nervous systems for Bodian silver staining. II. Modified solutions. (In preparation)

Gregory, G. E.: Alcoholic Bouin fixation of insect nervous systems for Bodian silver staining. III. A shortened single impregnation method. (In preparation)

Gregory, G. E., Greenway, A. R., Lord, K. A.: Alcoholic Bouin fixation of insect nervous systems for Bodian silver staining. I. Composition of 'aged' fixative. (In preparation)

Griffiths, G. W.: Transport of glial cell acid phosphatase by endoplasmic reticulum into damaged axons. J. Cell Sci. *36:*361–389 (1979)

Griffiths, G., Boschek, B.: Rapid degeneration of visual fibers following retinal lesions in the dipteran compound eye. Neurosci. Lett. *3:*253–258 (1976)

Guillery, R. W.: Light- and electron-microscopical studies of normal and degenerating axons. In: *Contemporary Research Methods in Neuroanatomy* (W. J. Nauta, S. O. E. Ebbesson, eds.), pp. 77–105. Springer-Verlag, Berlin, New York, 1970

Günther, J., Schürmann, F. W.: Zur Feinstruktur des dorsalen Riesenfasersystems im Bauchmark des Regenwurms. II. Synaptische Beziehungen der proximalen Riesenfaserkollateralen. Z. Zellforsch. Mikrosk. Anat. *139:*369–396 (1973)

Guthrie, D. M.: The regeneration of motor axons in an insect. J. Insect Physiol. *13:*1593–1611 (1967)

Hagmann, L. E.: A method for injecting insect tracheae permanently. Stain Technol. *15:*115–118 (1940)

Håkanson, R., Sjöberg, A.-K., Sundler, F.: Formaldehyde-induced fluorescence of peptides with N-terminal 3,4-dihydroxyphenylalanine of 5-hydroxytryptophan. Histochemie *28:*367–371 (1971)

Hake, T.: Studies on the reactions of OsO_4 and $KMnO_4$ with amino acids, peptides and proteins. Lab. Invest. *14:*1208–1212 (1965)

Halmi, N. S.: Differentiation of two types of basophils in the adenohypophysis of the rat and the mouse. Stain Technol. *27*:61–64 (1952)

Hamberger, B., Malmfors, T., Sachs, C.: Standardization of paraformaldehyde and of certain procedures for the histochemical demonstration of catecholamines. J. Histochem. Cytochem. *13*:147p (1965)

Haug, F. M. S.: Heavy metals in the brain. Ergeb. Entwicklungsmech. Org. Anat. Entwicklungsgesch. *47*, Fasc.4:10–71 (1973)

Hausen, K.: Functional characterization and anatomical identification of motion sensitive neurons in the lobula plate of the blowfly *Calliphora erythrocephala*. Z. Naturforsch., Teil C *31*:629–633 (1976a)

Hausen, K.: Struktur, Funktion und Konnektivität bewegungsempfindlicher Interneurone im dritten optischen Neuropil der Schmeissfliege *Calliphora erythrocephala*. Ph.D. Thesis, University of Tübingen (1976b)

Hausen, K.: Signal processing in the insect eye. In: *Function and Formation of Neural Systems* (G. Stent, ed.), Dahlem Konferenzen, Berlin, 1977

Hausen, K., Strausfeld, N. J.: Sexually dimorphic interneuron arrangements in the fly visual system. Proc. R. Soc. Lond. Ser. B *208*:57–71 (1980)

Hayat, M. A.: *Principle and Techniques of Electron Microscopy*, Vol. 1. Van Nostrand-Reinhold, Princeton, New Jersey, 1970

Hayat, M. A.: *Positive Staining for Electron Microscopy*. Van Nostrand-Reinhold, New York, 1975

Heimer, L.: Selective silver-impregnation of degenerating axoplasm. In: *Contemporary Research Methods in Neuroanatomy* (W. J. H. Nauta, S. O. E. Ebbesson, eds.), pp. 106–131. Springer-Verlag, Berlin, New York, 1970

Hengstenberg, R.: Das Augenmuskelsystem der Stubenfliege *Musca domestica*. I. Analyse der "clock-spikes" und ihrer Quellen. Kybernetik *2*:56–77 (1971)

Hengstenberg, R.: The effect of pattern movement on the impulse activity of the cervical connective of *Drosophila melanogaster*. Z. Naturforsch., Teil C *28*:593–597 (1973)

Hengstenberg, R.: Spike responses of "non-spiking" visual interneurons. Nature (London) *270*:338–340 (1977)

Herdman, P. J., Taylor, J. J.: Suppression of connective tissue impregnation in a silver technique for demonstrating nerve fibres. Stain Technol. *50*:37–42 (1975)

Herkenham, M., Nauta, W. J. H.: Afferent connections of the habenular nuclei in the rat. A horseradish peroxidase study, with a note on the fiber-of-passage problem. J. Comp. Neurol. *173*:123–145 (1977)

Herlant, M.: L'hypophyse et le système hypothalamohypophysaire du Pangolin (*Manis tricuspis* Raf. et *Manis tetradactyla* L.). Arch. Anat. Microsc. Morphol. Exp. *47*:1–24 (1958)

Herman, M. M., Miquel, J., Johnson, M.: Insect brain as a model for the study of aging. Age-related changes in *Drosophila melanogaster*. Acta Neuropathol. *19*:167–183 (1971)

Hess, A.: Experimental anatomical studies of pathways in the severed central nerve cord of the cockroach. J. Morphol. *103*:479–502 (1958a)

Hess, A.: The fine structure of nerve cells and fibers, neuroglia, and sheaths of the ganglion chain in the cockroach (*Periplaneta americana*). J. Biophys. Biochem. Cytol. *4*:731–742 (1958b)

Hess, A.: The fine structure of degenerating nerve fibers, their sheaths, and their terminations in the central nerve cord of the cockroach (*Periplaneta americana*). J. Biophys. Biochem. Cytol. *7:*339–344 (1960)

Highnam, K. C.: The histology of the neurosecretory system of the adult female desert locust, *Schistocerca gregaria*. J. Microsc. Sci. *102:*27–38 (1961)

Hirsch, J. C., Fedorko, M. E.: Ultrastructure of human leucocytes after simultaneous fixation with glutaraldehyde and osmium tetroxide and 'postfixation' in uranyl acetate. *J. Cell Biol. 38:*615–627 (1968)

Hökfelt, T.: A modification of the histochemical fluorescence method for the demonstration of catecholamines and 5-hydroxytryptamine, using Araldite as embedding medium. J. Histochem. Cytochem. *13:*518–519 (1965)

Hökfelt, T.: In vitro studies on central and peripheral monoamine neurons at the ultrastructural level. Z. Zellforsch. Mikrosk. Anat. *91:*1–74 (1968)

Holmes, W.: A new method for the impregnation of nerve axons in mounted paraffin sections. J. Pathol. Bacteriol. *54:*132–136 (1942)

Holmes, W.: Silver staining of nerve axons in paraffin sections. Anat. Rec. *86:*157–187 (1943)

Holmes, W.: The peripheral nerve biopsy. In: *Recent Advances in Clinical Pathology* (S. C. Dyke, ed.), Chapter 38, pp. 402–417. Churchill, London, 1947

Holmgren, E.: Studier öfuer hodens och Körtelartade hudorganens morphology: hos skandinaviska makrolepidopterlarver. Kongl. Svenska Vertenskaps-Akademiens Handlinger. Stockholm, *27:*3–82 (1895)

Holmgren, E.: Zur Kenntnis des Hauptnervensystems der Arthropoden. Anat. Anz. *12:*449–457 (1896)

Honegger, H. W.: Interommatidial hair receptor axons extending into the ventral nerve cord in the cricket *Gryllus campestris*. Cell Tissue Res. *182:*281–285 (1977)

Honegger, H.-W., Schürmann, F. W.: Cobalt sulphide staining of optic fibres in the brain of the cricket, *Gryllus campestris*. Cell Tissue Res. *159:*213–225 (1975)

Horridge, G. A., Burrows, M.: Synapses upon motoneurons of locusts during retrograde degeneration. Phil. Tans. R. Soc. Lond. Ser. B *269:*95–108 (1974)

Hoyer, H.: Über die Anwendung des Formaldehyds in der histologischen Technik. Verh. Anat. Ges. Anat. Anz. Ergenzungsheft *9:*236–238 (1894)

Hoyle, G.: Cellular mechanism underlying behaviour-neuroethology. Adv. Insect Physiol. *5:*349–444 (1970)

Hoyle, G.: The dorsal, unpaired median neurons of the locust metathoracic ganglion. J. Neurobiol. *9:*43–57 (1978)

Hoyle, G., Burrows, M.: Neural mechanisms underlying behavior in the locust *Schistocerca gregaria*. 1. Physiology of identified motoneurons in the metathoracic ganglion. J. Neurobiol. *4:*3–41 (1973)

Hoyle, G., Dagan, D., Moberly, B., Colquhoun, W.: Dorsal unpaired median insect neurons make neurosecretory endings on skeletal muscle. J. Exp. Zool. *187:*159–165 (1974)

Huber, P.: Aldehyd-Thionin: ein Farbeagens mit erweiterten Anwendungsmöglichkeiten. Mikroskopie *18:*317–338 (1963)

Hunt, S. P., Streit, P., Künzle, H., Cuénod, M.: Characterization of the pigeon isthmo-tectal pathway by selective uptake and retrograde movement of radioactive compounds and by Golgi-like horseradish peroxidase labeling. Brain Res. *129:*197–212 (1977)

Iles, J. F.: Organisation of motoneurones in the prothoracic ganglion of the cockroach *Periplaneta americana* (L). Phil. Trans. R. Soc. Lond. Ser. B *276:* 205–219 (1976a)

Iles, J. F.: Central terminations of muscle afferents on motoneurones in the cat spinal cord. J. Physiol. (London) *262:*91–117 (1976b)

Iles, J. F., Mulloney, B.: Procion yellow staining of cockroach motor neurones without the use of microelectrodes. Brain Res. *30:*397–400 (1971)

Ittycheriah, P. I.: Resorcin-fuchsin and in situ demonstration of neurosecretory material in insects. Folia Entomol. Mex. *28:*47–49 (1974)

Ittycheriah, P. I., Marks, E. P.: Performic acid-Resorcin Fuschsin: A technique for in the in situ demonstration of neurosecretory material in insects. Ann. Entomol. Soc. Am. *64:*762–765 (1971)

James, T. H.: Mechanism of photographic development. I. The general effect of oxidation on the development process and the nature of the induction period. J. Phys. Chem. *43:*701–719 (1939a)

James, T. H.: Mechanism of photographic development. II. Development by hydroquinone. J. Phys. Chem. *44:*42–57 (1939b)

James, T. H.: The reduction of silver ions by hydroquinone. J. Chem. Soc. *61:*648–652 (1939c)

Johansson, A. S.: Relation of nutrition to endocrine-reproductive functions in the milkweed bug *Oncopeltus fasciatus* (Dallas) (Heteroptera, Lygaeidae). Nytt Mag. Zool. *7:*1–132 (1958)

Jones, D. G.: The fine structure of the synaptic membrane adhesions on Octopus synaptosomes. Z. Zellforsch. Mikrosk. Anat. *88:*457–469 (1968)

Jonsson, G.: Quantitation of fluorescence of biogenic monoamines. Prog. Histochem. Cytochem. *2/4:*1–36 (1971)

Joslyn, M. A., Marsh, G. L.: The role of peroxidase in the deterioration of frozen fruits and vegetables. Science *78:*174–175 (1933)

Kalmring, K., Römer, H., Rehbein, H. G.: Connections of acoustic neurones within the CNS of grasshoppers. Naturwissenschaften *61:*454–455 (1974)

Kanaseki, T., Kadota, K.: The "vesicle in a basket." J. Cell Biol. *42:*202–220 (1969)

Kaneko, C. R. S., Kater, S. B.: Intracellular staining techniques in gastropod molluscs. In: *Intracellular Staining in Neurobiology* (S. B. Kater, C. Nicholson, eds.), pp. 151–156. Springer-Verlag, Berlin, New York, 1973

Kater, S. B., Nicholson, C., eds.: *Intracellular Staining in Neurobiology.* Springer-Verlag, Berlin, New York, 1973

Kater, S. B., Nicholson, C., Davis, W. J.: A guide to intracellular staining techniques. In: *Intracellular Staining in Neurobiology* (S. B. Kater, C. Nicholson, eds.), pp. 307–325. Springer-Verlag, Berlin, New York, 1973

Karelin, Yu. A.: On structural organisation of the descending pathways involved in initiation of the flight in the locust *Locusta migratoria*. Zh. Evol. Biokhim. Fiziol. *10:*526–528 (1974)

Karnovsky, M. J.: A formaldehyde-glutaraldehyde fixation of high osmolarity for use in electron microscopy. J. Cell Biol. *27:*137A (1965)

Katz, B.: *Nerve, Muscle, and Synapse.* McGraw-Hill, New York, 1966

Kawana, E., Akert, K., Sandri, C.: Zinc iodide osmium-tetroxide impregnation of nerve terminals in the spinal cord. Brain Res. *16:*325–331 (1969)

Kay, D.: *Techniques for Electron Microscopy.* Blackwell, Oxford, 1965

Keefer, D. A.: HRP as a retrogradely transported detailed dendritic marker. Brain Res. *140:*15 (1978)

Keefer, D. A., Spatz, W. B., Misgeld, U.: Golgi-like staining of neocortical neurons using retrogradely transported horseradish peroxidase. Neurosci. Lett. *3:*233–237 (1976)

Kemali, M.: A modification of the rapid Golgi method. Stain Technol. *51:*169–172 (1976)

Kenyon, F. C.: The meaning and structure of the so-called "mushroom bodies" of the hexapod brain. Am. Nat. *30:*643–650 (1896a)

Kenyon, F. C.: The brain of the bee. A preliminary contribution to the morphology of the nervous system of the arthropoda. J. Comp. Neurol. *6:*133–210 (1896b)

Kerkut, G. A.: Catecholamines in invertebrates. Br. Med. Bull. *29:*100–104 (1973)

Kien, J.: A preliminary report on cobalt sulphide staining of locust visual interneurons through extracellular electrodes. Brain Res. *109:*158–164 (1976)

Kind, T. W.: The neurosecretory system of some Lepidoptera in relation to diapause and metamorphosis. In: *Neurosecretory Elements and Their Significance in Organisms* (N. L. Gerbilsky, A. L. Polenov, eds.), pp. 178–183. Moscow-Leningrad, 1964 (in Russian)

Kind, T. W.: The functional morphology of the insect neurosecretory system during active development and under different types of diapause. In: *Photoperiodic Adaptations in Insects and Acari* (A. S. Danilevsky, ed.), pp. 153–191. Leningrad University, Leningrad, 1968 (in Russian with English summary)

Kirkham, J. S., Goodman, L. J., Chappell, R. L.: Identification of cobalt in processes of stained neurones using X-ray energy spectra in the electron microscope. Brain Res. *85:*33–37 (1975)

Kirschfeld, K.: Die Projektion der optischen Umwelt auf das Raster der Rhabdomere im Komplexauge von Musca. Exp. Brain Res. *3:*248–270 (1967)

Kirschfeld, K.: Aufnahme und Verarbeitung optischer Daten in Komplexauge von Insekten. Naturwissenschaften *58:*201–209 (1971)

Klemm, N.: Monoaminhaltige Strukturen im Zentralnervensystem der Trichoptera (Insecta) Teil I. Z. Zellforsch. Mikrosk. Anat. *92:*487–502 (1968a)

Klemm, N.: Monoaminerge Zellelemente im stomatogastrischen Nervensystem der Trichopteren (Insecta) Z. Naturforsch., Teil B *23:*1297–1280 (1968b)

Klemm, N.: Monoaminhaltige Strukuren in Zentralnervensystem der Trichoptera (Insecta) Teil II. Z. Zellforsch. Mikrosk. Anat. *117:*537–558 (1971a)

Klemm, N.: Monoaminhaltige Zellelemente im stomatogastrischen Nervensystem und in den Corpora cardiaca von *Schistocerca gregaria* Forsk. (Insecta, Orthoptera). Z. Naturforsch., Teil B *26:*1085–1086 (1971b)

Klemm, N.: Monoamine-containing nervous fibres in foregut and salivary gland of the desert locust, *Schistocerca gregaria* Forskål (Orthoptera, Acrididae). Comp. Biochem. Physiol. A *43:*207–211 (1972)

Klemm, N.: Vergleichende-histochemische Untersuchungen über die Verteilung monoaminhaltiger Strukturen in Oberschlundganglion von Angehörigen verschiedener Insektenordnungen. Entomol. Ger. *1*:21–49 (1974)

Klemm, N.: Histochemistry of putative transmitter substances in the insect brain. Prog. Neurobiol. *7*:99–169 (1976).

Klemm, N.: Biogenic monoamines in the stomatogastric nervous system of members of several insect orders. Ent. Gen. (1980, in press)

Klemm, N., Axelsson, S.: Determination of dopamine, noradrenaline and 5-hydroxytryptamine in the cerebral ganglion of the desert locust, *Schistocerca gregaria* Forsk. (Insecta, Orthoptera). Brain Res. *57*:289–298 (1973)

Klemm, N., Björklund, A.: Identification of dopamine and noradrenaline in the nervous structures of the insect brain. Brain Res. *26*:459–464 (1971)

Klemm, N., Falck, B.: Monoamines in the pars intercerebralis-corpus cardiacum-complex in locusts. Gen. Comp. Endocrinol. *34*:180–192 (1978)

Klemm, N., Schneider, G.: Selective uptake of indolamine into nervous fibers in the brain of the desert locust, *Schistocerca gregria* Froskål (Insecta). A fluorescence and electron microscopic investigation. Comp. Biochem. Physiol. C *50*:177–182 (1975)

Knyazeva, N. I.: [Рецепторы конечностей азиатской саранчи (*Locusta migratoria* L.), участвующие в рефлексе запуска полета. В кн.: Современные проблемы структуры и функции нервной системы насекомых. (Тр.Всерос. Энтомол.общ. 53). Ленинград, 132–147]. The receptors of the extremities in *Locusta migratoria* L. involved in the reflex initiating flight. Present problems of the structure and the function of the nervous system in insects. Tr. Vseross Entomol. Obshch. *53*:132–147 (1969)

Knyazeva, N. I.: Morphofunctional specialization of the neurons of the II-nd type in *Locusta migratoria*. Zh. Evol. Biokhim. Fiziol. *6*:81–87 (1970a)

Knyazeva, N. I.: [Проприоцепторы крылового аппарата, регулирующие полет азиатской саранчи. *Locusta migratoria* L. Энтомол.обозр. *49*:517–522.] The proprioceptors of the wing apparatus, regulating the flight of *Locusta migratoria* L. Entomol. Obozr. *49*:517–522 (1970b)

Knyazeva, N. I.: [Характер иннервации мультиполярными .нейронами мышц и связанных с ними тканей у .азиатской сарамчи *Locusta migratoria* L. (Orthoptera, Acrididae) Энтомол.обзр. *55*:36–40.] The character of the innervation of the muscles and the tissue attached by multipolar neurons in *Locusta migratoria* L. (Orthoptera, Acrididae). Entomol. Obozr. *55*:36–40 (1976)

Knyazeva, N. I., Fudalewicz-Niemczy, K. W., Rosciszewska, M.: Proprioceptors of the house-cricket (*Gryllus domesticus* L.) (Ortoptera). Acta Biol. Cracov. Ser. Zool. *18*:33–44 (1975)

Kolb, H.: Organization of the outer plexiform layer of the primate retina: Electron microscopy of Golgi-impregnated cells. Phil. Trans. R. Soc. Lond. Ser. B *258*:261–283 (1970)

Kolmogorova, E. J.: Structure of the central parts of the nervous system in *Opistorchis felineus*. Zool. Zh. *38*:1627–1633 (1959)

Kolmogorova, E. J.: [Нервная система метацеркария лнчинки *Opisthorchis felineus*.] [Сб.научных работ .кафедры гистол. и змбриол. Пермского мед.инст.: .43–49.] *The Nervous System of the Metacercaria of the Larva Opisthorchis felineus*, pp. 43–49. Collection of Scientific Works of the Department of Histology and Embryology at the Medical Institute of Perm, 1960.

Konečný, M., Pličzka, Z.: Über die Möglichkeiten der Anwendung des Aldehyd-Fuchsins (Gomori) in der Histochemie. Acta Histochem. *5:*247–260 (1958)

Kopsch, F.: Erfahrungen über die Verwendung der Formaldehyde bei der Chromsilber-Imprägnation. Anat. Anz. *11:*727–729 (1896)

Kraehenbuhl, J. P., Calardy, R. E., Jamieson, J. D.: Preparation and characterization of an immuno-electron microscope tracer consisting of a heme-octapeptide coupled to Fab. J. Exp. Med. *139:*208–223 (1974)

Krasne, F. B., Lee, S. H.: Survival of functional synapses on crustacean neurons lacking cell bodies. Brain Res. *121:*43–57 (1977)

Kravitz, E. A., Stretton, A. O. W., Alvarez, J., Furshpan, E. J.: Determination of neuronal geometry using an intracellular dye injection technique. Fed. Proc., Fed. Am. Soc. Exp. Biol. *27:*749 (1968)

Kristensson, K., Olsson, Y.: Retrograde axonal transport of protein. Brain Res. *29:*363–365 (1971)

Kristensson, K., Olsson, Y.: Retrograde transport of horseradish peroxidase in transected axons. 3. Entry into injured axons and subsequent localization in perikaryon. Brain Res. *115:*201–213 (1976)

Kristensson, K., Olsson, Y., Sjöstrand, J.: Axonal uptake and retrograde transport of exogenous proteins in the hypoglossal nerve. Brain Res. *29:*399–406 (1971)

Kuffler, S. W., Nicholls, J. G.: *From Neuron to Brain.* Sinauer Assoc., Sunderland, Massachusetts, 1976

Lachi, P.: La formalina como mezzo de fissazione in sostituzione all'acido osmico nel metodo de Ramón y Cajal. Anat. Anz. *10:*790 (1895)

Lafon-Cazal, M., Arluison, M.: Localization of monoamines in the corpora cardica and the hypocerebral ganglion of locusts. Cell Tissue Res. *172:*517–527 (1976)

Lagutenko, Yu. P.: [К морфолгии элементов ассоциативного .аппарата брюшного мозга медицинской пиявки. Вопросы .нейрогистологии, биофизики парабиоза и биофизической .нейродинамики. Тр.Ленингр.о-ва естествоисп. *75:*.18–24.] On morphology of the associative apparatus of the abdominal brain of the medical leech. Questions of physical neurodynamics. Tr. Leningr. O-va. Estestvoispyt. *75:*18–24 (1974). Works of the Society of Naturalists, Leningrad

Lagutenko, Yu. P.: [К морфологии дорсальной и вентральной .ассоциативых областей нейропиля брюшного мозга .медицинской пиявки. Эволюция межнейрональных отношений. .Тр.Ленингр.о-ва естествоисп *77:*49–63.] On morphology of the dorsal and ventral associative areas of the neuropil of the abdominal brain of the medical leech. The evolution of interneuronal relations. Tr. Leningr. O-va. Estestvoispyt. *77:*49–63 (1975a)

Lagutenko, Yu. P.: [Особенности межнейрональных связей в заднем медиальном ядре ганглия брюшной цепочки .медицинской пиявки. Эволюция межнейрональных отношений. .Тр.Ленингр.о-ва естествоисп. *77:*63–69.] Characteristics of the interneuronal relations in the dorsomedial nucleus of the ganglion of the abdominal chain in the medical leech.

The evolution of interneuronal relations. Tr. Leningr. O-va. Estestvoispyt. *77*:63–69 (1975b)

Lagutenko, Yu. P.: [Сравнительная нейрофизиология и нейрохимия. Ленинград, Наука: 24–32.] Architectonics and connections of paired associative neurons in the abdominal nervous chain of the leech *Hirudo medicinalis*. *Comparative Neurophysiology and Neurochemistry,* pp. 24–32. Nauka, Leningrad, 1976

Lagutenko, Yu. P.: Эволюционно-морфологический анализ афферентной системы в брюшном мозге медицинской пиявки. Сб. Туловищный мозг членистоногих и червей. Тр. Биолого-Почвенного института ДВНЦ СССР. Изд. "Наука" Ленинград, 179–194, 1977

Lagutenko, G. P.: Структура нейропиля брюшного мозга полихет. Арх. Анат.Гистол.Эмбриол. *74*:77–84 (1978)

Lam, D. M. K., Steinman, L.: The uptake of [³H]aminobutyric acid in the goldfish retina. Proc. Natl. Acad. Sci. U.S.A. *68*:2777–2781 (1971)

Lamparter, H. E.: Die strukturelle Organisation des Prothorakalganglion bei der Waldameise (*Formica lugubris*). Z. Zellforsch. Mikrosk. Anat. *74*:198–231 (1966)

Lamparter, H. E., Akert, K., Sandri, C.: Wallersche Degeneration im Zentralnervensystem der Ameise. Elektronenmikroskopische Untersuchungen am Prothorakalganglion von *Formica lugubris* Zett. Schweiz. Arch. Neurol., Neurochir. Psychiatr. *100*:337–354 (1967)

Lamparter, H. E., Akert, K., Sandri, C.: Localization of primary sensory afferents in the prothoracic ganglion of the wood ant (*Formica lugubris* Zett.): A combined light and electron microscopic study of secondary degeneration. J. Comp. Neurol. *137*:367–376 (1969)

Lamparter, H. E., Steiger, U., Sandir, C., Akert, K.: Zum Feinbau der Synapsen im Zentralnervensystem der Insekten. Z. Zellforsch. Mikrosk. Anat. *99*:435 –442 (1969)

Landolt, A. M.: Elektronenmikroskopische Untersuchungen an der Perikaryenschicht der Corpora pedunculata der Waldameise (*Formica lugubris* Zett.) mit besonderen Berücksichtigung der Neuron-Glia- Beziehung. Z. Zellforsch. Mikrosk. Anat. *66*:701–736 (1965)

Landolt, A. M., Ris, H.: Electron microscopic studies on somasomatic interneuronal junctions in the corpus pedunculatum of the wood ant (*Formica lugubris* Zett.). J. Cell Biol. *28*:391–403 (1966)

Lane, N. J.: The organization of insect nervous systems. In: *Insect Neurobiology* (J. E. Treherne, ed.), pp. 1–74. North-Holland Publ., Amsterdam, 1974

Lane, N. J., Swales, L. S.: Interrelationships between golgi, GERL and synaptic vesicles in the nerve cells of insect and gastropod ganglia. J. Cell Sci. *22*: 435–453 (1976)

LaVail, J. H., LaVail, M. M.: Retrograde axonal transport in the central nervous system. Science *176*:1416–1417 (1972)

LaVail, J. H., LaVail, M. M.: The retrograde intra-axonal transport of horseradish peroxidase in the chick visual system. J. Comp. Physiol. *157*:303–352 (1974)

LaVail, J. H., Winston, K. R., Tish, A.: A method based on retrograde axonal transport of protein for identification of cell bodies of origin of axons terminating within the CNS. Brain Res. *58*:470–477 (1973)

Lázár, G.: Application of cobalt-filling technique to show retinal projections in the frog. Neuroscience *3*:725–737 (1978)

Lenard, J., Singer, J. S.: Alteration of the conformation of proteins in red blood cell membranes and in solution by fixatives used in electron microscopy. J. Cell Biol. *37*:117 (1968)

Levi-Montalcini, R., Chen, R. S.: Selective outgrowth of nerve fibers in vitro from embryonic ganglia of *Periplaneta americana*. Arch. Ital. Biol. *109*:307–337 (1971)

Lewis, P. R., Knight, D. P.: Staining methods for sectioned material. In: *Practical Methods in Electron Microscopy* (A. M. Glauert, ed.), Vol. 5, Part 1. Elsevier, New York, 1977

Liesegang, R. E.: Die Kolloidchemie der histologischen Silberfärbung. Kolloidchem. Betr. *3*:1–46 (1911)

Liesegang, R. E.: Histologische Versilberungen. Z. Wiss. Mikrosk. Tech. *45*, 3:273–279 (1928)

Liesegang, R. E.: See original references in Naturwiss. Wochenschr. *11*:353 (1896) and Phot. Arch. *21*:221 (1896) See also Stern, K. H. The Liesegang phenomenon. Chem. Rev. *57*, 79–99 (1954)

Liesegang, R. E., Rieder, W.: Versuche mit einer Keimmethode zum Nachweis von Silver in Gewebschnitten. Z. Mikrosk. Wiss. Tech. *38*:334–338 (1911)

Lillie, R. D.: *Histopathologic Technic and Practical Histochemistry*. McGraw-Hill, New York, 1965

Lindsay, R. D., ed.: "Computer Analysis of Neuronal Structures." Plenum, New York, 1977

Lindvall, O., Björklund, A., Sensson, L.-Å.: Fluorophore formation from catecholamines and related compounds in the glyoxylic acid fluorescence histochemical method. Histochemie *39*:197–227 (1974)

Lindvall, O., Björklund, A., Falck, B., Sensson, L.-Å.: Combined formaldehyde and glyoxylic acid reactions. I. New possibilities for microspectrofluorometric differentiation between phenylethylamines, indolylethylamines and their precursor amino acids. Histochemistry *46*:27–52 (1975)

Linossier, M. G.: Contribution a l'étude des ferments oxydants. Sur la peroxydase du pus. Compt. Rend. Hebd. Séances Acad. Sci. Paris *50*:373–375 (1898)

Llinas, R., Nicholson, C.: Electrophysiological properties of dendrite and somata in alligator Purkinje cells. J. Neurophysiol. *34*:532–551 (1971)

Locke, M., Collins, J. V.: The structure and formation of protein granules in the fat body of an insect. J. Cell Biol. *26*:857–884 (1965)

Loew, O.: Catalase, a new enzyme of general occurrence. U.S.D.A. (Washington D.C.) Rep. *68*:1–47 (1901)

Luft, J. H.: Improvements in epoxy resin embedding methods. J. Biophys. Biochem. Cytol. *9*:409–414 (1961)

Luft, J. H.: The structure and properties of the cell surface coat. Int. Rev. Cytol. *45*:291–381 (1976)

Luft, J. H., Wood, R. L.: The extraction of tissue protein during and after fixation with osmium tetroxide in various buffer systems. J. Cell Biol. *19*:46A (1963)

Lund, R. D., Collett, T. S.: A survey of reduced silver techniques for demonstrating neuronal degeneration in insects. J. Exp. Zool. *167*:391–410 (1968)

MacFarland, W. E., Davenport, H. A.: Staining paraffin sections with protargol. 6. Impregnation and differentiation of nerve fibers in adrenal glands of mammals. Stain Technol. *16:*53–58 (1941)

McClean, M., Edwards, J. S.: Target discrimination in regenerating insect sensory nerve. J. Embryol. Exp. Morphol. *36:*19–39 (1976)

McGuire, S. R., Opel, H.: Resorcin fuchsin staining of neurosecretory cells. Stain Technol. *44:*235–237 (1969)

McKenzie, J. D., Vogt, B. A.: An instrument for light microscope analysis of three-dimensional neuronal morphology. Brain Res. *111:*411–415 (1976)

McMillan, P. N., Luftig, R. B.: Preservation of erythrocyte ghost ultrastructure achieved by various fixatives. Proc. Natl. Acad. Sci., *70:*3060–3064 (1973)

Magalhaes-Castro, H. H., Dolabela de Lima, A., Saraiva, P. E. S., Magalhaes-Castro, B.: HRP labeling of cat tectotectal cells. Brain Res. *148:*1–13 (1978)

Maillet, M.: La technique de Champy á l'osmium ioduré de potassium et la modification de Maillet à l'osmium ioduré de zinc. Trab. Inst. Cajal Invest. Biol. *54:*1–36 (1962)

Malmgren, L., Olsson, Y.: A sensitive method for histochemical demonstration of horseradish peroxidase in neurons following retrograde axonal transport. Brain Res. *148:*279–294 (1978)

Mann, F. G., Saunders, B. C.: *Practical Organic Chemistry.* Longmans, Green, New York, 1960

Manton, I.: The possible significance of some details of flagellar bases in plants. J. R. Microsc. Soc. *82:*279–285 (1964)

Mark, R. F., Marotte, L. R., Johnstone, J. R.: Reinnervated eye muscles do not respond to impulses in foreign nerves. Science *170:*193–194 (1970)

Martin, M. R., Mason, C. A.: The seventh cranial nerve of the rat. Visualization of efferent and afferent pathways by cobalt precipitation. Brain Res. *121:*21–41 (1977)

Martin, R., Barlow, J., Miralto, A.: Application of the zinc iodide-osmium tetroxide impregnation of synaptic vesicles in cephalopod nerves. Brain Res. *15:*1–16 (1969)

Mason, A.: Morphology of leech Retzius cells demonstrated by intracellular injection of horseradish peroxidase. Comp. Biochem. Physiol. *A61:*213–216 (1978)

Mason, C. A.: New features of the brain-retrocerebral neuroendocrine complex of the locust *Schistocerca vaga* (Scudder). Z. Zellforsch. Mikrosk. Anat. *141:*19–32 (1973)

Mason, C. A.: Delineation of the rat visual system by the axonal iontophoresis-cobalt sulfide precipitation technique. Brain Res. *85:*287–293 (1975)

Mason, C. A., Lincoln, D. W.: Visualization of the retino-hypothalamic projection in the rat by cobalt precipitation. Cell Tissue Res. *168:*117–131 (1976)

Mason, C. A., Nishioka, R. S.: The use of the cobalt chloride-ammonium sulfide precipitation technique for the delineation of invertebrate and vertebrate neurosecretory systems. *Neurosecretion, Final Neuroendocrine Pathway, 6th Int. Symp. Neurosecretion, 1973,* pp. 48–58 (1974)

Mason, C. A., Sparrow, N., Lincoln, D. W.: Structural features of the retionohypothalamic projection in the rat during normal development. Brain Res. *132:*141–148 (1977)

Matus, A. I.: Ultrastructure of the superior cervical ganglion fixed with zinc iodide and osmium tetroxide. Brain Res. *17:*195–203 (1970)

Mees, C. E. K., James, T. H.: *The Theory of the Photographic Process,* 3rd ed. Macmillan, New York, 1966

Melamed, J., Trujillo-Cenoz, O.: Electron microscopic observations on the reactional changes occurring in insect nerve fibres after transection. Z. Zellforsch. Mikrosk. Anat. *59:*851–856 (1963)

Mendenhall, B., Murphey, R. K.: The morphology of cricket giant interneurones. J. Neurobiol. *5:*565–580 (1974)

Meola, S. M.: Sensitive paraldehyde-fuchsin technique for neurosecretory system of mosquitoes. Trans. Am. Microsc. Soc. *89:*66–71 (1970)

Mesulam, M. M.: The blue reaction product in horseradish peroxidase histochemistry: Incubation parameters and visibility. J. Histochem. Cytochem. *24:* 1273–1280 (1976)

Mesulam, M., van Hoesen, G. W., Pandya, D. N., Geschwind, N.: Limbic and sensory connections of the inferior parietal lobe (area PG) in the rhesus monkey: A study with a new method for horseradish peroxidase histochemistry. Brain Res. *136:*393–414 (1977)

Milburn, N. S., Bentley, D. R.: On the dendritic topology and activation of cockroach giant interneurons. J. Insect Physiol. *17:*607–623 (1971)

Millonig, G.: Advantages of a phosphate buffer for OsO_4 solutions in fixation. J. Appl. Phys. *32:*1637 (1961)

Mittenthal, J. E., Wine, J. J.: Connectivity patterns of crayfish giant interneurons: Visualization of synaptic regions with cobalt dye. Science *179:*182–184 (1973)

Mobbs, P. G.: Golgi staining of material containing cobalt-filled profiles in the insect CNS. Brain Res. *105:*563–566 (1976)

Mobbs, P. G.: Some aspects of the structure and development of the locust ocellus. Ph.D. Thesis, University of London (1978)

Mollenhauer, H. H.: Plastic embedding mixtures for use in electron microscopy. Stain Technol. *39:*111–114 (1964)

Monti, R.: Ricerche microscopiche sul sistema nervoso degli insetti. Boll. Sci. Pavia *15:*105–122 (1893)

Monti, R.: Ricerche microscopiche sul sistema nervoso degli insetti. Boll. Sci. Pavia *16:*6–17 (1894)

Moskowitz, N.: Silver protein staining of nerve fibers in celloidin sections. Stain Technol. *42:*221–225 (1967)

Muller, K. J., McMahan, U. J.: The shapes of sensory and motor neurons and the distribution of their synapses in ganglia of the leech: A study using intracellular injection of horseradish peroxidase. Proc. R. Soc. Lond. Ser. B *194:* 481–499 (1976)

Mulloney, B.: Microelectrode injection, axonal iontophoresis, and the structure of neurons. In: *Intracellular Staining in Neurobiology* (S. B. Kater, C. Nicholson, eds.), pp. 99–113. Springer-Verlag, Berlin, New York, 1973

Murdock, L. L.: Catecholamines in arthropods: A review. Comp. Gen. Pharmacol. *2:*254–274 (1971)

Murphey, R. K.: Characterization of an insect neuron which cannot be visualized *in situ.* In: *Intracellular Staining in Neurobiology* (S. B. Kater, C. Nicholson, eds.), pp. 135–150. Springer-Verlag, Berlin, New York, 1973

Murphey, R. K., Mendenhall, B., Palka, J., Edwards, J. S.: Deafferentation slows the growth of specific dendrites of identified giant interneurons. J. Comp. Neurol. *159*:407–418 (1975)

Muskó, I. B., Zs.-Nagy, I., Deák, G.: Fluorescence microscopy and microspectrofluorometry of monoamines in the brain of *Locusta migratoria migratorioides* R. F. (Insecta, Orthoptera) with special regard on the protocerebrum. Annl. Biol. Tihany *40*:85–94 (1973)

Nakata, T. S., Nishijima, S., Akai, M.: Combined staining by thiolacetic acid and the Bielschowsky method. Stain Technol. *46*:151–154 (1971)

Nauta, W. J. H., Ebbesson, S. O. E., eds.: *Contemporary Research Methods in Neuroanatomy.* Springer-Verlag, Berlin, New York, 1970

Nauta, W. J. H., Pritz, M. B., Lasek, R. J.: Afferents to the cat caudoputamen studied with horseradish peroxidase. An evaluation of a retrograde neuroanatomical research method. Brain Res. *67*:219–238 (1974)

Neville, A. C.: Motor unit distribution of the dorsal longitudinal flight muscles in locusts. J. Exp. Biol. *40*:123–136 (1963)

Newmywaka, G. A.: [Иннервация кишки у дождевого червя (*Allolobophora caliginosa*). ДАН СССР, *55*:533–536.] The innervation of the ceacum in the rainworm (*Allolobophora caliginosa*). Dokl. Akad. Nauk SSSR *55*:533–536 (1947a)

Newmywaka, G. A.: [Брюшной мозг дождевго червя (*Allolobophora caliginosa*). ДАН СССР, *58*:1483–1486.] The abdominal brain of the rainworm (*Allolobophora caliginosa*). Dokl. Akad. Nauk SSSR *58*:1483–1486 (1947b)

Newmywaka, G. A.: [Материалы по сравнителной гистологии .нервной системы. Брюшной мозг дождевого червя. В сб.: .Памяти А.А. Заварзина, Москва-Ленинград: 27–53.] Material for comparative histology of the nervous system. The abdominal brain of the rainworm. In the collection: *In Memory of Zavarzin,* pp. 27–53. Moscow-Leningrad, 1948

Newmywaka, G. A.: [Об иннервации выделительного аппарата .у аннелид. ДАН СССР, *87*:1063–1066.] On the innervation of the excretory apparatus of annelids. Dokl. Akad. Nauk SSSR *87*:1063–1066 (1952)

Newmywaka, G. A.: [Об участии вегетативных нервов в .иннервадии соматических мышц у *Allolobophora caliginosa*. ДАН СССР, *10*:855–857.] The participation of the vegetative nerves in the innervation of the somatic muscles in *Allolobophora caliginosa*. Dokl. Akad. Nauk SSSR *110*:885–857 (1956)

Newmywaka, G. A.: [Нервная система дождевого червя. Москва-Ленинград.] *The Nervous System of the Rainworm.* Moscow-Leningrad, 1966

Nicholson, C., Kater, S. B.: The development of intracellular staining. In: *Intracellular Staining in Neurobiology* (S. B. Kater, C. Nicholson, eds.), pp. 1–19. Springer-Verlag, Berlin, New York, 1973

Nickel, E., Waser, P. G.: An electron microscopic study of denervated motor endplates after zinc iodide-osmium impregnation. Brain Res. *13*:168–176 (1969)

Nijhout, H. F.: Axonal pathways in the brain-retrocerebral neuroendocrine complex of *Manduca sexta* (L.) (Lepidoptera: Sphingidae). Int. J. Insect Morphol. Embryol. *4*:529–538 (1975)

Nishimura, K., Fujita, T., Nakajima, M.: Catabolism of tryptamine by cockroach head enzyme preparations. Pestic. Biochem. Physiol. *5*:557–565 (1975)

Obermayer, M.: Brain structure: Modular organization in the brain. Plates 3 and 4 in: *European Molecular Biology Laboratory Research Reports 1978*, pp. 10–14 (1978)

Obermayer, M., Strausfeld, N. J.: Increasing Golgi impregnation by diffusion of CrO_4^{2-} from a micropipette. (in preparation)

van Orden, L., Burke, J. P., Geyer, M., Lodeon, F. V.: Localization of depletion-sensitive and depletion-resistant norepinephrine storage sites in autonomic ganglia. J. Pharmacol. Exp. Ther. *174*:56–71 (1970)

Orlov, J.: [К вопросу о чувствительной иннервации мышц у насекомых. Изв.Биол.н-иссл.ин-та при Пермск.ин-те, *1*:10–18.] On the sensory innervation of the muscles in insects. Izv. Biol. Nanchno-Issled. Inst. Biol. Stn. Permsk. Gos. Univ. *1*:10–18 (1922)

Orlov, J.: Die Innervation des Darmes der Insekten (Larven von Lamellicorniern). Z. Wiss. Zool. *122*:452–502 (1924)

Orlov, J.: Über den histologischen Bau der Ganglien des Mundmagennervensystems der Insekten. Z. Mikrosk. Anat. Forsch. *2*:39–110 (1925)

Orlov, J.: Das Magenganglion des Flußkrebses. Z. Mikrosk.-anat. Forsch. *8*:73–96 (1927)

Orlov, J.: Über den histologischen Bau des Ganglion des Mundmagennervensystems der Crustaceen. Z. Zellforsch. Mikrosk. Anat. *8*:493–541 (1929)

Orlov, J.: Über die Innervation des Schlundes bei Arthropoda. Z. Mikrosk.-Anat. Forsch *20*:1–2, 98–106 (1930)

Osborne, M. P.: The fine structure of synapses and tight junctions in the central nervous system of the blowfly larva. J. Insect Physiol. *12*:1503–1512 (1966)

Osborne, M. P.: The ultrastructure of nerve-muscle synapses. In: *Insect Muscle* (P. N. R. Usherwood, ed.), pp. 151–205. Academic Press, New York, 1975

O'Shea, M., Rowell, C. H. F.: A spike-transmitting electrical synapse between visual interneurones in the locust movement detector system. J. Comp. Physiol. *97*:143–158 (1975)

O'Shea, M., Williams, J. L. D.: The anatomy and output connection of a locust visual interneurone; the lobula giant movement detector (LGMD) neurone. J. Comp. Physiol. *91*:257–266 (1974)

O'Shea, M., Rowell, C. H. F., Williams, J. L. D.: The anatomy of a locust visual interneurone: The descending contralateral movement detector. J. Exp. Biol. *60*:1–12 (1974)

Paget, G. E., Eccleston, E.: Aldehyde-thionin: A stain having similar properties to aldehyde-fuchsin. Stain Technol. *34*:223–226 (1959)

Palade, G. E.: Intracellular localization of acid phosphatase. J. Exp. Med. *94*:535–548 (1951)

Palay, S. L., Chan-Palay, V.: A guide to the synaptic analysis of the neuropil. Cold Spring Harbor Symp. Quant. Biol. *40*:1–16 (1975)

Palka, J., Edwards, J. S.: The cerci and abdominal giant fibres of the house cricket, *Acheta domesticus*. II. Regeneration and effects of chronic deprivation. Proc. R. Soc. Lond. Ser. B *185*:105–121 (1974)

Palka, J., Levine, R., Schubiger, M.: The cercus-to-giant interneuron system of crickets. I. Some attributes of the sensory cells. J. Comp. Physiol. *119*:267–283 (1977)

Palmgren, A.: A rapid method for selective silver staining of nerve fibres and nerve endings in mounted paraffin sections. Acta Zool. (Stockholm) *29*:377–392 (1948)

Pan, K. C., Goodman, L. J.: Ocellar projections within the central nervous system of the worker honey bee, *Apis mellifera*. Cell Tissue Res. *176*:505–527 (1977)

Panov, A. A.: Neurosecretory cells in the abdominal ganglia of Orthoptera (Insecta). Entomol. Obozr. *43*:789–800 (1964)

Panov, A. A.: Neurosecretion in the brain of the sun pest, *Eurygaster integriceps* (Heteroptera, Scutelleridae). Zool. Zh. *48*:1640–1651 (1969) (in Russian with English summary)

Panov, A. A.: Composition of the medial neurosecretory cell group in the brain of several bugs. Zool. Zh. *51*:46–56 (1972) (in Russian with English summary)

Panov, A. A.: Composition of the medial neurosecretory cells of the pars intercerebralis in *Calliphora* and *Lucilia* adults. Zool. Anz. *196*:23–27 (1976)

Panov, A. A.: Neurosecretory cells in the pars intercerebralis of Orthoptera. Zool. Anz., *201(1/2)*:49–63 (1978)

Panov, A. A., Davydova, E. D.: Medial neurosecretory cells in the brain of Mecoptera and Neuropteroidea (Insecta). Zool. Anz. *197*:345–365 (1976)

Panov, A. A., Kind, T. W.: Neurosecretory cell system in the brain of Lepidoptera (Insecta). Dokl. Akad. Nauk SSSR *153*:1186–1189 (1963)

Panov, A. A., Melnikova, E. Ju.: Structure of the neurosecretory system in Lepidoptera. II. Light and electron microscopy of the medial neurosecretory cells in larval brain of *Hyphantria cunea* Drury (Lepidoptera). Gen. Comp. Endocrinol. *23*:361–375 (1974)

Pantin, C. F. A.: *Notes on Microscopical Technique for Zoologists*. Cambridge Univ. Press, London, New York, 1946

Pareto, A.: Die zentrale Verteilung der Fühlerafferenz bei Arbeiterinnen der Honigbiene, *Apis mellifera* L. Z. Zellforsch. Mikrosk. Anat. *131*:109–140 (1972)

Partanen, S.: Simultaneous fluorescence histochemical demonstration of catecholamines and tryptophyl-peptides in endocrine cells. Histochemistry *43*:295–303 (1975)

Paul, K. G.: Die Isolierung von Meerrettichperoxidase. Acta Chem. Scand. *12*:1312–1318 (1958)

Payton, B. W.: Histological staining properties of Procion yellow M4RS. J. Cell Biol. *45*:659–662 (1970)

Payton, B. W., Bennett, M. V. L., Pappas, G. D.: Permeability and structure of junctional membranes at an electrotonic synapse. Science *166*:1641–1643 (1969)

Peachey, L. D.: Thin sections. I. A study of section thickness and physical distortion produced during microtomy. J. Biophys. Biochem. Cytol. *4*:233–242 (1958)

Pearl, G. S., Anderson, K. V.: Use of cobalt impregnation as a method of identifying mammalian neural pathways. Physiol. Behav. *15*:619–622 (1975)

Pearson, A. A., O'Neill, S. L.: A silver-gelatin method for staining nerve fibers. Anat. Rec. *95*:297–301 (1946)

Pearson, K. G., Fourtner, C. R.: Identification of the somata of common inhibitory motoneurons in the metathoracic ganglion of the cockroach. Can. J. Zool. *51*:859–866 (1973)

Pearson, K. G., Fourtner, C. R.: Nonspiking interneurons in the walking system of the cockroach. J. Neurophysiol. *38*:33–52 (1975)

Pearson, L.: The corpora pedunculata of *Sphinx ligustri* L. and other Lepidoptera:

An anatomical study. Phil. Trans. R. Soc. Lond. Ser. B *259:*477–516 (1971)

Pease, D. C.: *Histological Techniques for Electron Microscopy.* Academic Press, New York, 1964

Peters, A.: Experiments on the mechanism of silver staining. I. Impregnation. Q. J. Microsc. Sci. *96:*84–102 (1955a)

Peters, A.: Experiments on the mechanism of silver staining. II. Development. Q. J. Microsc. Sci. *96:*103–115 (1955b)

Peters, A.: Experiments on the mechanism of silver staining. III. Quantitative studies. Q. J. Microsc. Sci. *96:*301–315 (1955c)

Peters, A.: Experiments on the mechanism of silver staining. IV. Electron microscope studies. Q. J. Microsc. Sci. *96:*317–322 (1955d)

Peters, A.: A general-purpose method of silver staining. Q. J. Microsc. Sci. *96:*323–328 (1955e)

Peters, A.: The fixation of central nervous tissue and the analysis of electron micrographs of the neuropil, with special reference to the cerebral cortex. In: *Contemporary Research Methods in Neuroanatomy* (W. J. H. Nauta, S. O. E. Ebbeson, eds.), pp. 58–76. Springer-Verlag, Berlin, New York, 1970

Peute, J., van de Kamer, J. C.: On the histochemical differences of aldehyd-fuchsin positive material in the fibres of the hypothalamo-hypophyseal tract of *Rana temporaria.* Z. Zellforsch. Mikrosk Anat. *83:*441–448 (1967)

Perrachia, C., Mittler, B. S.: Fixation by means of glutaraldehyde -hydrogen peroxide reaction products. J. Cell Biol. *53:*234–238 (1972)

Pfenninger, K. H.: The cytochemistry of synaptic densities. I. An analysis of the bismuth iodide impregnation method. J. Ultrastruct. Res. *34:*103–122 (1971a)

Pfenninger, K. H.: The cytochemistry of synaptic densities. II. Proteineceous components and mechanisms of synaptic connectivity. J. Ultrastruct. Res. *35:*451–475 (1971b)

Pfenninger, K. H.: *Synaptic Morphology and Cytochemistry Progress in Histochemistry and Cytochemistry,* Vol. 5, No. 1. Fischer, Stuttgart, 1973

Pfenninger, K. H., Sandri, C., Akert, K., Engster, C. H.: Contribution to the problem of structural organization of the presynaptic area. Brain Res.*12:*10–18 (1969)

Pflugfelder, O.: Vergleichende anatomische, experimentelle und embryologische Untersuchungen über das Nervensystem und die Sinnesorgane der Rhynchaten. Zoologica (N.Y.) *34:*1–102 (1937)

Pierantoni, R.: An observation on the giant fiber posterior optic tract in the fly. Kybernetik *5:*157–163. Leipzig, 1973

Pierantoni, R.: A look into the cock-pit of the fly. The architecture of the lobula plate. Cell Tissue Res. *171:*101–122 (1976)

Pihl, E., Falkmer, S.: Trials to modify the sulfide-silver method for ultrastructural tissue localization of heavy metals. Acta Histochem. *27:*33–41 (1967)

Pipa, R. L.: A cytochemical study of neurosecretory and other neuroplasmic inclusions in *Periplaneta americana.* Gen. Comp. Endocrinol. *2:*44–52 (1962)

Pipa, R. L., Cook, E. F., Richards, A. G.: Studies on the hexapod nervous system. II. The histology of the thoracic ganglia of the adult cockroach, *Periplaneta americana* (L.). J. Comp. Neurol. *113:*401–433 (1959)

Pitman, R. M.: Block intensification of neurones stained with cobalt sulphide: A method for destaining and enhanced contrast. J. Exp. Biol. *78*:295–297 (1979)

Pitman, R. M., Tweedle, C. D., Cohen, M. J.: Branching of central neurons: Intracellular cobalt injection for light and electron microscopy. Science *176*:412–414 (1972)

Pitman, R. M., Tweedle, C. D., Cohen, M. J.: The form of nerve cells: Determination by cobalt impregnation. In: *Intracellular Staining in Neurobiology* (S. B. Kater, C. Nicholson, eds.), pp. 83–97. Springer-Verlag, Berlin, New York, 1973

Ploem, J. S.: A new microscopic method for the visualization of blue formaldehyde-induced catecholamine fluorescence. Arch. Int. Pharmacodyn. Ther. *182*:421–424 (1969)

Plotnikova, S. I.: [К сравнительной морфологии вегетативной .нервной системы. Рецепторные клетки и эффекторные нервные .окончания в пищеварительном тракте личинки .стрекозы *Aeschna* sp. Докл.АН СССР, Новая серия, *68*:923–926.] Comparative morphology of the vegetative nervous system. Receptor cells and effector nerve endings in the digestive tract of the locust larva *Aeschna* sp. Dokl. Akad. Nauk SSSR [N.S.] *68*:923–926 (1949)

Plotnikova, S. I.: [Иннервация кишечника азиатской .саранчи *Locusta migratoria* L. (Orthoptera, Acrididae) .Энтомол.обозр., *46*:122–126.] The innervation of the intestine in *Locusta migratoria* L. (Orthoptera, Acrididae). Entomol. Obozr. *46*:122–126 (1967a)

Plotnikova, S. I.: The structure of the sympathethic nervous system of insects. Proc. Int. Symp. on Neurobiol. Invertebr., pp. 59–68 (1967b)

Plotnikova, S. I.: Effector neurons with several axons in the ventral nerve cord of *Locusta migratoris*. Zh. Evol. Biokhim. Fiziol. *5*:339–341 (1969)

Plotnikova, S. I.: [О чувствительном нейропиле ганглиев .брюшной цепочки Locusta migratoria L. Эволюция межнейрональных отношений. Тр.Ленингр.о-ва естествоисп. *77*:5–21.] On the sensory neuropil of the ganglia of the abdominal nervous chain in *Locusta migratoria* L. The evolution of interneuronal relations. Tr. Leningr. o-va. estestvoispyt. *77*:5–21 (1975)

Plotnikova, S. I.: [Эффекторные нейроны торакальных ганглиев .туловищного мозга азиатской саранчи. Туловищный мозг .членистоногих и червей. Тр.Биолого-Почвенного института ДВНЦ АН СССР. Изд. Наука (впечати).] *Effector Neurons of the Thoracic Ganglia of the Truncal Brain of Locusta migratoria L. The Truncal Brain of Arthropods and Worms.* Works of the Institute of Biology and Soil DVNC of the Academy of Sciences of the USSR, "Nauka" Leningrad, 136–167 (1977)

Plotnikova, S. I.: Дорсальные коннективные волокна грудных ганглиев азиатской саранчи. Сб. Морфологические основы функциональной эволюции. Ленинград. Изд. "Наука", Ленинград, 24–29, 1978

Plotnikova, S. I.: Структурная организация центральной нервной системы насекомых. Изд. "Наука". Ленинград (книга выходит из печати в сентябре 1979 года), 1979

Pollard, T. D., Ito, S.: Cytoplasmic filaments of *Amoeba proteus*. J. Cell Biol. *46*:267–289 (1970)

Polley, E. H.: Silver staining of nerve tissue with a new silver proteinate. Anat. Rec. *125:*509–519 (1956)

Porter, R. W., Davenport, H. A.: Golgi's dichromate-silver method I. Effects of embedding. 2. Experiments with modifications. Stain Technol. *24:*117–126 (1949)

Power, M. E.: The brain of *Drosophila melanogaster.* J. Morphol. *72:*517–559 (1943)

Power, M. E.: The antennal centers and their connections within the brain of *Drosophila melanogaster.* J. Comp. Neurol. *85:*485–517 (1946)

Power, M. E.: The thoracico-abdominal nervous system of an adult insect, *Drosophila melanogaster.* J. Comp. Neurol. *88:*347–409 (1948)

Prior, D. J., Fuller, P. M.: The use of a cobalt iontophoresis technique for identification of the mesencephalic trigeminal nucleus. Brain Res. *64:*472–475 (1973)

Prosser, C. L.: *Comparative Animal Physiology.* Saunders, Philadelphia, Pennsylvania, 1973

Purves, D., McMahan, U. J.: Procion yellow as a marker for electron microscopic examination of functionally identified nerve cells. In: *Intracellular Staining in Neurobiology* (S. B. Kater, C. Nicholson, eds.), pp. 71–82. Springer-Verlag, Berlin, New York, 1973

Querido, A.: Gold intoxication of nervous elements on the permeability of the , blood-brain barrier. Acta Psychiatr. Neurol. *22:*1–2:97–151 (1948)

Quicke, D. L. J., Brace, R. C.: Differential staining of cobalt and nickel-filled neurons using rubeanic acid. J. Microscopy *115:*161–163 (1979)

Raabe, M.: Mise en évidence, chez les Insectes d'ordres variés, d'éléments neurosécréteurs tritocérébraux. C. R. Hebd. Seances Acad. Sci. *257:*1171–1173 (1963)

Raabe, M.: Etude des phénomènes de neurosécrétion au niveau de la chaîne nerveuse ventrale des Phasmides. Bull. Soc. Zool. Fr. *90:*631–654 (1965)

Raabe, M.: Neurosécrétion dans la chaîne nerveuse ventrale des Insectes et organes neurohaemaux métamériques. Arch. Zool. Exp. Gen. *112:*679–694 (1971)

Raabe, M., Monjo, D.: Recherches histologiques et histochimiques sur la neurosécrétion chez le Phasme, *Clitumnus extradentatus:* Les neurosécrétions de type C. C. R. Hebd. Seances Acad. Sci., Ser. D *270:*2021–2024 (1970)

Rademakers, L. M. P. M.: Identification of a secretomotor centre in the brain of *Locusta migratoria,* controlling secretory activity of the adipokinetic hormone producing cells of the corpus cardiacum. Cell Tissue Res. *184:*381–395 (1977)

Radl, E.: *Neue Lehre vom Zentralen Nervensystem.* Engelmann, Leipzig, 1912

Ramade, F., L'Hermite, P.: Mise en évidence de neurones adrénergiques par la microscopie de fluorescence dans la système nerveux de *Calliphora erythrocephala* Meig., et *Musca domestica* L. C. R. Hebd. Seances Acad. Sci. Ser. D *272:*3313–3317 (1971)

Ramon-Moliner, E.: A chlorate-formaldehyde modification of the Golgi method. Stain Technol. *32:*105–116 (1957)

Ramon-Moliner, E.: A source of error in the study of Golgi-stained material. Arch. Ital. Biol. *105:*139–148 (1967)

Ramon-Moliner, E.: The Golgi-Cox technique. In: *Contemporary Research Methods in Neuroanatomy* (W. J. H. Nauta, S. O. E. Ebbesson, eds.) pp. 32–55. Springer-Verlag, Berlin, New York, 1970

Rastad, J., Jankowska, E., Westman, J.: Arborization of initial axon collaterals of spinocervical tract cells stained intracellularly with horseradish peroxidase. Brain Res. *135:*1–10 (1977)

Rath, O.: Zur Kenntnis der Hauptsinnesorgane und des sensiblen Nervensystems der Arthropoden. Z. Wiss. Zool. *61:*499–539 (1896)

Ravetto, C.: Alcian blue-alcian yellow: A new method for the identification of different acid groups. J. Histochem. Cytochem. *12:*44–45 (1964)

Rees, D., Usherwood, P. N. R.: Fine structure of normal and degenerating motor axons and nerve-muscle synapses in the locust, *Schistocerca gregaria.* Comp. Biochem. Physiol. A *43:*83–101 (1972)

Rehbein, H. G.: Experimentell-anatomische Untersuchungen über den Verlauf der Tympanalnervenfasern im Bauchmark von Feldheuschrecken, Laubheuschrecken und Grillen. Verh. Dtsch. Zool. Ges. *66:*184–189 (1973)

Rehbein, H.: Auditory neurons in the ventral cord of the locust: Morphological and functional properties. J. Comp. Physiol. *110:*233–250 (1976)

Rehbein, H., Kalmring, K., Römer, H.: Structure and function of acoustic neurons in the thoracic ventral nerve cord of *Locusta migratoria* (Acrididae). J. Comp. Physiol. *95:*263–280 (1974)

Rehm, M.: Sekretionsperioden neurosekretorischer Zellen im Gehirn von Ephestia kühniella. Z. Naturforsch., Teil B *5:*167–169 (1950)

Reichardt, W., Poggio, T.: Visual control of orientation behavior in the fly. Quart. Rev. Biophys. *9:*311–438 (1976)

Reid, N.: Ultramicrotomy. In: *Practical Methods in Electron Microscopy* (A. M. Glauert, ed.), Vol. 3, Part 3. Am. Elsevier, New York, 1975

Reinecke, M., Walther, C.: Aspects of turnover and biogenesis of synaptic vesicles at locust neuromuscular junctions as revealed by zinc iodide-osmium tetroxide (ZIO) reacting with intravesicular SH-groups. J. Cell Biol. *78:*839–855 (1978)

Remler, M., Selverston, A., Kennedy, D.: Lateral giant fibers of crayfish: Location of somata by dye injection. Science *162:*281–283 (1968)

Reynolds, E. S.: The use of lead citrate at high pH as an electron opaque stain in electron microscopy. J. Cell Biol. *17:*208–212 (1963)

Ribi, W.: The neurons of the first optic ganglion of the bee, *Apis mellifera.* Ergeb. Anat. Entwicklungsgesch. *50,* 4:1–43 (1975)

Ribi, W., Berg, G. J.: Light and electron microscopic structure of Golgi-stained neurons in the vertebrate (new rapid Golgi procedure). Cell Tissue Res. *205,* 1–10 (1980)

Richards, F. M., Knowles, J. R.: Glutaraldehyde as a protein crosslinking reagent. J. Mol. Biol. *37:*231 (1968)

Ritzèn, M.: Cytological identification and quantitation of biogenic monoamines. A microspectrofluorometric and autoradiographic study. Thesis, University of Stockholm (1967)

Robertson, H. A.: The innervation of the salivary gland of the moth, *Manduca sexta:* Evidence that dopamine is the transmitter. J. Exp. Biol. *63:*413–419 (1975)

Robertson, H. A.: Octopamine, dopamine and noradrenaline content of the brain of the locust, *Schistocerca gregria.* Experientia *32:*552–553 (1976)

Rock, M. K., Blankenship, J. E., Lebeda, F. J.: Penis retractor muscle of *Aplysia:* Excitatory motor neurons. J. Neurobiol. *8:*569–579 (1977)

Rogoff, W. M.: The Bodian technic and the mosquito nervous system. Stain Technol. *21:*59–61 (1946)

Rogosina, M.: Über das periphere Nervensystem der Aeschna-Larve. Zellforsch. Mikrosk. Anat. *6:*732–758 (1928)

Romanes, G. J.: The staining of nerve fibres in paraffin sections with silver. J. Anat. *84:*104–115 (1950)

Romeis, B.: Taschenbuch der mikroskopischen Tecknik. 13. Aufl., Verlag von R. Oldendburg, München, Berlin (1932)

Rosa, C. G.: Preparation and use of aldehyde fuchsin stain in the dry form. Stain Technol. *28:*299–302 (1953)

Ross, S. B.: Structural requirements for uptake into catecholamine neurons. In: *The Mechanism of Neural and Extraneural Transport of Catecholamines* (D. M. Panton, ed.), pp. 67–93. Raven, New York, 1976

Ross, S. B., Rengi, A. L.: Tricyclic antidepressant agents: I. Comparison of the uptake of ^{3}H-noradrenaline and ^{14}C-5-hydroxytryptamine in slices and crude synaptosome preparations of mid-brain–hypothalamus region of the rat brain. Acta Pharmacol. Toxicol. *36:*382–394 (1975)

Rowell, C. H. F.: A general method for silvering invertebrate central nervous systems. Q. J. Microsc. Sci. *104:*81–87 (1963)

Rowell, C. H. F., Dorey, A. E.: The number and size of axons in the thoracic connectives of the desert locust, *Schistocerca gregaria* Forsk. Z. Zellforsch. Mikrosk. Anat. *83:*288–294 (1967)

Rowell, C. H. F., O'Shea, M., Williams, J. L. D.: The neuronal basis of a sensory analyzer, the acridid movement detector system. IV. The preference for small field stimuli. J. Exp. Biol. *68:*157–185 (1977)

Rutschke, E., Richter, D., Thomas, H.: Autoradiographische Untersuchungen zum Einbau von ^{3}H-Dopamin und ^{3}H-5-hydroxytryptamin in das Gehirn von *Periplaneta americana* L. Zool. Jahrb., Abt. Anat. Ontog. Tiere *95:*439–447 (1976)

Sabatini, D. D., Bensch, K., Barnett, R. J.: Cytochemistry and electron microscopy. The preservation of cellular ultrastructure and enzymatic activity by aldehyde fixation. J. Cell Biol. *17:*19–58 (1963)

Samuel, E. P.: Impregnation and development in silver staining. J. Anat. *87:*268–276 (1953a)

Samuel, E. P.: The mechanism of silver staining. J. Anat. *87:*278–286 (1953b)

Samuel, E. P.: Towards controllable silver staining. Anat. Rec. *116:*511–519 (1953c)

Samuel, E. P.: Gold toning. Stain Technol. *28:*225–229 (1953d)

Sánchez y Sánchez, D.: Datos para el conocimiento histogenico de los centros opticos de los insectos. Evolución de algunos elementos retinianos del *Pieris brassicae* L. Trab. Lab. Invest. Biol. Univ. Madrid *14:*189–231 (1916)

Sánchez y Sánchez, D.: L'histogenèse dans les centres nerveux des insectes pendant les métamorphoses. Trab. Lab. Invest. Biol. Univ. Madrid *23:*29–52 (1925)

Sánchez y Sánchez, D.: Contribution á la connaissance de la structure des corps fongiformes (calices) et de leurs pèdicules chez la blatte commune (*Stylopyga* (*Blatta*) *orientalis*). Trab. Lab. Invest. Biol. Univ. Madrid *28:*149–185 (1933)

Sánchez y Sánchez, D.: Contribution à l'étude de l'origine et de neurologie chez les insectes. Trab. Lab. Invest. Biol. Univ. Madrid *30:*299–353 (1935)

Sánchez y Sánchez, D.: Sur le centre antenno-moteur ou antennaire postérieur de l'abeille. Trav. Lab. Rech. Biol. Univ. Madrid *31:*245–269 (1937)

Sánchez y Sánchez, D.: Contribution à la connaissance des centres nerveux des insectes. Nouveaux apports sur la structure du cerveau des abeilles (*Apis mellifica*). Trab. Inst. Cajal Invest. Biol. *32,* Part I:123–210; Part II:165–236 (1940)

Sandeman, D. C.: Integrative properties of a reflex motoneuron in the brain of the crab, *Carcinus maenas*. Z. Vergl. Physiol. *64:*450–464 (1969)

Sandeman, D. C., Okajima, A.: Statocyst induced eye movements in the crab *Scylla serrata:* III. The anatomical projections of sensory and motor neurones and the responses of the motor neurones. J. Exp. Biol. *59:*17–38 (1973)

Sakharov, D. A.: Cellular aspects of invertebrate neuropharmacology. Annu. Rev. Pharmacol. *10:*335–352 (1970)

Sakharova, A. V., Sakharov, D. A.: Visualization of intraneuronal monoamines by treatment with formalin solutions. Prog. Brain Res. *34:*11–25 (1971)

Satir, P., Gilula, N. B.: The fine structure of membranes and intercellular communication in insects. Annu. Rev. Entomol. *18:*143–166 (1973)

Saunders, B. C., Holmes-Siedle, A. G., Stark, B. P.: *Peroxidase: The Properties and Uses of a Versatile Enzyme and Related Catalysts.* Butterworths, London, 1964

Scharrer, E., Scharrer, B.: Neurosecretion. In: *Handbuch der mikroskopischen Anatomie* (W. von Möllendorf, ed.), Vol. 6, pp. 453–1066. Springer-Verlag, Berlin, New York, 1954

Scharrer, B., Wurzelmann, S.: 1974. Observations on synaptoid vesicles in insect neurons. Zool. Jahrb., Abt. Allg. Zool Physiol. *78:*387–396 (1974)

Schiebler, T. H.: Darstellung der B-Zellen in Pankreasinseln und von Neurosekret mit Pseudoisocyanin. Naturwissenschaften *45:*214 (1958)

Schooneveld, H.: Structural aspects of neurosecretory and corpus allatum activity in the adult colorado beetle, *Leptinotarsa decemlineata* Say., as a function of daylength. Neth. J. Zool. *20:*151–237 (1970)

Schürmann, F.-W.: Über die Struktur der Pilzkörper des Insektenhirns. I. Synapsen im Pedunculus. Z. Zellforsch. Mikrosk. Anat. *103:*363–381 (1970)

Schürmann, F.-W.: Synaptic contacts of association fibres in the brain of the bee. Brain Res. *26:*169–176 (1971)

Schürmann, F.-W.: Über die Struktur der Pilzkörper des Insektenhirns. II. Synaptische Schaltungen im Alpha-Lobus des Heimchens *Acheta domesticus* L. Z. Zellforsch. Mikrosk. Anat. *127:*240–257 (1972)

Schürmann, F.-W.: Über die Struktur der Pilzkörper des Insektenhirns. III. Die Anatomie der Nervenfasern in den Corpora pedunculata bei *Acheta domes-*

the rat and their differentiation from mitochondria by the peroxidase method. J. Biophys. Biochem. Cytol. *5:*193 (1959)

Straus, W.: Factors affecting the cytochemical reaction of peroxidase with benzidine and the stability of the blue reaction product. J. Histochem. Cytochem. *12:*462 (1964a)

Straus, W.: Factors affecting the state of injected horseradish peroxidase in animal tissues and procedures for the study of phagosomes and phago-lysosomes. J. Histochem. Cytochem. *12:*470 (1964b)

Straus, W.: Improved staining for peroxidase with benzidine and improved double staining immunoperoxidase procedures. J. Histochem. Cytochem. *20:*272–278 (1972)

Strausfeld, N. J.: Golgi studies on insects. II. The optic lobes of Diptera. Phil. Trans. R. Soc. Lond. Ser. B. *258:*135–223 (1970)

Strausfeld, N. J.: The organisation of the insect visual system (light microscopy). I. Projections and arrangements of neurones in the lamina ganglionaris of Diptera. Z. Zellforsch. Mikrosk. Anat. *121:*377–441 (1971)

Strausfeld, N. J.: *Atlas of an Insect Brain.* Springer-Verlag, Berlin, New York, 1976

Strausfeld, N. J.: Male and female visual neurons in dipterous insects. Nature (Lond.): *283,* 381–383 (1980)

Strausfeld, N. J., Blest, A. D.: Golgi studies on insects. I. The optic lobes of Lepidoptera. Phil. Trans. R. Soc. Lond. Ser. B *258:*81–134 (1970)

Strausfeld, N. J., Campos-Ortega, J. A.: L3, the third second-order neuron of the first visual ganglion in the "neural superposition" eye of *Musca domestica.* Z. Zellforsch. *139:*397–403 (1973a)

Strausfeld, N. J., Campos-Ortega, J. A.: The L4 monopolar neuron: A substrate for lateral interaction in the visual system of the fly *Musca domestica.* Brain Res. *59:*97–117 (1973b)

Strausfeld, N. J., Campos-Ortega, J. A.: Vision in insects: Pathways possibly underlying neural adaptation and lateral inhibition. Science *195:*894–897 (1977)

Strausfeld, N. J., Hausen, K.: The resolution of neuronal assemblies after cobalt injection into neuropil. Proc. R. Soc. Lond. Ser. B *199:*463–476 (1977)

Strausfeld, N. J., Obermayer, M.: Resolution of intraneuronal and transsynaptic migration of cobalt in the insect visual and central nervous systems. J. Comp. Physiol. *110:*1–12 (1976)

Stretton, A. O. W., Kravitz, E. A.: Neuronal geometry: Determination with a technique of intracellular dye injection. Science *162:*132–134 (1968)

Stretton, A. O. W., Kravitz, E. A.: Intracellular dye injection: The selection of Procion yellow and its application in preliminary studies of neuronal geometry in the lobster nervous system. In: *Intracellular Staining in Neurobiology* (S. B. Kater, C. Nicholson, eds.), pp. 21–40. Springer-Verlag, Berlin, New York, 1973

Strong, O. S.: Notes on neurological methods. Anat. Anz. *10:*494 (1895)

Sulston, J., Dew, M., Brenner, S.: Dopaminergic neurons in the nematode *Caenorhabditis elegans.* J. Comp. Neurol. *163:*215–226 (1975)

Sumner, B. E. H.: A histochemical study of aldehyde fuchsin staining. J. R. Microsc. Soc. [3] *84:*329–338 (1965)

Svidersky, V. L.: [Нейрофизиология полета саранчи. Л. 1–215.] *Neurophysiology of Locust Flight.* Leningrad. 1–215, 1973

Svidersky, V. L.: Afferent mechanisms of the maintenance of flight in locusta. In: *Neurobiology of Invertebrates* (J. Salánki, ed.), pp. 479–485. Plenum, New York, 1967

Svidersky, V. L., Knyazeva, N. I.: [Центральное преобразование импульсов, поступающих от рецепторов головы к нейронам крыловых мышц саранчи. ДАН СССР, *183:*486–489.]The central transformation of impulses which are sent from the receptors of the head to the neurons of the wing muscles of locusts. Dokl. Akad. Nauk SSSR *183:*486–489 (1968)

Svidersky, V. L., Varanka, I.: Responses of descending interneurons to stimulation of wind head receptors in locusts. Neurophysiology *5:*602–610 (1973)

Székely, G.: The morphology of motor neurons and dorsal root fibers in the frog's spinal cord. Brain Res. *103:*275–290 (1976)

Székely, G., Gallyas, F.: Intensification of cobaltous sulphide precipitate in frog nervous tissue. Acta Biol. Acad. Sci. Hung. *26:*175–188 (1975)

Székely, G., Kosaras, B.: Dendro-dendritic contacts between frog motoneurons shown with the cobalt labelling technique. Brain Res. *108:*194–198 (1976)

Tandan, B. K., Dogra, G. S.: An in situ study of cystine and cysteine-rich neurosecretory cells in the brain of *Sarcophaga ruficornia* (Fabricius 1794) (Diptera, Cyclorrhapha). Proc. R. Entomol. Soc. Lond. Ser. A *41:*57–66 (1966)

Tasaki, I., Tsukahara, Y., Ito, S., Wayner, M. J., Yu, W. Y.: A simple, direct and rapid method for filling microelectrodes. Physiol. Behav. *3:*1009–1010 (1968)

Taylor, H. M., Truman, J. W.: Metamorphosis of the abdominal ganglia of the tobacco hornworm, *Manduca sexta.* J. Comp. Physiol. *90:*367–388 (1974)

Theorell, H.: Crystalline peroxidase. Enzymologia *10:*250–252 (1942)

Thomsen, E.: Effect of removal of neurosecretory cells in the brain of adult *Calliphora erythrocephala* Meig. Nature (London) *161:*439 (1948)

Thomsen, E., Lea, A. O.: Control of the medial neurosecretory cells by the corpus allatum in *Calliphora erythrocephala.* Gen. Comp. Endocrinol. *12:*51–57 (1968)

Thomsen, E., Thomsen, M.: Darkfield microscopy of living neurosecretory cells. Experientia *10:*206 (1954)

Timm, F.: Zur Histochemie des Ammonshorngebietes. Z. Zellforsch. Mikrosk. Anat. *48:*548–555 (1958a)

Timm, F.: Zur Histochemie der Schwermetalle. Das Sulfid-Silber-Verfahren. Dtsch. Z. Gesamte Gerichtl. Med. *46:*706–711 (1958b)

Toh, Y., Kuwabara, M.: Synaptic organization of the fleshfly ocellus. J. Neurocytol. *4:*271–287 (1975)

Tranzer, J. P.: A new amine storing compartment in adrenergic axons. Nature (London), New Biol. *237:*57–58 (1972)

Tranzer, J. P., Thoenen, H.: An electron microscopic study of selective, acute degeneration of sympathetic nerve terminals after administration of 6-hydroxydopamine. Experientia *24:*155–156 (1968)

Tranzer, J. P., Thoenen, H., Snipes, R. L., Richards, J. G.: Recent developments on the ultrastructural aspect of adrenergic nerve endings in various experimental conditions. Prog. Brain Res. *31:*33–46 (1969)

Trujillo-Cenóz, O.: Some aspects of the structural organization of the intermediate retina of dipterans. J. Ultrastruct. Res. *13:*1–33 (1965)

Trujillo-Cenóz, O.: Some aspects of the structural organization of the medulla in muscoid flies. J. Ultrastruct. Res. *27:*535–553 (1969)

Trujillo-Cenóz, O., Melamed, J.: The development of the retina-lamina complex in muscoid flies. J. Ultrastruct. Res. *42:*554–581 (1973)

Truman, J. W., Reiss, S. E.: Dendritic reorganization of an identified motoneuron during metamorphosis of the tobacco hornworm moth. Science *192:*477–479 (1976)

Tsvileneva, V. A.: The nerve cord of *Aeschna.* 1. The structure of the thoracic ganglia. Izv. Akad. Nauk SSSR, Ser. Biol. *2:*91–128 (1950)

Tsvileneva, V. A.: The nerve cord of *Aeschna.* Izv. Akad. Nauk SSSR, Ser. Biol. *2:*66–116 (1951)

Tsvileneva, V. A.: The nerve structure of the ixodid ganglion. Zool. Jahrb., Abt. Anat. Ontog. Tiere *81:*576–602 (1964)

Tsvileneva, V. A.: *On the Evolution of Arthropod Nerve Cord.* "Nauka," Leningrad, 199 pages (1970)

Tsvileneva, V. A.: On the contacts of nervous elements in the central nervous system of a crayfish. Zh. Evol. Biokhim. Fiziol. *8:*78–85 (1972)

Tsvileneva, V. A., Fedosova, T. V.: [Об аппарате гигантских элементов в центральной нервной системе речного рака. Эволюция межнейрональных отношений. Тр.Ленингр.о-ва естествоиспыт. *77:*41–49.] On the apparatus of gigantic elements in the central nervous system of the crayfish. The evolution of interneuronal connections. Tr. Leningr. O-va. Estestvoispyt. *77:*41–49 (1975)

Tsvileneva, V. A., Fedosova, T. A., Titova, V. A.: Architechtonics of the nervous elements in crayfish thoracic ganglia. Zool. Jahrb., Abt. Anat. Ontog. Tiere *96,* 5:1–44 (1976)

Tsvileneva, V. A., Titova, V. A., Fedosova, T. V.: Architectonics of the nerve elements in the first thoracic ganglion of the crayfish. Tr. Leningr. O-va. Estestvoispyt. *77:*31–40 (1975)

Tweedle, C. D., Pitman, R. M., Cohen, M. J.: Dendritic stability of insect central neurons subjected to axotomy and de-afferentation. Brain Res. *60:*471–476 (1973)

Tyrer, N. M., Altman, J. S.: Spatial distribution of synapses onto thoracic motor neurones in locusts. Cell Tissue Res. *166:*389–398 (1976)

Tyrer, N. M., Altman, J. S.: Motor and sensory flight neurons in a locust demonstrated using cobalt chloride. J. Comp. Neurol. *157:*117–138 (1974)

Tyrer, N. M., Bell, E. M.: The intensification of cobalt-filled neurone profiles using a modification of Timm's sulphide-silver method. Brain Res. *73:*151–155 (1974)

Tyrer, N. M., Bacon, J. P., Davies, C. A.: Primary sensory projections from the wind-sensitive head hairs of the locust, *Schistocerca gregaria.* I. Distribution in the CNS. Cell Tissue Res. *203:*79–92

Ungewitter, L. H.: A urea silver nitrate method for nerve endings and nerve fibres. Stain Technol. *26:*73–76 (1951)

Usherwood, P. N. R.: Response of insect muscle to denervation. I. Resting potential changes. J. Insect Physiol. *9:*247–255 (1963a)

Usherwood, P. N. R.: Response of insect muscle to denervation. II. Changes in neuromuscular transmission. J. Insect Physiol. *9*:811–825 (1963b)

Usherwood, P. N. R.: Nerve-muscle transmission. In: *Insect Neurobiology* (J. E. Treherne, ed.), pp.245–305. North-Holland Publ., Amsterdam, 1974

Usherwood, P. N. R., Rees, D.: Quantitative studies of the spatial distribution of synapic vesicles within normal and degenerating motor axons of the locust, *Schistocerca gregaria*. Comp. Biochem. Physiol. A *43*:103–118 (1972)

Usherwood, P. N. R., Cochrane, D. G., Rees, D.: Changes in structural, physiological and pharmacological properties of insect excitatory nerve-muscle synapses after motor nerve section. Nature (London) *218*:589–591 (1968)

Valverde, F.: The Golgi method. A tool for comparative structural analyses. In: *Contemporary Research Methods in Neuroanatomy* (W. J. H. Nauta, S. O. E. Ebbesson, eds.), pp. 12–3ʳ. Springer-Verlag, Berlin, New York, 1970

Van der Loos, H.: Une combinaison de deux vieilles méthodes histologiques pour le système nerveux central. Monatsschr. Psychiatr. Neurol. *132*:330–334 (1956)

Vanegas, H., Holländer, H., Distel, H.: Early stages of uptake and transport of horseradish peroxidase by cortical structures, and its use for the study of local neurons and their processes. J. Comp. Neurol. *177*:193–212 (1978)

Van Orden, L. S., III: Principles of fluorescence microscopy and photomicrography with applications to Procion yellow. In: *Intracellular Staining in Neurobiology* (S. B. Kater, C. Nicholson, eds.), pp. 61–69. Springer-Verlag, Berlin, New York, 1973

Varanka, I., Sversky, V. L.: Functional characteristics of the interneurons of wind-sensitive hair-receptors on the head in *Locusta migratoria* L. I. Interneurons with excitatory responses. Comp. Biochem. Physiol. A *48*:411–426 (1974a)

Varanka, I., Sviderksy, V. L.: Functional characteristics of the interneurons of wind-sensitive hair receptors on the head in *Locusta migratoria* L. II. Interneurons with inhibitory responses. Comp. Biochem. Physiol. A *48*:427–438 (1974b)

Vaughan, P. F. T., Neuhoff, V.: The metabolism of tyrosine, tyramine and L-3,4-dihydroxyphenylalanine by cerebral and thoracic ganglia of the locust, *Schistocerca gregaria*. Brain Res. *117*:175–180 (1976)

Vial, J. D.: The early changes in the axoplasm during Wallerian degeneration. J. Biophys. Biochem. Cytol. *4*:551–556 (1958)

Voigt, G. E.: Untersuchungen mit der Sulfidsilbermethode an menschlichen und tierischen Bauchspeicheldrüsen (unter besonderer Berücksichtigung des Diabetes mellitus und experimenteller Metallvergiftungen). Virchows Arch. Pathol. Anat. Physiol. *332*:295–323 (1959)

Vowles, D. M.: The structure and connexions of the corpora pedunculata in bees and ants. Q. J. Microsc. Sci. *96*:239–255 (1955)

Vrensen, G., De Groot, D.: Quantitative stereology of synapses: A critical investigation. Brain Res. *58*:25–35 (1973)

Vrensen, G., De Groot, D.: Osmium-zinc-iodide staining and the quantitative study of central synapses. Brain Res. *74*:131–142 (1974)

Waller, A.: Experiments on the section of the glossopharyngeal and hyperglossal nerves of the frog, and observations of the alterations produced thereby in the structure of their primitive fibres. Phil. Trans. R. Soc. Lond. Ser. B *140:*423–469 (1850)

Wang, J. I., Mahler, H. R.: Topography of the synaptosomal membrane. J. Cell Biol. *71:*639–658 (1976)

Wässle, H., Riemann, H. J.: The mosaic of nerve cells in the mammalian retina. Proc. R. Soc. Lond. Ser. B *200:*441–461 (1978)

Watson, M. L.: Staining of tissue sections for electron microscopy with heavy metals. J. Biophys. Biochem. Cytol. *4:*475–478 (1958)

Watson, S. J., Barchas, J. D.: Catecholamine histofluorescence using cryostat sectioning and glyoxylic acid in unperfused frozen brain. Histochem. J. *9:*183–196 (1977)

Weast, R. C., ed.: *Handbook of Chemistry and Physics,* 56th ed. CRC Press, Cleveland, Ohio, 1975

Whaley, W. M., Govindachari, T. R.: The Pictet-Spengler synthesis of tetrahydroisoquinolines and related compounds. Org. React. *6:*151–206 (1951)

Wigglesworth, V. B.: A new method for injecting the tracheae and tracheoles of insects. Q. J. Microsc. Sci. *91:*217–224 (1950)

Willmore, L. J., Fuller, P. M., Butler, A. B., Bass, N. H.: Neuronal compartmentation of ionic cobalt in rat cerebral cortex during initiation of epileptifrom activity. Exp. Neurol. *47:*280–289 (1975)

Wine, J. J.: Invertebrate central neurons: Orthograde degeneration and retrograde changes after axonotomy. Exp. Neurol. *38:*157–169 (1973)

Wine, J. J., Mittenthal, J. E., Kennedy, D.: The structure of tonic flexor motoneurons in crayfish abdominal ganglia. J. Comp. Physiol. *93:*315–335 (1974)

Wine, J. J.: Neuronal organization of crayfish escape response behavior. J. Neurophysiol. *40:*1078–1097 (1977)

Wirtz, R. A., Hopkins, T. L.: Tyrosine and phenylalanine concentrations in the cockroaches *Leucophaea maderae* (F.) and *Periplaneta americana* (L.) in relation to the cuticle formation and ecdysis. Comp. Biochem. Physiol. A *56:*263–266 (1977)

Webster, A. H., Halpern, J.: Homogeneous catalytic activation of molecular hydrogen by silver salts. J. Phys. Chem. *60:*280–285 (1956)

Webster, A. H., Halpern, J.: Homogeneous catalytic activation of molecular hydrogen in aqueous solution by silver salts. J. Phys. Chem. *61:*1239–1248 (1957)

Weiss, M. J.: A reduced silver staining method applicable to dense neuropiles, neuroendocrine organs, and other structures in insects. Brain Res. *39:*268–273 (1972)

Weiss, M. J.: Golgi-Cox procedure with aldehyde prefixation (manuscript). Personal communication (1972)

Welsh, J. H.: Catecholamines in invertebrates. In: *Handbuch der experimentellen Pharmakologie* (H. Blascko, E. Muscholl, eds.), Vol. 33, pp. 79–109. Springer-Verlag, Berlin, New York, 1972

Wendelaar Bonga, S. E.: Ultrastructure and histochemistry of neurosecretory cells and neurohaemal areas in the pond snail *Lymnea stagnalis* (L.). Z. Zellforsch. Mikrosk. Anat. *108:*190–224 (1970)

Weyer, F.: Über drüsenartige Nervenzellen im Gehirn der Honigbiene, *Apis melliphica* L. Zool. Anz. *112:*137–141 (1935)

Wolman, M.: Studies of the impregnation of nervous tissue elements. I. Impregnation of axons and myelin. Q. J. Microsc. Sci. *96:*329–336 (1955a)

Wolman, M.: Studies of the impregnation of nervous tissue elements. II. The nature of the compounds responsible for the impregnation of axons; practical considerations. Q. J. Microsc. Sci. *96:*337–341 (1955b)

Wood, J. G., Barrnett, R. J.: Histochemical demonstration of norepinephrine at a fine structural level. J. Histochem. *12:*197–209 (1964)

Wood, M. R., Pfenninger, K. H., Cohen, M. J.: Two types of presynaptic configurations in insect central synapses: An ultrastructural analysis. Brain Res. *130:*25–45 (1977)

Yamasaki, T., Narahashi, T.: The effects of potassium and sodium ions on the resting and action potentials of the cockroach giant axon. J. Insect Physiol. *3:*146–158 (1959)

Yau, K.-W.: Physiological properties and receptive fields of mechanosensory neurones in the head ganglion of the leech: Comparison with homologous cells in segmental ganglia. J. Physiol. (London) *263:*489–512 (1976a)

Yau, K.-W.: Receptive fields, geometry and conduction block of sensory neurones in the central nervous system of the leech. J. Physiol. (London) *263:*513–538 (1976b)

Young, D.: Specifity and regeneration in insect motor neurons. In: *Developmental Neurobiology of Arthropods* (D. Young, ed.), pp. 179–202. Cambridge Univ. Press, London, New York, 1973

Zawarzin, A. A.: Histologische Studien über Insekten. Das Herz der Aeschnalarven. Z. Wiss. Zool. *97*, 3:481–510 (1911)

Zawarzin, A. A.: Histologische Studien über Insekten. 2. Das sensible Nervensystem der Aeschnalarven. Z. Wiss. Zool. *100*, 2:245–286 (1912a)

Zawarzin, A. A.: Histologische Studien über Insekten. 3. Über das sensible Nervensystem der Larven von *Melolontha vulgaris*. Z. Wiss. Zool. *100*, 3:447–458 (1912b)

Zawarzin, A. A.: Histologische Studien über Insekten. 4. Die optischen Ganglien der Aeschnalarven. Z. Wiss. Zool. *108*, 2:175–257 (1913)

Zawarzin, A. A.: Über die histologische Beschaffenheit des unpaaren ventralen Nervs der Insekten (Histologische Studien über Insekten, 5). Z. Wiss. Zool. *122*, 1:97–115 (1924a)

Zawarzin, A. A.: Histologische Studien über Insekten. 6. Zur Morphologie der Nervenzentren. Das Bauchmark der Insekten. Ein Beitrag zur vergleichenden Histologie. Z. Wiss. Zool. *122*, 3–4:323–424 (1924b)

Zawarzin, A. A.: Der Parallelismus der Strukturen als ein Grundprinzip der Morphologie. Z. Wiss. Zool. *124*, 1:118–212 (1925)

Zawarzin, A. A.: *Outlines on Evolutionary Histology of the Nervous System.* Leningrad, Medgis, 1941 (reprinted: *Selected Works,* Vol. III. Akad. Nauk SSSR, Moskva-Leningrad, 1950)

Zettler, F., Järvilehto, M.: Decrement-free conduction of graded potentials along the axon of a monopolar neuron. Z. Vergl. Physiol. *75:*402–421 (1971)

Zon, L.: The physical chemistry of silver staining. Stain Technol. *11:*53–67 (1936)